Springer Tracts in Electrical and Electronics Engineering

Springer Tracts in Electrical and Electronics Engineering (STEEE) publishes the latest developments in Electrical and Electronics Engineering - quickly, informally and with high quality. The intent is to cover all the main branches of electrical and electronics engineering, both theoretical and applied, including:

- Signal, Speech and Image Processing
- Speech and Audio Processing
- Image Processing
- Human-Machine Interfaces
- Digital and Analog Signal Processing
- Microwaves, RF Engineering and Optical Communications
- Electronics and Microelectronics, Instrumentation
- Electronic Circuits and Systems
- Embedded Systems
- Electronics Design and Verification
- Cyber-Physical Systems
- Electrical Power Engineering
- Power Electronics
- Photovoltaics
- Energy Grids and Networks
- Electrical Machines
- Control, Robotics, Automation
- Robotic Engineering
- Mechatronics
- Control and Systems Theory
- Automation
- Communications Engineering, Networks
- Wireless and Mobile Communication
- Internet of Things
- Computer Networks

Within the scope of the series are monographs, professional books or graduate textbooks, edited volumes as well as outstanding PhD theses and books purposely devoted to support education in electrical and electronics engineering at graduate and post-graduate levels.

Review Process

The proposal for each volume is reviewed by the main editor and/or the advisory board. The books of this series are reviewed in a single blind peer review process.

Ethics Statement for this series can be found in the Springer standard guidelines here https://www.springer.com/us/authors-editors/journal-author/journal-author-helpdesk/before-you-start/before-you-start/1330#c14214

Manish Kumar
Editor

Computational Analysis of Perovskite Solar Cells

Predicting the Future Energy

Springer

Editor
Manish Kumar
Department of Physics
Atma Ram Sanatan Dharma College, University of Delhi
New Delhi, Delhi, India

ISSN 2731-4200 ISSN 2731-4219 (electronic)
Springer Tracts in Electrical and Electronics Engineering
ISBN 978-981-95-6520-7 ISBN 978-981-95-6521-4 (eBook)
https://doi.org/10.1007/978-981-95-6521-4

This Springer imprint is published by the registered company Springer Nature Singapore Pte Ltd.
The registered company address is: 152 Beach Road, #21-01/04 Gateway East, Singapore 189721, Singapore

Contents

About the Editor

Manish Kumar obtained his M.Tech. degree in Ceramic Engineering from IIT(BHU), Varanasi and Ph.D. degree in Physics from Institute of Science, Department of Physics, Banaras Hindu University, Varanasi, India. At present, he is serving as Associate Professor at Department of Physics, ARSD College, University of Delhi, New Delhi, India since last more than 13 years. His research area focuses on environmental friendly multifunctional materials for energy and memory devices, materials for green energy solutions, multiferroics, magnetoelectrics, composites, perovskite materials for photovoltaic applications, supercapacitors, white light emitting mate-rials, solar simulation via different software and DFT calculation. Recently, he was listed among top 2% scientists worldwide by Stanford University, USA, Ranking-2021 & 2025 and published more than 155 research publications in international journals of high repute, two authored books, six edited books (published/in press) in Springer Nature, 38 international book chapters and one Indian Patent with h-index of 34, i10-index of 94 including more than 3600 citations as per Google-Scholar. He has delivered the talks in a number of national and international conferences. He is also a member of various research committees of different universities. In addition, he has completed two national research projects and received the best research award from his institute. He is guiding/ guided number of Ph.D., M.Tech./M.Sc. and B.Sc./B.Tech. students as supervisor and/ cosupervisor. Dr. Kumar is serving and/ served as an Editor in number of highly reputed journals such as, Scientific Reports (Nature), Materials Today Communica-tions (Elsevier), ECS Sensors Plus (IOP, ECS), High Performance Polymers (SAGE), Materials Today: Proceedings (Elsevier) and Springer Nature Books. He has orga-nized one national conference (RAFM-2020) as co-convener and three international conferences (RAFM-2022, RAFM-2024 and APGC-2025) as convener.

Chapter 1
Introduction of Perovskite Solar Cells on Computational Analysis

Shweta Dhakla, Parvesh K. Deendyal, Anjali Kumari, Tushar Nayak, Ankur Taya, Sarvesh Kumar, and Manish K. Kashyap

Abstract Over the decades, hybrid halide perovskite (HHP)-based solar cells have emerged as the groundbreaking technology in the domain of renewable energy sector owing to their low fabrication cost and impressive power conversion efficiency (PCE) of 27.3% at present. Despite of several remarkable properties of HHP material like tunable band gap, low exciton binding energy, high charge carrier mobility, high diffusion length, facile and low-cost-based fabrication process, these perovskite solar cells (PSCs) are still far from the door of photovoltaic market. Therefore, there is a need to explore and analyze the PSCs from the computational perspective along with the experimental approach. Computational analysis has the potential to significantly advance the development of perovskite-based photovoltaics, allowing the prediction of material properties, optimization of device designs, and the investigation of new perovskite compositions. This chapter deals with the importance of computational analysis in PSC research along with the comprehensive discussion on various computational approaches including Density Functional Theory (DFT), molecular dynamics (MD) simulations, and machine learning (ML). Various device simulation and modeling software such as SCAPS-1D, COMSOL, and AMPS are further discussed in this chapter. In addition, future directions for further research

S. Dhakla · A. Kumari · T. Nayak · M. K. Kashyap (✉)
Renewable Energy Laboratory, School of Physical Sciences, Jawaharlal Nehru University, New Delhi, India
e-mail: mkkashyap@jnu.ac.in

P. K. Deendyal
Renewable Energy Laboratory, School of Physical Sciences, Jawaharlal Nehru University, New Delhi, India

Government Polytechnic for Women, Faridabad, Haryana, India

A. Taya
Department of Physics, Mukand Lal National College, Yamunanagar, Haryana, India

S. Kumar
Inter-University Accelerator Centre (IUAC), New Delhi, India

Department of Physics & Astrophysics, University of Delhi, Delhi, India

M. Kumar (ed.), *Computational Analysis of Perovskite Solar Cells*, Springer Tracts in Electrical and Electronics Engineering,
https://doi.org/10.1007/978-981-95-6521-4_1

and development in the field of PSCs have also presented. The computational approach is expected to play a significant role in shaping the future of sustainable energy technologies and realizing the commercial potential of PSCs.

Keywords Perovskite · Computational · Photovoltaics · DFT · Machine learning

Introduction

The growing global energy requirement and dwindling reserves of non-renewable fossil fuels as well as their negative environmental impact are driving the move toward reliable energy technologies. Out of all the renewable energy sources, solar energy is considered the most feasible lasting solution. Particularly, perovskite materials have attracted huge interest for solar power applications due to their incredible properties like tunable band gap, high absorption coefficient, excellent carrier mobility, low exciton binding energy, and long carrier diffusion lengths along with the low cost and simple manufacturing processes (Dou et al. 2014; Kim et al. 2015; Castelli et al. 2014; Deendyal et al. 2024) as shown in Fig. 1.1. Over the past 15 years, perovskite-based solar cells have achieved unprecedented advancements in power conversion efficiency (PCE), jumping from 3.8% in 2009 (Kojima et al. 2009) to a highest certified value of 27.3% (NREL n.d.; Green et al. 2026) at present. This tremendous rise in efficiency is unparalleled by any other photovoltaic technology and also in contrast to the gradual development of commercial silicon solar cells, which took nearly 50 years to achieve a PCE of 27.9 % (Green et al. 2026).

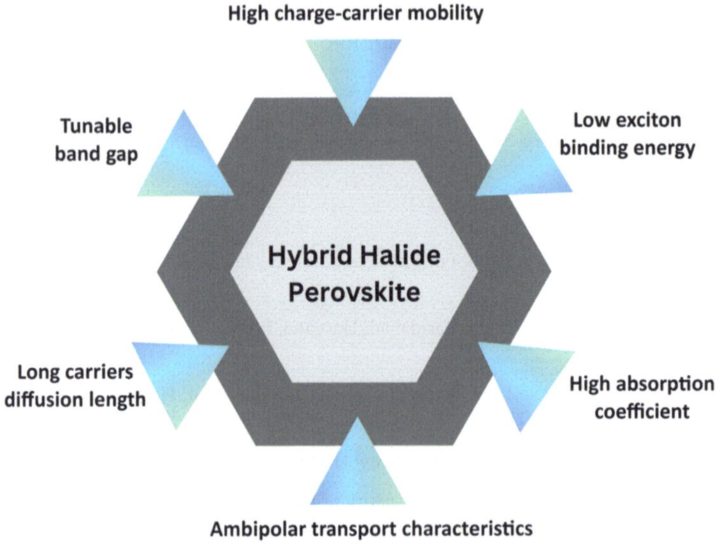

Fig. 1.1 Various properties of hybrid halide perovskites, suitable for photovoltaics

In 2009, Miyasaka and his colleagues employed $MAPbX_3$ ($X = I$, Br) perovskite nanocrystals as light absorbers in dye-sensitized solar cells (DSSCs) with liquid electrolytes (Kojima et al. 2009). The photovoltaic cell was based on the DSSC architecture, utilizing an iodide/bromide-based perovskite nanocrystalline particles efficiently sensitize on TiO_2 as n-type semiconductors for visible-light conversion. Since the PSCs were neither so efficient (3.8%/3.13% for the $MAPbI_3/MAPbBr_3$) nor so stable in a liquid electrolyte configuration. This laid the foundation for advancing more efficient next-generation solar cells. The PCE of $MAPbI_3$-based solar cell was eventually advanced to 6.54% in $MAPbI_3$-based quantum dot synthesized solar cells (Im et al. 2011) by Park and colleagues in 2011, but the identical device based on N719 dye displayed a significantly lower PCE (3.89%) (Im et al. 2011). However, these devices based on liquid electrolyte still required more modifications as the acquired PCE was inadequate, thus earned little attention from photovoltaic community in the same domain. In 2012, several groups investigated the potential of "Spiro-OMeTAD," a solid-state hole transporting layer (HTL) in place of liquid electrolyte in device architecture. Several milestones are significant in the roadmap of solid-state PSC device; Park et al. (Kim et al. 2012) developed a solid-state device and reported the improvement in the fabricated PSC, yielding a PCE of 9.7%. At the same time, Lee et al. (2012) fabricated a PSC device with Spiro-OMeTAD as HTL and a mixed halide perovskite ($MAPbI_{3-x}Cl_x$) as a light harvester on top of the Al_2O_3 layer, demonstrating an increment in PCE value of 10.9%. In the mid of 2013, various innovations took place for the fabrication of solar cells with high PCEs (>12%) (Burschka et al. 2013; Liu et al. 2013; Green et al. 2015). Intense efforts were made in the following years to improve the PCE, including band gap adjustment, solvent engineering, interfacial engineering, additives engineering and an increase in charge carrier generation capabilities. In 2015, Ahn et al. (2015) fabricated PSCs using the Lewis base adduct of lead (II) iodide with $MAPbI_3$ as an absorber and found the additional rise in PCE to 19.7%. It is worth to mention that a certified PCE of 23.7% for fabricated device using $FAPbI_3$ perovskite doped with methylene diammonium dichloride ($MDACl_2$) was reported by Min et al. (2019) in 2019. With consistent efforts in same direction, the champion PSC device (NREL n.d.; Green et al. 2026) has currently exhibited the following photovoltaic parameters: high PCE = 27.3%, open-circuit voltage (V_{oc}) = 1.20 V, short-circuit current density (J_{sc}) = 26.64 mA/cm^2 and fill-factor (FF) = 85.4%.

Perovskites are materials represented by the formula ABX_3, where A and B are cations of different sizes (A being larger than B) and X is an anion. The crystal structure of perovskite is depicted in Fig. 1.2. In perovskite structure, A-site constitutes either organic cation (MA^+, FA^+) or inorganic cation (Cs^+, Rb^+), whereas B- and X-site constitute divalent cation (Pb^{2+}, Sn^{2+}) and halide anion (Cl^-, I^-, Br^-), respectively. These perovskites have several significant characteristics, including high solar absorption, ease of fabrication and low non-radiative carrier recombination rates and the ability to capitalize on over 20 years of development of related organic and dye-sensitized solar cells.

The crystallographic stability and structure can be predicted by calculating tolerance factor "t" and octahedra factor "μ", where t is defined as the ratio of the

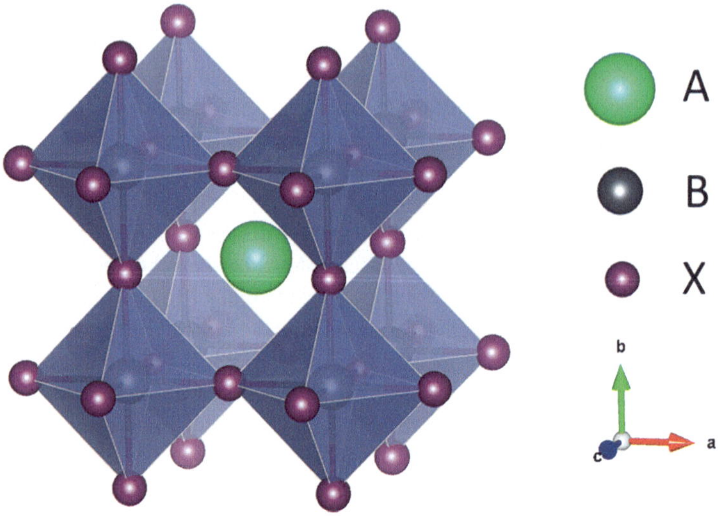

Fig. 1.2 Schematic of perovskite ABX_3 type crystal structure

distance *A-X* to the distance *B-X* in a model of idealized solid-sphere. R_A, R_B, R_X are the ionic radii of *A*, *B*, *X*, respectively.

$$t = \frac{R_A + R_X}{\sqrt{2}\left(R_B + R_X\right)} \tag{1.1}$$

and

$$\mu = \frac{R_B}{R_X} \tag{1.2}$$

Generally, for halide perovskites (X = F, Cl, Br, I), we have $0.81 < t < 1.11$ and $0.44 < \mu < 0.90$ (Green et al. 2014). If "t" lies in between 0.89 and 1.0, then the structure is likely to be cubic; for lower "t" value, the structure likely to be less symmetric, i.e., tetragonal or orthorhombic (Green et al. 2014). Nevertheless, for any particular perovskite, transitions between these structures upon heating are frequent, with the high-temperature phase typically being cubic.

In past few years, the efficiency of PSCs has been reached to 26.7% (NREL n.d.; Green et al. 2026). However, the poor long-term device stability of PSCs is still a big issue, which led to a question that whether promising advances could be carried from the research laboratories to industry for practical applications. Therefore, this long-term stability issue needs to be solved for PSCs. According to multiple reports,

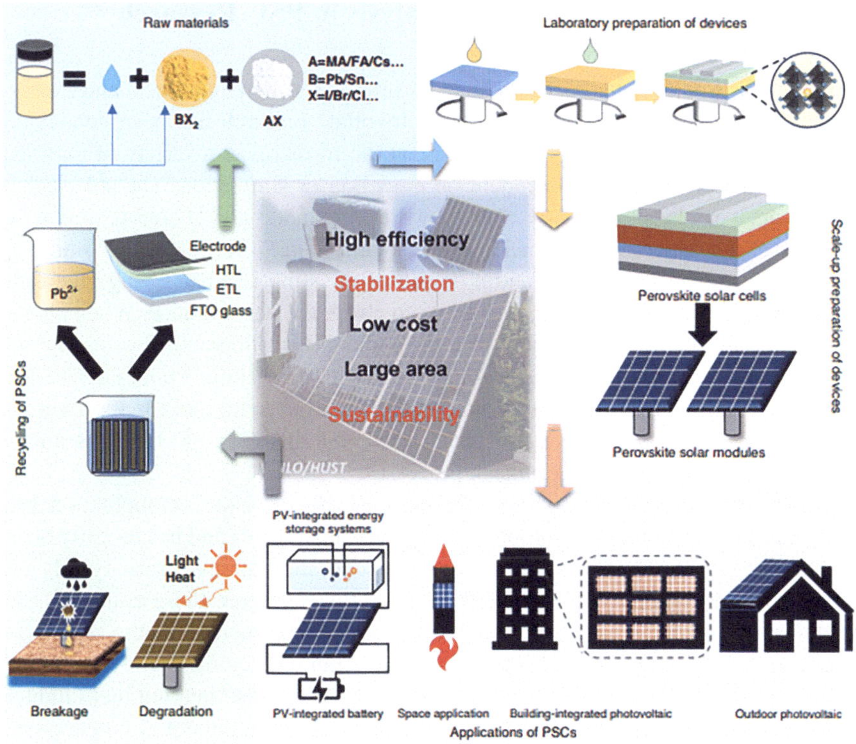

Fig. 1.3 The life cycle of perovskite materials, reproduced with permission from (Yang et al. 2024)

moisture, UV light, oxygen, solution processing, and thermal stress are major factors influencing the stability of PSCs. Sometimes rapid degradations occur when devices are exposed to those environmental factors (Yang et al. 2016; Niu et al. 2015; Rong et al. 2015; Leijtens et al. 2015). Niu et al. (2015) concluded that oxygen, together with moisture, could lead to the irreversible degradation of $CH_3NH_3PbI_3$ (CH_3NH_3 = MA). The $TiO_2/CH_3NH_3PbI_3$ film was subjected to air with 60% humidity at 35°C for 18 h, resulting in a significant reduction in absorption between 530 and 800 nm. Moreover, during fabrication, deployment and disposal process, lead causes toxicity issues. The complete life cycle of perovskite material is highlighted in Fig. 1.3.

With the incredible advancements in a less time span, PSCs are still far from photovoltaic market. Hence, there is a need to investigate and analyze the PSC area thoroughly from both experimental and theoretical fronts. This chapter aims to provide an in-depth exploration of how computational approaches might be advantageous in PSC research as elucidated in the following sections.

Importance of Computational Analysis in PSC Research

Research has traditionally been conducted primarily through theoretical and experimental methods. Now, the majority of photovoltaic research has been conducted using a combination of both, with a focus on the former. For a variety of reasons, modern research is mostly dependent on modeling or simulation. The solar cell's performance can be significantly altered by different materials. Therefore, it is wise to simulate the material performance before synthesizing it for real uses. Similarly, the modifications in the architectural design of the solar cell can change the efficiency, so it is better to simulate the device to check the performance. When physical investigation and measurement of a phenomenon are impractical or easily not possible, the modeling is typically employed. While modeling cannot achieve the exact precision of physical measurements, it serves as a useful tool for approximating real-world behavior, which can provide valuable insights into the phenomena and deepen our understanding of such processes.

Due to its steady advancement over the past three decades, computation has emerged as a crucial tool for comprehending defects in solids and has recently been extensively employed in PSC research. Solar cell simulations primarily rely on semiconductor equations to analyze their behavior. However, these equations are nonlinear and challenging to solve analytically, making numerical methods the preferred approach. Modeling and simulations are commonly employed in the second and third generation solar cells to boost efficiency while also making them lighter and less expensive. Advancements in materials, device architecture, contact systems, and characterization techniques have led to significant improvements in solar cell efficiency. Further enhancement can be achieved through modeling and analysis. Numerical modeling makes the fundamentals of solar cells easier to comprehend and reveals the key factors impacting their photovoltaic performance.

For PSCs, the electronic properties of perovskite materials such as band gap tuning and defect states, are investigated via Density Functional Theory (DFT) approach (Das et al. 2022; Eya and Dzade 2023). Using computational methods, we can predict the stability and performance by substituting ion in the perovskite structure. By simulating charge transport, modeling of optical absorption spectra, studying recombination processes through computational methods, we can get idea about efficiency of photovoltaic device. Computational techniques and analysis of different lattice configurations give us insight into ion mobility and its impact on device hysteresis and longevity. Molecular dynamics (MD), a widely utilized computational tool, helps in scalability and manufacturing by simulating perovskite crystallization and film formation during fabrication. Using machine learning (ML) models, we can accelerate the discovery of new perovskite structure by analyzing larger database. These approaches based on the data-driven models which suggest optimal processing conditions, reducing experimental trails and errors which after all reduce the experimental costs.

Key Computational Methods in PSC Research

In PSC research, several computational methods such as DFT, MD simulations, and ML models play a pivotal role in understanding material properties, optimizing device performance and predicting new materials, respectively. Collectively, these computational approaches contribute to a comprehensive understanding of PSCs, facilitating the design of more durable and efficient device architectures. All the aforementioned key approaches for PSC research are discussed in details as follows:

Density Functional Theory (DFT)

In the light of theoretical approach, numerous first-principles techniques can be utilized to predict the desired material attributes. For determining the perovskite's properties, ab initio techniques can be utilized to resolve the many-body problem through materials modeling. These techniques follow the bottom-up approach and can be solved directly using the many-body Schrödinger equation, depending on first-principles quantum mechanical approach including few approximations. No information derived from experimental data and curve fitting procedures are required in these calculations.

The DFT is a computational modeling approach based on quantum mechanics that can be employed to investigate the ground state properties of both existent and computationally designed materials that seem to have great promise and even have not been synthesized experimentally yet. Various features such as structural, electronic, optical, and magnetic of many-body systems can be easily determined via DFT approach, utilizing a potential acting on system's electrons. This DFT potential includes an "external potential" (V_{ext}) dictated by the system's elemental composition and structure as well as an "effective potential" (V_{eff}) resulting from interelectronic interactions. The DFT, as the name implies, is primarily based on two factors: (1) electron density ($p(r)$) and (2) functional, in which any property is a functional of $p(r)$. In condensed matter physics and computational materials science, it is a well-suited and accomplished concept for complicated material simulation as well as a highly effective method for calculating the properties of metal, semiconductor, and insulator.

The DFT is a powerful computational tool extensively employed in PSCs research to better understand and optimize its material properties. It permits extensive investigations into the band structure and density of states (DOS), providing valuable insights into the electronic properties and band gap of perovskite materials (Taya et al. 2019; Taya et al. 2022), which are vital for efficient light absorption. The DFT is primarily concerned with determining the ground state electronic properties of perovskites. However, predicting and understanding the dynamic behavior of electrons, atoms, and molecules in HHPs necessitates approaches that extend beyond DFT. In such situation, time-dependent density functional theory (TD-DFT) comes into picture (Deendyal et al. 2024; Runge and Gross 1984). The quantum mechanical framework of TD-DFT is critical for understanding the dynamics that govern light interaction with the various layers of PSCs. In TD-DFT, the exchange-correlation functionals are based not only on the spatial density of electrons but also on their time-dependent density.

In the late 1920s and 1930s, Thomas, Fermi, and Dirac (Thomas 1927; Fermi and Lincei 2011; Dirac 1930) pioneered the development of methods for calculating the ground state properties of many-particle systems using electron density. Slater (1974) in 1951 proposed to represent the exchange-correlation (XC) potential by a function proportional to the 1/3 power of the electron density. The DFT became a fully promised theory in 1964, following the formulation of the Hohenberg and Kohn theorems (Kohn and Sham 1965), which explicitly confirmed the use of density as a fundamental quantity. Furthermore, notable progress has been made in the DFT domain to incorporate some necessary modifications, allowing it to be effectively applied to the cases of atoms, molecules, bulk as well as low dimensional systems (metals, semiconductors, and insulators), surfaces, dopant-based systems, etc. For a wide number of solid-state systems, DFT results are relatively satisfactory w.r.t. experimental counterparts.

Molecular Dynamics for Stability and Aging Analysis

Molecular dynamics (MD) is a well-established method (Mattoni et al. 2016; Marx 2000) to study the thermodynamics and statistical behavior of complex materials that can be particularly appropriate for soft crystals as is the case of hybrid perovskites. The ab initio molecular dynamics (AIMD), a computational approach that integrates quantum mechanics and molecular dynamics, is an effective tool for studying the atomic-scale dynamics of PSCs (Car and Parrinello 1985). This technique is used to model the growth of perovskite films from vapors or solutions, yielding information about the parameters that influence high-quality formation in PSCs. AIMD offers a detailed evaluation of the impact of moisture, temperature, and pressure on different types of electron and hole transport layers (ETL/HTL) and the perovskite absorber layer used in PSCs. It also provides insights into ion migration and material degradation, which are crucial for understanding the performance and longevity of PSCs under real-world operating conditions. Additionally, AIMD facilitates the design and optimization of more efficient and stable perovskite materials for PSCs by modeling different perovskite structures and compositions. For instance, Mosconi et al. (2014) investigated the hybrid perovskites and analyzed the structural and electronic properties of organic-halide lead perovskites by a combined AIMD calculations and IR-spectroscopy. Using the dipole-dipole correlation function, the MD was used to model infrared spectra. Their analysis of the rotational motion of MA cations revealed that the molecule can rotate within the inorganic framework within a few picoseconds. Additionally, fluctuations in the electronic band gap ranging from 0.1 to 0.2 eV were observed, which increased with temperature during ion dynamics, emphasizing the interaction between the material's structural and electronic properties.

Machine Learning in Material Property Prediction

The research in the domain of PSCs has made outstanding breakthroughs with a dramatically increase in PCE from 3.8% to 27.3% in last 15 years (Kojima et al. 2009; NREL n.d.), which could only have been achievable with the combined efforts of researchers worldwide. Recent years have seen a great deal of research into the effects of environmental stressors like temperature, humidity, and oxygen on the stability of solar cells and the degradation behavior of materials under which they will be used or stored (Li et al. 2020; Liu et al. 2024; Cheng et al. 2024; Odabaşı and Yıldırım 2020). One of the active areas of research in the same direction is to computationally model or screen materials (typically perovskites) to aid in the discovery of new material with more stability and suitable band gap (Odabaşı and Yıldırım 2020; Srivastava et al. 2023), which are frequently followed by experimental work to synthesize and test the chosen materials. The Machine Learning (ML) and data-driven approaches have recently gained unprecedented interests among materials scientists due to their potential in materials design and discovery (Raccuglia et al. 2016; Tao et al. 2021). The ML is an interdisciplinary field that integrates statistics, computer science, mathematics, and engineering to produce a key area of artificial intelligence (Ahmadi et al. 2021). Due to the abundance of theoretically and experimentally derived data on the properties, structure, and behavior of perovskite materials, ML provides a variety of novel approaches for understanding, optimizing, and accelerating the development of PSC. Materials informatics is employed to develop databases, models, and algorithms that enable data-driven discovery and design of perovskites. The potential of ML algorithms to assess the perovskite attributes based on structural and compositional parameters makes it easier to choose the appropriate perovskite for solar energy conversion. Algorithms in various machine learning models can efficiently screen and prioritize customized perovskite materials with desirable properties, significantly minimizing the time and resources needed for experimental synthesis and testing. This highlights the growing importance of machine learning as a crucial tool in PSC material discovery and design.

ML, a class of artificial intelligence, aims to develop computer programs that utilize statistics and algorithms to learn from large datasets. It can assist in finding patterns in data that are invisible to the human eye, building prediction models and creating algorithms to be used in further research (Yılmaz and Yıldırım 2021). Machine learning is systematically implemented through a structured process that includes: (1) defining research objectives, (2) constructing the dataset, (3) selecting descriptors to characterize the data, (4) identifying suitable ML techniques, (5) developing and optimizing ML models, and (6) assessing and implementing the findings. The general workflow of ML in perovskite materials and related applications is depicted in Fig. 1.4.

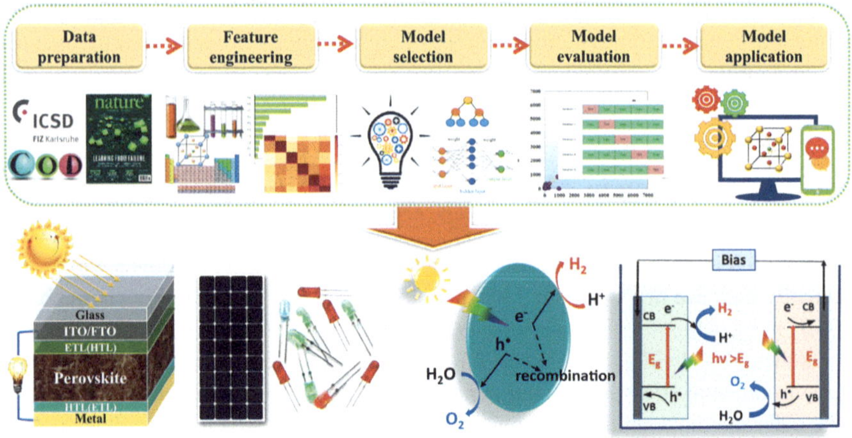

Fig. 1.4 Workflow of ML in perovskite materials and related applications (adapted from (Tao et al. 2021) with permission)

Device Simulation and Modeling Software of PSCs

Numerical simulation is the most powerful tool for developing a highly effective solar cell, obtaining solar cell parameters, and optimizing various geometrical and technological aspects of solar cells (Dhakla et al. 2024; Deendyal et al. 2023a; Dhakla 2023; Deendyal et al. 2023b; Singh et al. 2023). It plays a significant role in demonstrating the physical mechanisms of solar cells and proposed physical model. In addition, the fabrication cost can be minimized by optimizing device characteristics using simulation tools. Meanwhile, this model is more helpful in understanding the influence of various device parameters on the solar cell performance without extensive experimental efforts. Various numerical simulation models that are majorly used in PV technology include the Solar Cells Capacitance Simulator (SCAPS), Analysis of Microelectronics and Photonic Structure (AMPS), COMSOL Multiphysics, Semi-conducting Emissive Thin Film Optics Simulator (SETFOS), Technology Computer-aid Design (TCAD), etc. The comprehensive overview of the most widely used simulation tools for different topologies of metal halide PSCs in the photovoltaic research domain is discussed as follows:

SCAPS-1D

Solar Cells Capacitance Simulator (SCAPS) is a one-dimensional numerical modeling tool equipped with advanced features to simulate solar cell devices to predict optoelectronic performance, developed at the Department of Electronics and

Information Systems (ELIS) of the University of Gent, Belgium (Burgelman et al. 2000). This program was originally designed for polycrystalline cell architectures based on cadmium telluride (CdTe) and copper indium selenide ($CuInSe_2$). With further development of the SCAPS-1D tool, its application was also expanded to c-Si, amorphous silicon (a-Si), and gallium arsenide (GaAs)-based solar cells, etc. The performance parameters of the device are obtained by solving three equations: (1) electron continuity equation, (2) Poisson's equation, (3) hole continuity equation. This tool allows the researchers to investigate the impact of doping concentration and thickness of the perovskite layer on the solar cells (Dhakla et al. 2024; Deendyal et al. 2023b). Other parameters affecting the performance of the solar cells include defect density, carrier diffusion length, and electron affinity of HTL and ETL layers (Haider et al. 2018). Some of the features that SCAPS-1D provides are given below (Burgelman et al. 2000):

1. Solar cells with maximum of seven conducting layers can be simulated.
2. It supports bulk along with interface defect definition.
3. A compound semiconductor with a distinct composition can be defined.
4. It supports current density-voltage (J-V), capacitance-voltage (C-V), and external quantum efficiency (EQE) analysis for the device.
5. Inter- and intra-band tunneling are supported.
6. It supports curve fitting.
7. It supports scripting features for defining problem and its execution.
8. It supports batch execution for parametric variation and optimization.

Several previous studies show productive implementation of SCAPS-1D on the optimization of PSCs for a PCE. In 2020, Lakhdar et al. (Lakhdar and Hima 2020) obtained an efficiency of 10.79% using SCAPS-1D to simulate Ge-based PSC, ITO/PEDOT:PSS/$MAGeI_3$/PCBM/Ag. Optimizing the parameters like ETL/HTL and absorber layer, diverse layer thickness, the PCE of the simulated PSC get increased to 11.16%. The best PCE was obtained by considering C60 as ETL. In 2021, Patel et al. (Patel 2021) numerically analyzed the device structure of FTO/TiO_2/$CH_3NH_3SnI_3$/Cu_2O/anode. The study includes the optimization of defect density, acceptor density, absorber layer thickness, and work function of anode materials. The PCE reached 18.36% when the thickness was optimized to 500 nm. In 2023, Deendyal et al. (2023b) studied dealt with the numerical simulation and optimization of (GA/FA/Cs)PbI_3 using SCAPS-1D software. The best combination of ETL and HTL was obtained out of 35 device structures. The elected PSC structure was (FTO/LBSO/(GA/FA/Cs)PbI_3/CBTS/Au) with optimized defect density/thickness of 1×10^{13} cm^{-3}/700 nm. By considering Ni as the back electrode and materials having a work function greater than 5 eV, the best PCE of 20.36% with FF of 80.17% was obtained. Subudhi and Punetha (2023) in 2023 used the SCAPS-1D tool to study various hybrid PSCs with different perovskite layers and analyzed their performance. The highest efficiency of 26.35% corresponding to cell configuration ITO/TIO_2-SnO_2/(FA/MA)PbI_3/CBTS/Metal was achieved.

AMPS

Analysis of Microelectronic and Photonic Structures (AMPS) is another useful tool created by the University of Illinois that allows the researchers a comprehensive framework for modeling and analyzing solar cells (Zhu et al. 1999). The device's behavior is characterized by solving Poisson's and electron/hole continuity equation. The 1D AMPS tool calculates the quasi-Fermi level and electrostatic potential throughout the device by solving three coupled nonlinear differential equations along with the specified boundary conditions. This software utilizes two different models: the density of states and the lifetime model. Some of the advanced features, that this software allows, are material optimizations, structural designs, defect mitigation, and performance prediction of the solar cells. While investigating the material defect on the device, Liu et al. (2014) with the help of the AMPS tool, devised a high-efficiency perovskite solar cell. As per the simulation studies on the device configuration of $TiO_2/CH_3NH_3PbI_3$/Spiro planar heterojunctions, they found that the efficiency of the device was highly determined by thickness/defect density of absorber layer, hole mobility and acceptor density of HTL and back contact work function. Under standard simulation conditions, an efficiency exceeding 20% was achieved. In 2022, Hajjiah et al. (2022) introduced bilayer all-perovskite solar cells through material crystallinity and device simulation using a 1D AMPS platform. This study utilized $CsPbI_3$ and $FAPbI_3$ perovskites in a bilayer structure and analyzed their density of states (DOS), bandstructure, and optical absorption through DFT approach. The optimum thickness achieved for the ETL/$CsPbI_3$/HTL was 30/200/80 nm. The PCE of the optimized structure was obtained to be 19.94% with FF of 74%. In 2024, Kerara et al. (2024) worked on numerical simulation of FTO/NiO/$CH_3NH_3PbI_3$/(ZnO or TiO_2)/Mn perovskite solar cell performed using AMPS-1D software. The device configuration consists of two ETL layers—zinc oxide and titanium oxide. The optimized thickness of 260 nm/300 nm was obtained for TiO_2/absorber layer. With the above mentioned optimization, the PCE of the device was obtained to be 28.2% with the optimized doping concentration of 10^{16} cm^{-3}.

COMSOL Multiphysics

COMSOL Multiphysics is a powerful tool that enables scientists to study, analyze, and optimize solar cells for higher efficiency. Simulation tools such as COSMOL Multiphysics can save time, resources, and manpower without physically constructing the solar cell. It integrates multiple aspects, including thermal, electrical, and optical behaviors, into a unified simulation, making it ideal for solar cell modeling. Additionally, it enables researchers to optimize the model in three dimensions. Furthermore, COMSOL provides a variety of modules, including Heat Transfer, Electromagnetic Waves, Semiconductor and Radio Frequency (RF) that can be utilized independently or in combination to enhance the precision of numerical analysis. The COSMOL comes up with the feature of a graphical and interactive display that helps in the intuitive display of the model. A key advantage of COMSOL is its ability to test each layer independently. This tool is based on the Finite Element Model (FEM) which allows solving complex geometries and finding material properties in PSCs, enabling accurate predictions of the device's performance. The COMSOL is also used to model the charge carrier dynamics which includes charge generation, recombination, and transport in perovskite layers. In 2020, Zandi et al. (2020a) investigated a CZTSSe thin-film solar cell with the device configuration of Mo/Mo(S, Se)$_2$/CZTSSe/CdS/ZnO/ITO/air layers. The influence of Mo(S, Se)$_2$ on the optical generation rate was calculated which includes coupling of thermal, optical, and electrical physics within COSMOL. The study concluded that incorporating Mo(S,Se)$_2$ enhanced the open-circuit voltage (V_{oc}) from 0.46V to 0.513V. Moreover, COMSOL was employed to investigate PSC numerically with reduced graphene oxide (RGO) contact (Zandi et al. 2020b). This was accomplished by combining the heat transfer module with the electromagnetic wave module in COMSOL. Typically, metallic contacts were employed for heat dissipation; however, RGO contacts demonstrated the enhanced heat dissipation capabilities and improved thermal stability in the PSC. In 2021, Makableh et al. (2021) examined an optoelectronic model with different ETL design structures. The results concluded that using a sandwich layer of TiO_2-ZnO-TiO_2 with a thickness of 10-50-20 nm achieved a PCE of 19.26% improving the performance of resulting device. Moreover, Beepat et al. (2025) used COMSOL multiphysics to simulate MXene-doped PSC of configuration TiO_2/$MAPbI_3$ + MXene/Spiro-OMeTAD. COMSOL multiphysics was utilized for the first time to simulate PSCs incorporating 2D $Ti_3C_2T_x$ MXene as a dopant. The study included the impact of thickness and concentration on the $MAPbI_3$ + MXene layer. With the help of COMSOL Multiphysics, the maximum efficiency corresponding to optimum thickness and concentration was found to be 19.87%.

Other Software: SILVACO TCAD and SETFOS

There are several computer programs other than SCAPS-1D, AMPS, and COMSOL Multiphysics that empower for accurate simulation of optoelectronic devices. SILVACO TCAD is a powerful device simulator that allows for multidimensional device simulation. Silvaco, a company based in the USA, developed this tool, which is widely utilized for simulating the optical, electrical, and thermal behavior of various devices. Its basic functionality includes three steps, namely (1) process simulation, (2) complex mesh generation, and (3) device simulation. The process simulator is utilized in various advanced fabrication techniques including dopant diffusion, oxidation, deposition, activation, and others. The process simulator is employed with different art fabrication processes which include oxidation, dopant diffusion, deposition, activation, etc. Complex mesh generation is equipped with device meshing features in all dimensions. To provide realistic device simulation, the device simulator comprises fundamental features such as optoelectronic, radiation effect, self-heating, and electrochemistry-based device simulations. The SILVACO TCAD was used to investigate the impact of 3D texturing in Si-solar cells to enhance optical absorption (Jahanshah et al. 2008). It was used to investigate the effect of surface texturing on GaAs solar cells (Jalil et al. 2008). It was also utilized to study the performance limits of perovskite solar cells (Kumar et al. 2022).

The precision of simulations relies on accuracy of the device model, which is ultimately derived from experimental data. The SETFOS is another powerful CPU-efficient simulator developed by Prof. Ruhstaller, Fluxim AG (Advanced modeling of solar cells and OLEDs n.d.). Developed using Java, this simulator is specifically designed for advancing thin-film optoelectronic technologies. Along with perovskite (Chiang et al. 2017), this simulator has been developed to study organic-based semiconductor solar cells (Bi et al. 2018) and thin-film devices (Lampande et al. 2016). This simulator incorporates advanced optics modules, drift-diffusion, absorption-emission to analyze various charge transport mechanisms and optical properties. The SETFOS tool's multi-target and multi-variable optimization capability enables researchers to explore device characteristics in a way that is not achievable through experimental methods. The reliability and flexibility of the tool significantly accelerate fundamental research in device physics. Bhattarai et al. (2021) used SETFOS software to conduct a detailed investigation on double graded $MAPbI_3$-based PSCs. The unique band gap combination in PSC at optimal thickness provided an impressive PCE of 19.20%. More recently, Dikhit and Kumar Tripathy (2024) investigated the azitidinium cation for its potential as an alternative for FA/MA ion in PSCs using combined SETFOS and DFT approaches. The champion device, i.e., ITO/Indium-gallium-zinc oxide (IGZO)/$AZSnCl_3$/CuI/Ag achieved an incredible PCE of 26.1%, with a J_{sc} and V_{oc} of 26.83 mA/cm^2 and 1.26 V, respectively.

Challenges and Future Perspectives

PSCs are at the forefront of energy research owing to their distinctive blend of facile solution processability, low materials cost, and the impressive device performance. These solar cells can achieve ~33% efficiency based on theoretical predicted Shockley-Queisser (SQ) limit which is as higher than the PCE of 27.3 % at present (NREL n.d.). There is still a more headway to make to achieve the most stable and efficient version of PSCs. Added to this, resources, manpower, and time can be balanced by taking into account the computational tool for PSC research. These tools can assist in investigating the physics phenomena in solar cells without constructing them physically. One of the significant challenges in the computational analysis of PSCs is the high computational cost of these methods, particularly for bigger systems or extended simulations, which restricts their wider applicability. Accurately simulating ionic dynamics, defect physics and halide segregation are main hurdles as these processes need time-dependent and computationally costly simulation techniques. Techniques like molecular dynamics and DFT frequently fail to describe the complex interactions seen in perovskite materials including ion migration, defect states and non-radiative recombination mechanisms which are essential for comprehending device performance. Another problem is integrating ML with experimental data, as a lack of high-quality and standardized datasets reduces the reliability and applicability of ML models. In addition, the lack of high-quality datasets impedes the development of robust ML models, and large-scale simulations for intricate compositions require significant computational resources. Looking ahead, advances in computational approaches offer intriguing solutions to these problems.

Hybrid approaches integrating DFT, MD, and ML are believed to enhance both the accuracy and efficiency of computer simulations. Computational restrictions will be addressed by developing more efficient algorithms and utilizing high-performance computing (HPC). With the use of advanced prediction models, the combination of AI with big data analytics offers enormous promise for material discovery, defect engineering and performance optimization. To make computational approaches much more useful, extensive, and standardized databases on perovskite materials and properties should be built. The collaborative databases and open platforms for sharing computational results will improve reproducibility and global research efforts. Real-world device performance can be predicted using comprehensive device modeling that takes into account the environmental variables *viz* temperature, humidity, and UV exposure. By tackling these challenges and embracing future directions, computational research in PSCs can overcome current constraints, opening the path for advancing the scalable, highly-efficient, and robust energy solutions.

Conclusion

Perovskite solar cells have gained unprecedented interest in solar energy sector owing to their rapid progress in PCE of 27.3% (till date) in last 15 years. Perovskite materials possess a unique blend of incredible properties like facile and low-cost fabrication methods, high absorption coefficient, tunable band gap, and many more, making them potential substitutes for traditional PV technologies. However, an extensive comprehension of the foundations of perovskite is essential for achieving the stable and high-performance PSCs. The computational analysis has become an indispensable tool for investigating their potential, providing valuable information about the electronic properties, device architecture, and stability. This chapter has explored the various computational approaches such as DFT, MD simulations, and ML which play a significant role in understanding material properties, optimizing device performance and predicting new materials, respectively. Furthermore, device simulation and modeling software including SCAPS-1D, COMSOL, AMPS, and others have been outlined which are commonly employed in PSCs research to model, analyze, and optimize their photovoltaic performance. Together, these computational approaches contribute in the comprehensive understanding and advancement of PSCs technology, accelerating the design of more efficient, stable, and cost-effective PSCs for next-generation energy solutions.

References

Advanced modeling of solar cells and OLEDs (n.d.). https://www.fluxim.com/setfos-intro

Ahmadi M, Ziatdinov M, Zhou Y, Lass EA, Kalinin SV (2021) Machine learning for high-throughput experimental exploration of metal halide perovskites. Joule 5(11):2797–2822. https://doi.org/10.1016/J.JOULE.2021.10.001

Ahn N, Son DY, Jang IH, Kang SM, Choi M, Park NG (2015) Highly reproducible perovskite solar cells with average efficiency of 18.3% and best efficiency of 19.7% fabricated via Lewis base adduct of lead(II) iodide. J Am Chem Soc 137(27):8696–8699. https://doi.org/10.1021/JACS.5B04930

Beepat KG, Sharma DP, Mahajan A, Pathak D, Kumar V (2025) MXene doped perovskite solar cell simulation for enhanced efficiency. https://www.comsol.jp/paper/download/1442892/mxene-doped-perovskite-solar-cell-simulation-for-enhanced-efficiency.pdf. Accessed 6 Jan 2025

Bhattarai S, Sharma A, Muchahary D, Gogoi D, Das TD (2021) Numerical simulation study for efficiency enhancement of doubly graded perovskite solar cell. Opt Mater 118:111285. https://doi.org/10.1016/J.OPTMAT.2021.111285

Bi S, Ouyang Z, Shaik S, Li D (2018) Effect of donor-acceptor vertical composition profile on performance of organic bulk heterojunction solar cells. Sci Rep 8(1):9574

Burgelman M, Nollet P, Degrave S (2000) Modelling polycrystalline semiconductor solar cells. Thin Solid Films 361:527–532. https://doi.org/10.1016/S0040-6090(99)00825-1

Burschka J et al (2013) Sequential deposition as a route to high-performance perovskite-sensitized solar cells. Nature 499:7458. https://doi.org/10.1038/nature12340

Car R, Parrinello M (1985) Unified approach for molecular dynamics and density-functional theory. Phys Rev Lett 55(22):2471–2474. https://doi.org/10.1103/PhysRevLett.55.2471

Castelli IE, García-Lastra JM, Thygesen KS, Jacobsen KW (2014) Bandgap calculations and trends of organometal halide perovskites. APL Mater 2(8):81514. https://doi.org/10.1063/1.4893495/120783

Cheng P, An Y, Jen AKY, Lei D (2024) New nanophotonics approaches for enhancing the efficiency and stability of perovskite solar cells. Adv Mater 36(17):2309459. https://doi.org/10.1002/ADMA.202309459

Chiang KM, Hsu BW, Chang YA, Yang L, Tsai WL, Lin HW (2017) Vacuum-deposited organometallic halide perovskite light-emitting devices. ACS Appl Mater Interfaces 9(46):40516–40522. https://doi.org/10.1021/ACSAMI.7B12805

Das T, Di Liberto G, Pacchioni G (2022) Density functional theory estimate of halide perovskite band gap: when spin orbit coupling helps. J Phys Chem C 126(4):2184–2198. https://doi.org/10.1021/ACS.JPCC.1C09594

Deendyal PK, Dhakla S, Singh H, Kumar S, Kashyap MK (2023a) Suitability of LBSO/CuI as an effective ETL/HTL for perovskite solar cells: a dry lab approach. Indian J Pure Appl Phys 61(11):931–933. https://doi.org/10.56042/IJPAP.V61I11.2441

Deendyal PK, Dhakla S, Singh H, Taya A, Kumar S, Kashyap MK (2023b) Design and optimization of a triple-cation perovskite solar cell for efficiency promotion using numerical simulations. ChemistrySelect 8(31):e202302252. https://doi.org/10.1002/SLCT.202302252

Deendyal PK et al (2024) Chronological evolution of stability in hybrid halide perovskite solar cells. Sol RRL 8(6):2300957. https://doi.org/10.1002/SOLR.202300957

Dhakla SDPSHMMKSKM (2023) Hole transport layer optimization for mixed halide perovskite based solar cells to achieve better photovoltaic performance. J Pure Appl Phys 61(10):851–853
Dhakla S, Deendyal PK, Singh H, Kumar S, Kashyap MK (2024) Optimization of photovoltaic environment for enhanced performance of GA0.053MA0.947PbI3-based perovskite solar cell. IEEE J Photovolt 14(1):85–92. https://doi.org/10.1109/JPHOTOV.2023.3319603
Dikhit A, Kumar Tripathy S (2024) Enhancing the solar cell efficiency with Azetidinium based perovskites: an investigation for device optimization using DFT and SETFOS. Sol Energy 284:113089. https://doi.org/10.1016/J.SOLENER.2024.113089
Dirac PAM (1930) Note on exchange phenomena in the Thomas atom. Math Proc Camb Philos Soc 26(3):376–385. https://doi.org/10.1017/S0305004100016108
Dou L et al (2014) Solution-processed hybrid perovskite photodetectors with high detectivity. Nat Commun 5(1):1–6. https://doi.org/10.1038/ncomms6404
Eya HI, Dzade NY (2023) Density functional theory insights into the structural, electronic, optical, surface, and band alignment properties of BaZrS3 chalcogenide perovskite for photovoltaics. ACS Appl Energy Mater 6(11):5729–5738. https://doi.org/10.1021/ACSAEM.3C00103
Fermi E, Lincei R (2011) Statistical method to determine some properties of atoms. Rend Lincei Sci Fisiche Nat 22(4):283–314
Green MA, Ho-Baillie A, Snaith HJ (2014) The emergence of perovskite solar cells. Nat Photonics 8(7):506–514. https://doi.org/10.1038/nphoton.2014.134
Green MA, Emery K, Hishikawa Y, Warta W, Dunlop ED (2015) Solar cell efficiency tables (version 46). Prog Photovolt 23(7):805–812
Green, M. A., Dunlop, E. D., Yoshita, M., Kopidakis, N., Bothe, K., Siefer, G., ... & Jiang, J. Y. (2026). Solar cell efficiency tables (version 67). Progress in Photovoltaics: Research and Applications, 0: 1–15. https://doi.org/10.1002/pip.70068
Haider SZ, Anwar H, Wang M (2018) A comprehensive device modelling of perovskite solar cell with inorganic copper iodide as hole transport material. Semicond Sci Technol 33(3):035001. https://doi.org/10.1088/1361-6641/AAA596
Hajjiah A, Gamal M, Kandas I, Gorji NE, Shehata N (2022) DFT and AMPS-1D simulation analysis of all-perovskite solar cells based on CsPbI3/FAPbI3 bilayer structure. Sol Energy Mater Sol Cells 248:112026. https://doi.org/10.1016/J.SOLMAT.2022.112026
Im JH, Lee CR, Lee JW, Park SW, Park NG (2011) 6.5% efficient perovskite quantum-dot-sensitized solar cell. Nanoscale 3(10):4088–4093. https://doi.org/10.1039/C1NR10867K
Jahanshah F et al (2008) An investigation of three-dimensional texturing in silicon solar cells for enhanced optical absorption. In: Conference record of the IEEE photovoltaic specialists conference. https://doi.org/10.1109/PVSC.2008.4922676
Jalil SM, Abdullah L, Ahmad I, Abdullah H (2008) The effect of surface texturing on GaAs solar cell using TCAD tools. In: IEEE international conference on semiconductor electronics, proceedings, ICSE, pp 280–283. https://doi.org/10.1109/SMELEC.2008.4770323
Kerara M, Naas A, Gueddim A, Meglali O (2024) Numerical study and optimization of a perovskite solar cell based on methylammonium lead iodide. Trans Electr Electron Mater 25(5):665–673. https://doi.org/10.1007/S42341-024-00549-W
Kim HS et al (2012) Lead iodide perovskite sensitized all-solid-state submicron thin film mesoscopic solar cell with efficiency exceeding 9%. Sci Rep 2(1):1–7. https://doi.org/10.1038/srep00591
Kim YH et al (2015) Multicolored organic/inorganic hybrid perovskite light-emitting diodes. Adv Mater 27(7):1248–1254. https://doi.org/10.1002/ADMA.201403751
Kohn W, Sham LJ (1965) Self-consistent equations including exchange and correlation effects. Phys Rev 140(4A). https://doi.org/10.1103/PHYSREV.140.A1133
Kojima A, Teshima K, Shirai Y, Miyasaka T (2009) Organometal halide perovskites as visible-light sensitizers for photovoltaic cells. J Am Chem Soc 131(17):6050–6051. https://doi.org/10.1021/JA809598R
Kumar A, Gupta U, Gupta N, Goyal AK (2022) IOP Conf Ser Mater Sci Eng. https://doi.org/10.1088/1757-899X/1225/1/012016/meta

Lakhdar N, Hima A (2020) Electron transport material effect on performance of perovskite solar cells based on CH3NH3GeI3. Opt Mater 99:109517. https://doi.org/10.1016/J.OPTMAT.2019.109517

Lampande R, Kim GW, Park MJ, Kang BY, Kwon JH (2016) Efficient light harvesting in inverted polymer solar cells using polymeric 2D-microstructures. Sol Energy Mater Sol Cells 151:162–168. https://doi.org/10.1016/J.SOLMAT.2016.03.005

Lee MM, Teuscher J, Miyasaka T, Murakami TN, Snaith HJ (2012) Efficient hybrid solar cells based on meso-superstructured organometal halide perovskites. Science 338(6107):643–647. https://doi.org/10.1126/SCIENCE.1228604

Leijtens T, Eperon GE, Noel NK, Habisreutinger SN, Petrozza A, Snaith HJ (2015) Stability of metal halide perovskite solar cells. Adv Energy Mater 5(20):1500963

Li N, Niu X, Chen Q, Zhou H (2020) Towards commercialization: the operational stability of perovskite solar cells. Chem Soc Rev 49(22):8235–8286. https://doi.org/10.1039/D0CS00573H

Liu M, Johnston MB, Snaith HJ (2013) Efficient planar heterojunction perovskite solar cells by vapour deposition. Nature 501:7467. https://doi.org/10.1038/nature12509

Liu F et al (2014) Numerical simulation: toward the design of high-efficiency planar perovskite solar cells. Appl Phys Lett 104(25):253508. https://doi.org/10.1063/1.4885367/940833

Liu C et al (2024) Two-dimensional perovskitoids enhance stability in perovskite solar cells. Nature 633(8029):2024. https://doi.org/10.1038/s41586-024-07764-8

Makableh YF, Hassan W, Awad IA, Aljaiuossi G (2021) Comprehensive electrical modeling analysis of heterojunction perovskite solar cells by using different electron transport nanostructured layers. Superlattice Microst 150:106777. https://doi.org/10.1016/J.SPMI.2020.106777

Marx JH (2000) Ab initio molecular dynamics: theory and implementation. Mod Methods Algorith Quant Chem 1(141):301–449

Mattoni A, Filippetti A, Caddeo C (2016) Modeling hybrid perovskites by molecular dynamics. J Phys Condens Matter 29(4):043001. https://doi.org/10.1088/1361-648X/29/4/043001

Min H et al (2019) Efficient, stable solar cells by using inherent bandgap of a-phase formamidinium lead iodide. Science 366(6466):749–753. https://doi.org/10.1126/SCIENCE.AAY7044

Mosconi E, Quarti C, Ivanovska T, Ruani G, De Angelis F (2014) Structural and electronic properties of organo-halide lead perovskites: a combined IR-spectroscopy and ab initio molecular dynamics investigation. Phys Chem Chem Phys 16(30):16137–16144. https://doi.org/10.1039/C4CP00569D

Niu G, Guo X, Wang L (2015) Review of recent progress in chemical stability of perovskite solar cells. J Mater Chem A Mater 3(17):8970–8980. https://doi.org/10.1039/C4TA04994B

NREL (n.d.) Best research-cell efficiency chart | photovoltaic research. https://www.nrel.gov/pv/cell-efficiency.html. Accessed: Dec 2025

Odabaşı Ç, Yıldırım R (2020) Machine learning analysis on stability of perovskite solar cells. Sol Energy Mater Sol Cells 205:110284. https://doi.org/10.1016/J.SOLMAT.2019.110284

Patel PK (2021) Device simulation of highly efficient eco-friendly CH3NH3SnI3 perovskite solar cell. Sci Rep 11(1):1–11. https://doi.org/10.1038/s41598-021-82817-w

Raccuglia P et al (2016) Machine-learning-assisted materials discovery using failed experiments. Nature 533(7601):73–76. https://doi.org/10.1038/nature17439

Rong Y, Liu L, Mei A, Li X, Han H (2015) Beyond efficiency: the challenge of stability in mesoscopic perovskite solar cells. Adv Energy Mater 5(20):1501066. https://doi.org/10.1002/AENM.201501066

Runge E, Gross EKU (1984) Density-functional theory for time-dependent systems. Phys Rev Lett 52(12):997–1000. https://doi.org/10.1103/PhysRevLett.52.997

Singh H, Dhakla S, Deendyal PK, Kumar A, Kumar S, Kashyap MK (2023) Photovoltaic performance of mixed cation K0.005MA0.995PbI3-based perovskite solar module. Appl Res 2(2):e202200077. https://doi.org/10.1002/APPL.202200077

Slater JC (1974) The self-consistent field for molecules and solids: quantum theory of molecules and solids, vol 4. McGraw Hill, New York

Srivastava M, Hering AR, An Y, Correa-Baena JP, Leite MS (2023) Machine learning enables prediction of halide perovskites' optical behavior with >90% accuracy. ACS Energy Lett 8(4):1716–1722. https://doi.org/10.1021/ACSENERGYLETT.2C02555
Subudhi P, Punetha D (2023) Pivotal avenue for hybrid electron transport layer-based perovskite solar cells with improved efficiency. Sci Rep 13(1):1–15. https://doi.org/10.1038/s41598-023-33419-1
Tao Q, Pengcheng X, Li M, Wencong L (2021) Machine learning for perovskite materials design and discovery. NPJ Comput Mater 7(1):23
Taya A, Rani P, Thakur J, Kashyap MK (2019) First principles study of structural, electronic and optical properties of Cs-doped CH3NH3PbI3 for photovoltaic applications. Vacuum 160:440–444. https://doi.org/10.1016/J.VACUUM.2018.12.008
Taya A, Kumar S, Hackett TA, Kashyap MK (2022) Optical absorption and stability enhancement in mixed lead, tin, and germanium hybrid halide perovskites for photovoltaic applications. Vacuum 201:111106. https://doi.org/10.1016/J.VACUUM.2022.111106
Thomas LH (1927) The calculation of atomic fields. Math Proc Camb Philos Soc 23(5):542–548. https://doi.org/10.1017/S0305004100011683
Yang L, Barrows AT, Lidzey DG, Wang T (2016) Recent progress and challenges of organometal halide perovskite solar cells. Rep Prog Phys 79(2):026501. https://doi.org/10.1088/0034-4885/79/2/026501
Yang C et al (2024) Achievements, challenges, and future prospects for industrialization of perovskite solar cells. Light Sci Appl 13(1):1–48. https://doi.org/10.1038/s41377-024-01461-x
Yılmaz B, Yıldırım R (2021) Critical review of machine learning applications in perovskite solar research. Nano Energy 80:105546. https://doi.org/10.1016/J.NANOEN.2020.105546
Zandi S, Saxena P, Razaghi M, Gorji NE (2020a) Simulation of CZTSSe thin-film solar cells in COMSOL: three-dimensional optical, electrical, and thermal models. IEEE J Photovolt 10(5):1503–1507. https://doi.org/10.1109/JPHOTOV.2020.2999881
Zandi S, Saxena P, Gorji NE (2020b) Numerical simulation of heat distribution in RGO-contacted perovskite solar cells using COMSOL. Sol Energy 197:105–110. https://doi.org/10.1016/J.SOLENER.2019.12.050
Zhu H, Kalkan AK, Hou J, Fonash SJ (1999) Applications of AMPS-1D for solar cell simulation. AIP Conf Proc 462(1):309–314. https://doi.org/10.1063/1.57978

Chapter 2
Generation of Defects in Perovskite Absorber

Piyali Maity

Abstract Perovskite materials play an important role as absorber in the field of photovoltaic applications due to their remarkable light absorption and charge transport properties. The defects generation in perovskite absorbers for fabrication of solar cells is a major research topic. However, the defects generated in these materials significantly impact their efficiency, stability, and performance. This chapter provides the detailed mechanisms of underlying defect generation in perovskite absorbers, emphasizing on types of defects and their formation processes. The defects have been categorized into—vacancies, interstitials, and grain boundaries. Their formation mechanisms including synthesis conditions, thermal processing, and environmental conditions will be described thoroughly. These defects impact the electronic properties, charge transport, and recombination processes. By employing advanced synthesis techniques, proper material selection, and defect control strategies, researchers can improve the performance and stability of perovskite-based photovoltaic devices. This chapter will highlight the in-depth understanding of optimizing perovskite materials for next-generation solar energy technologies and emphasizes the importance of computational approaches in understanding and mitigating defect-related challenges in perovskite absorbers. Numerical analysis such as density functional theory (DFT), machine learning (ML), SCAPS (solar cell capacitance simulator), and SILVACO ATLAS play as significant tools in simulation and help to understand the defect dynamics.

Keywords Light absorption · Charge transport · Grain boundaries · Photovoltaics · Solar energy

P. Maity (✉)
City University of New York, New York, NY, USA

M. Kumar (ed.), *Computational Analysis of Perovskite Solar Cells*, Springer Tracts in Electrical and Electronics Engineering,
https://doi.org/10.1007/978-981-95-6521-4_2

Introduction

Perovskite solar cells (PSCs) have experienced significant advancements in recent years, characterized by their high-power conversion efficiencies, cost-effective manufacturing processes, and straightforward fabrication methods (Akbulatov et al. 2017; Barbé et al. 2017; Behrouznejad et al. 2016; Nayak et al. 2019). Nevertheless, the existence of defects within the perovskite absorber layers poses a considerable challenge to achieving both optimal efficiency and long-term stability. This chapter delves into the formation of defects in perovskite absorbers, exploring their various types, origins, and the computational modeling techniques employed to study them (Kojima et al. 2009; Srivastava et al. 2021). Furthermore, it addresses the influence of these defects on the operational performance of PSCs. By utilizing computational methodologies, researchers are able to anticipate defect behavior and develop strategies for their mitigation, thereby enhancing the efficiency and durability of PSCs (Jung et al. 2019; Li et al. 2016; Kim et al. 2017; Lee and Ebong 2017; Feurer et al. 2019).

Perovskite solar cells have shown remarkable promise in the field of solar energy conversion, achieving power conversion efficiencies comparable to those of conventional silicon-based solar cells (Wang et al. 2018; Ono et al. 2020; Simmons and Taylor 1971; Ran et al. 2018; Tian et al. 2020; Elumalai et al. 2016; Prasanna et al. 2023). However, despite their rapid advancements and potential, a major challenge to attaining long-term stability and peak efficiency lies in the defects present within the perovskite absorber layer. These defects may originate from various stages of fabrication, processing, or environmental influences, significantly affecting both the material characteristics and the performance of the resulting solar devices (Xie et al. 2021; Liu et al. 2021). This chapter delves into the mechanisms behind defect generation in perovskite absorbers, outlining the common types of defects, the contributing factors to their emergence, and their effects on the performance and stability of PSCs. Additionally, it examines the computational methods and models employed to analyze defects and forecast their behavior in perovskite materials. Lastly, we will explore approaches aimed at reducing defect formation to enhance the overall efficiency and longevity of PSCs (Kojima et al. 2009; Ball and Petrozza 2016; Duan and Uddin 2022; Jeng et al. 2014).

Perovskite materials, particularly metal halide perovskites (MHPs), have emerged as fundamental components in the development of next-generation photovoltaic technologies. These materials possess an ABX_3 crystal structure, where A typically represents a cation (such as methylammonium, formamidine, or cesium), B denotes a metal cation (such as lead or tin), and X signifies a halide anion (including iodine, bromine, or chlorine) (Toshniwal et al. 2017; D'Innocenzo et al. 2014; Oga et al. 2014). Perovskite solar cells have demonstrated remarkable efficiencies, exceeding 25%, thereby positioning them as viable competitors to traditional silicon solar cells. Nevertheless, despite these significant advancements, the presence of defects within the perovskite absorber layer continues to pose challenges to the performance and stability of perovskite solar cells (Zhang et al. 2017; Lee et al. 2012; Yang et al. 2017).

Defects present in perovskite materials can have a profound impact on charge transport, recombination processes, and overall device efficacy. It is essential to comprehend the mechanisms behind the formation of these defects, their influence on the properties of the material, and strategies for their mitigation to advance the development of efficient and stable perovskite solar cells. This chapter examines the various categories of defects, their origins during the processing of materials, and their contribution to the decline in PSC performance. Furthermore, the chapter will investigate the significance of computational modeling in forecasting defect behavior and informing defect engineering practices (Kojima et al. 2009; Jung et al. 2019; Snaith 2013; Xiao et al. 2015; Tvingstedt et al. 2015).

Perovskite materials, especially metal halide perovskites (MHPs), have attracted considerable interest in the realm of photovoltaic (PV) technology due to their exceptional optoelectronic characteristics, which include high absorption coefficients, adjustable bandgaps, and straightforward fabrication processes. Since the initial introduction of perovskite solar cells in 2009, the power conversion efficiency of these devices has experienced a remarkable surge, escalating from below 4% to over 25%. This advancement has allowed them to surpass traditional thin-film technologies and come close to the efficiency levels of silicon solar cells (Li et al. 2016; Kim et al. 2017; Xiao et al. 2015; Tvingstedt et al. 2015; Agiorgousis et al. 2014; Homes et al. 2001).

Nevertheless, despite these significant improvements, defects within perovskite materials pose a substantial challenge to enhancing the performance and stability of perovskite solar cells (PSCs). Such defects may emerge during the synthesis of materials, the fabrication of devices, or even during their operational phases. These imperfections can generate trap states that act as recombination centers, consequently diminishing the lifetime of charge carriers and the overall efficiency of the device. Moreover, defects can adversely affect the long-term stability of PSCs, highlighting the necessity for a deeper understanding of their origins, types, and impacts (Feurer et al. 2019; Simmons and Taylor 1971; Tvingstedt et al. 2015; Agiorgousis et al. 2014; Homes et al. 2001; Wright et al. 2016; Viswanath et al. 1998).

This chapter aims to provide a comprehensive overview of defect generation in perovskite absorbers, concentrating on their classifications, sources, and the influence they exert on the performance of PSCs. Additionally, we will examine computational techniques employed to predict defect formation and behavior in perovskites, emphasizing how these methodologies can facilitate defect engineering to enhance the performance of PSCs.

Moreover, computational techniques provide strong toolkit and a pathway to understand the fundamentals and to mitigate the defects in PSCs. Unique properties can be investigated and explored by these computational tools. Such as Density Functional Theory (DFT) helps to investigate the atomic level information and to understand the fundamentals for defect formation and passivation. Machine learning stimulates the prediction of defects and discovery of various materials. This also provides large-scale cost-effective insights even better than DFT alone. SCAPS (Solar Cell Capacitance Simulator) and SILVACO ATLAS helps in simulation of device level effects of defects by solar cell efficiency optimization under ambient atmosphere. In conclusion, DFT and ML are used in selection of materials and defect studies, SCAPS and ATLAS are used for the performance evaluation of solar cell designs under the presence of defects in it. These techniques can be combined for the prediction and optimization of the future energy potential of PSCs. This can also address challenges which are related to efficiency, stability, and defect management. Studying defects are challenging due to the complex structure, anharmonicity, and softness of crystal in case of halide perovskites. Hence, this chapter focuses to address the unresolved issues and proposes future research direction by demonstrating the current research on computational studies on defects. This also highlights the theoretical studies should be more explored to establish strong links with experimental investigations for providing insights to the scientific and industrial field. This way advanced interdisciplinary approach will be developed to understand the properties of defects which will help to design modern strategies for improving perovskite solar cell devices performance (Guo et al. 2024). Large-scale device fabrication is required for commercial manufacturing of PSCs from laboratory scale. Sheet-to-sheet and roll-to-roll deposition techniques are used for the development of required industrial scale devices. For this manufacturing, laser ablation, drying, post-treatment, and industry-friendly solvents are necessary (Galagan 2018).

Perovskite Structure and Types of Defects: Crystal Structure and Defect Formation in Perovskites

The structure of the perovskite crystals must be discussed first in order to comprehend the causes and effects of faults in perovskite materials. The perovskite crystal structure is prone to flaws during both manufacture and operation because it is extremely sensitive to external influences like temperature, pressure, and ion migration. The defect properties, defect densities, and defects-related stability in perovskite films are controlled by the methodologies, such as, growth of large size crystals, grain boundary and surface passivation, photo-curing method, and modification of the substrates. Metal halide perovskites have high-defect-tolerance limit, but still, it is challenging that there are deep defects in perovskite thin films which hinder the photoconversion efficiency of perovskite solar cells approaching the Schockley-Queisser limit (Agiorgousis et al. 2014; Kim et al. 2014). Some defects cause electronic states in band gap. They act as Shockley-Read-Hall (SRH) nonradiative recombination centers. They create a substantial decrease in lifetimes of charge carriers and photovoltaic performance (Stranks 2017). The trap-assisted recombination can be reduced to reach higher voltage limit for the higher photocurrent by avoiding the recombination (Tress 2017). Time-domain ab initio simulations can give the atomistic description of the quantum dynamics of electron and hole trapping and recombination. This helps to generate deep insights into the mechanisms of charge and energy losses and shows path for the development of high-performance perovskite solar cells (Li et al. 2019).

Crystal Structure and the Role of Defects

A three-dimensional network of corner-sharing octahedra makes up the perovskite structure, with the halide anions (X-site) occupying the vertices and the metal cation (B-site) situated at the center of each octahedron. The interstitial area between the octahedra is occupied by the A-site cation. With variations in the A, B, and X ions impacting the material's electrical, optical, and stability properties, this adaptable structure permits a wide range of compositions. Figure 2.1 represents the perovskite crystal unit cell structure and structure of a generic Perovskite crystal.

Perovskite materials generally have the formula ABX_3, where "A" is a cation (such as methylammonium (MA), formamidinium (FA), or cesium (Cs)), "B" is a metal cation (like lead (Pb) or tin (Sn)), and "X" is a halide anion (usually iodide (I), bromide (Br), or chloride (Cl)). The A cation fills the gaps between the corner-sharing BX_6 octahedra that make up the crystal structure of perovskites (Tvingstedt et al. 2015; Agiorgousis et al. 2014; Homes et al. 2001; Wright et al. 2016; Viswanath et al. 1998; Asghar et al. 2024).

Because of its great flexibility, the perovskite structure permits a variety of compositions and substitutions in both the A and B sites. But because of its flexibility, the material is also more likely to develop flaws, particularly when processed improperly or when exposed to outside pressures like light or temperature changes.

Several things can lead to defects in the perovskite lattice (Asghar et al. 2024; Malek et al. 2020), such as: (a) Nonstoichiometric composition: Interstitials or vacancies result from changes in the ratio of A, B, or X ions. (b) Processing conditions: Defects during material deposition can be introduced by temperature, humidity, and solvent selection. (c) Ion migration: Since perovskites are ionic materials,

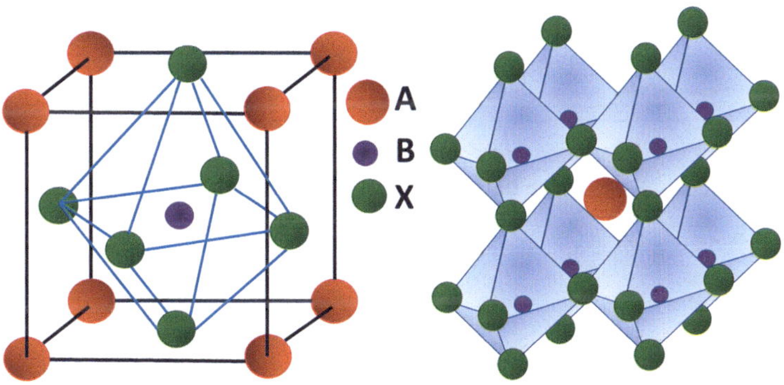

Fig. 2.1 (**a**) Perovskite crystal unit cell structure (**b**) structure of a generic perovskite crystal

defects may arise from ion migration in response to thermal excitation or applied electrical fields.

Types of Defects in Perovskite Absorbers

Dimensionality is a useful tool for classifying defects in perovskite materials. Point, line, and extended defects are a few of these. Depending on their dimensionality and type, defects in perovskite absorbers can be divided into multiple groups. Point defects, line defects, and extended defects are the main categories. The electronic structure of the material and the functionality of PSCs can be significantly impacted by these defects, which can occur during the synthesis, deposition, or operation of the perovskite layer. Figure 2.2 shows different types of defects in perovskite crystal surface. Defects concentrations can be defined as following equation:

$$N\left[X^q\right] = N_{\text{site}}.\exp\left[\frac{-E^f\left[X^q\right]}{k_B T}\right]$$

where N_{site} represents the number of available sites per volume. These sites can be occupied by the defects X. $E^f[X^q]$ represents the defect formation energy with q charge state. k_B represents Boltzmann constant and growth temperature is T (Guo et al. 2024; Zhang et al. 2020).

Point Defects

Point defects occur when individual atoms or ions are moved or eliminated from their regular locations within the lattice. In case of halide-based perovskites, point defects exist in different oxidation states. These defects are controlled by chemical

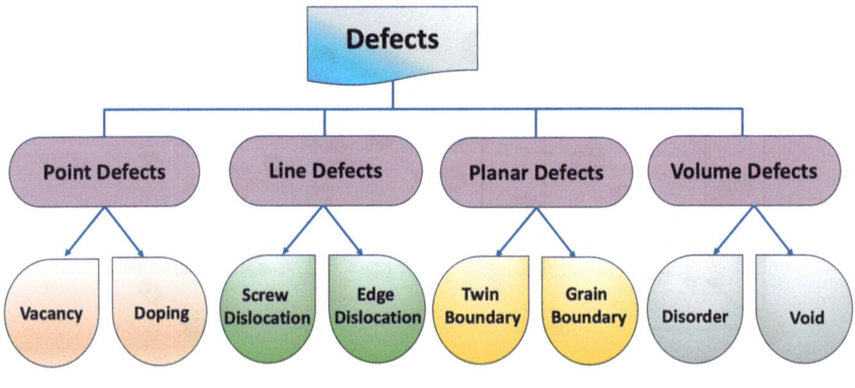

Fig. 2.2 Different types of defects in perovskite crystal surface

or electrical doping, and altered by flowing charges under ambient operating conditions. The electron–hole recombination dynamics are affected by the oxidation state. The most basic kind of defect is called point defect, which is caused by the movement or removal of a single atom or ion from its usual place within the lattice. Among these flaws are:

Vacancies: Absence of ions or atoms in the crystal lattice. Ions missing from the crystal structure are called vacancies. The A-site (such as methylammonium or formamidinium ions), the B-site (such as lead or tin cations), or the X-site (such as halide anions) of perovskites can all have vacancies. Because they have the potential to function as deep-level trap states, halide vacancies (like iodide vacancies) can trap charge carriers and cause recombination losses (Malek et al. 2020; Yu et al. 2018).

Interstitials: In general, atoms or ions that occupy sites in the lattice are unoccupied. When an atom or ion occupies an abnormal position in the lattice, interstitial defects generate. Localized strains can be introduced by these defects in the crystal and gives effect in the electronic properties of materials.

Substitutional defects: The substitutional defects are formed by replacing atoms of one element by atoms of another element within the lattice of the crystal. These defects occur when one type of atom or ion replaces another in the lattice. Such as lead (Pb) can substitute tin (Sn) at the B-site or methylammonium for formamidinium at the A-site. Overall stability and electronic properties of the material are influenced by these substitutional defects (D'Innocenzo et al. 2014; Snaith 2013; Agiorgousis et al. 2014; Malek et al. 2020).

Antisite defects: The exchange of positions between two ions of different types is termed as antisite defect, such as a halide ion occupying the B-site or a metal cation occupying the A-site.

Vacancies are particularly significant in perovskite materials. For instance, halide vacancies (such as iodide vacancies) can lead to the formation of deep-level states within the bandgap, which can act as recombination centers for charge carriers. Similarly, metal cation vacancies, such as lead vacancies, can also create trap states that reduce charge mobility (Lee et al. 2012; Viswanath et al. 1998).

Line and Planar Defects

Periodicity of the crystal along specific directions is disrupted by the line and planar defects. Misalignment of atomic planes along a particular direction is caused by line defects, which is also known as dislocation. When atoms or ions are displaced from their regular position, these defects occur. Dislocations propagate throughout the material and create areas of local strain which affects the structure and electronic properties of perovskites (Oga et al. 2014; Tvingstedt et al. 2015; Yu et al. 2018).

These include: (1) Dislocations: When the atomic planes are misaligned, local distortions in the crystal occurred. (2) Grain boundaries: Grain boundaries are the interfaces between grains in polycrystalline perovskite films. These can be a source of recombination centers that reduce the efficiency of perovskite solar cells. In particular, they are crucial in polycrystalline perovskite films, as the quality and size of the grains can prominently affect the electronic properties of materials. Charge carriers trapped by grain boundaries can lower the overall efficiency of solar cells.

Extended Defects

Extended defects affect large regions of the crystal lattice which are less localized compared to point defects. These defects include:

Stacking faults: Errors in the stacking sequence of crystal layers in the perovskite.

Twin Boundaries: Interfaces of two regions of the crystal where orientations mirrored creates twin boundaries. Twin boundaries play a role in modifying the overall electronic structure and properties of perovskite materials even if the extended defects are less common.

Identification of Defects

Optical characterization techniques such as steady-state photoluminescence (PL), time-resolved PL, and ultrafast femtosecond transient absorption spectroscopy are used to detect the defect states and their spatial position. Imaging techniques such as confocal fluorescence microscopy and scanning electron microscopy (SEM) help to investigate the local fluorescence lifetime imaging in real space at nanometer scale. This helps to modify the perovskite layer by evaluating their defect state density and effects on this layer carrier recombination. Noninvasive and label-free nonlinear optical methods such as femtosecond transient absorption microscopy and time-resolved-PL microscopy quantify and map the domain distributions of electronic excited states in perovskites. Scanning near-field optical microscopy (SNOM) and atomic force microscopy (AFM) estimate the origin of the radiative recombination center of photoexcited charge carriers in the perovskite layer. Also, computational quantum mechanical modeling method such as density functional theory (DFT), Machine Learning (ML), ATLAS help to determine defects properties of perovskites (Mahjabin et al. 2020).

Mechanism of Defect Generation and Dynamics in Perovskite Absorbers

The defects in perovskite materials are generated by various factors, such as deposition process, choice of precursors, and external conditions. Understanding the origins of these defects is important for the strategies to mitigate the formation of defects (Toshniwal et al. 2017; D'Innocenzo et al. 2014; Oga et al. 2014; Zhang et al. 2017; Tvingstedt et al. 2015; Agiorgousis et al. 2014; Homes et al. 2001; Wright et al. 2016).

Synthesis Conditions

Defects can be created during the synthesis of perovskite materials. The chemical composition of the precursors, the temperature during the synthesis and deposition, and the choice of solvents influence the creation of defects.

Nonstoichiometric composition: An alteration in the A, B, or X elements of perovskite materials during preparation of precursors can lead to vacancies or point defects in the prepared film such as an excess of lead in the solution of precursors can form the lead vacancies in the perovskite crystal materials.

Solvents: The choice of solvents during the deposition process also plays a significant role in defect creation. Solvents with high vapor pressure may evaporate too quickly during the deposition, resulting in incomplete crystallization and the formation of defects.

Annealing temperature: The degree of crystallinity and defect density are affected by the annealing temperature while the formation process of perovskite film.

Low-temperature annealing can cause the incomplete crystallization. High-temperature annealing can affect the creation of larger grain boundaries and defects due to thermal stresses.

Ion Migration and Defect Dynamics

Ion migration is the one of the important factors for the defect creation in perovskite materials. Perovskites are ionic materials. Ions can migrate in the lattice under certain conditions which creates defects such as vacancies or interstitials. Ion migration can occur due to the following reasons (Kojima et al. 2009; Oga et al. 2014).

Electric fields: Halides and metal cations can migrate through the lattice under applied external electric fields. This leads to the formation of vacancy at the A or B sites of perovskites. The materials are degraded by this migration with time, specifically in devices which needs to be operated for long term.

Thermal excitation: Ions can gain the optimum energy to migrate through the crystal lattice at higher temperature. This thermal ion migration gives result in defect formation, such as halide vacancies or metal cation interstitials. This gives effect in the performance, efficiency, and stability of the perovskite-based solar cells.

Environmental Condition

Environmental conditions such as light, oxidation, and moisture in air also contribute to the formation of defects in perovskite materials as absorbers.

Light and oxygen: Light- and oxygen-induced degradation cause perovskite materials to break down which gives formation in defects. These affect the electronic properties and are often driven by photogenerated charges or reactive oxygen derivatives. This leads to the halide vacancies formation or other defects.

Moisture: Humidity in air leads to the degradation of perovskite materials. Water causes the perovskites to degrade into non-perovskite phases by introducing defects in the material (Toshniwal et al. 2017).

Computational Methods for Defect Analysis in Perovskites

Theoretical computational tools are used to study defects in perovskite materials in recent research to help researchers understand defect formation and predict the effects of defects on various properties of materials. This guides the development of defect-mitigation strategies (Ball and Petrozza 2016; Guo et al. 2024; Malek et al. 2020; Yu et al. 2018).

Computational techniques help to study the defect generation and their behavior in perovskite absorbers. These methods provide insights into the formation energies of defects, their electronic properties, and their impact on material performance. Computational techniques have become a necessary tool to understand and mitigate defects in perovskites. Understanding and simulating the generation of defects in perovskite absorbers for perovskite solar cells (PSCs) are studied by advanced computational techniques such as Density Functional Theory (DFT), Machine Learning (ML), SCAPS simulation software, and SILVACO ATLAS. Detailed investigation of the material properties, defect behavior, and performance optimization of PSCs are performed by these tools. Below the function of these methods and how they are utilized in perovskite solar cell research are discussed in detail.

Density Functional Theory (DFT)

Density Functional Theory (DFT) is a quantum mechanical modeling approach. This is used to calculate the electronic structure of materials as well as properties of defects. DFT helps to calculate the formation energies of point defects, charge states, and interaction with charge carriers. This theoretical method gives information regarding the local electronic structure around defects. This helps to predict their impact on the optoelectronic properties of materials (Kojima et al. 2009; Xie et al. 2021; Duan and Uddin 2022; Jeng et al. 2014; Toshniwal et al. 2017; D'Innocenzo et al. 2014).

Analysis in DFT in defect studies include: (a) Calculation of energies for defect formation: Helps to calculate the energy for forming specific required defects under different conditions. (b) Analysis of electronic structure: Helps to determine the energy levels introduced by defects within the bandgap. These defects act as recombination centers. (c) Analysis of charge state: Determines how defects impact the distribution of charge in the material. These defects affect charge carrier dynamics.

Calculation of Defect Formation Energy

DFT calculation plays an important role in calculating the defect formation energy. This energy is required to create defect into an ideal crystal lattice. A defect is more likely to occur while material processing or under operational conditions when the formation energy is low (Elumalai et al. 2016; Xie et al. 2021). DFT allows for the calculation of formation energies of different types of defects such as vacancies, interstitials, and antisite defects in the perovskites. By simulating various defect configurations, the relative stability of these defects can be predicted. Moreover, their impact on the electronic properties of materials can be studied. These impacts charge transport, recombination rates, and overall cell efficiency. The formation energy depends on the environment conditions of the film during fabrication process such as concentrations of precursors, ratio of precursors, solvents, temperature, doping, etc. The formation energy can be expressed by the following equation:

$$E_f = E_v - E_{sc} + \mu_c$$

where $E_v$and E_{sc} represent the total energy of the vacancy defects and defect-free super cell structure, respectively. μc denotes the chemical potential of the material.

Effect of Defects on Electronic Properties

DFT calculations are capable of predicting how defects impact the electronic structure of perovskites. Localized states are introduced by defects in the bandgap that can trap charge carriers. This leads to higher recombination and lower charge mobility. These localized states can be deep or shallow. Deep states are specifically harmful for device performance. In the context of perovskite solar cells, DFT is particularly useful for: (1) defect formation and energetics, (2) electronic states and defect levels, (3) defect passivation. DFT helps to identify midgap states and defect-induced energy levels. These can trap charge carriers. These trapped charge carriers act as recombination centers which affects the performance of the PSCs. Defect passivation mechanisms can be explored, where the interaction of certain molecules or ions with defects lowers their harmful effects, which is crucial in optimizing material stability and device efficiency.

There are studies, which used DFT to model the formation of lead vacancies (Pb-vac) in a methylammonium lead iodide ($MAPbI_3$) perovskites. Then it helped determining the favorable Pb-vacancy formation energy and analyzed how the presence of vacancies forms defect states in the electronic structure which leads to recombination and reduces the efficiency. DFT-calculated spectrum is used to study the performance of the cell and the effect of perovskite absorber thickness, absorber dopant concentration, and defect densities at the interface of the electron transport layer (ETL). The DFT results demonstrated the agreement with the experiments and confirmed the superior performance of the $MAPbI_3$ absorber over $MAPbBr_3$ and $MAPbCl_3$. The optimum thickness, dopant concentration, and interface defect density were also investigated (Laali et al. 2020).

Molecular Dynamics (MD) Simulations

Molecular dynamics (MD) simulations are computational tool to study defect formation and migration in perovskite materials. MD simulations model the movement of atoms and ions with time. This model provides insights into the defect formation dynamics under specific conditions. By simulating at the atomic scale, researchers determine how defects evolve due to external temperature, external stress, and ion migration.

Ion Migration and Defect Evolution

MD simulation is used to understand how ions migrate within the lattice in perovskite. This leads to the formation of vacancies or interstitials. Ion migration impacts the instability of perovskites. It can result in phase transitions or the formation of harmful defects that degrade the performance of devices.

Effects of Temperature

Some perovskites are highly sensitive to temperature variation. Influence of temperature on defect formation can be estimated by the MD simulations. This technique is able to determine the defect states characteristics at high temperature indicating their thermal stability as perovskite solar cells. Grain boundary (GB) in perovskite crystal accelerates the nonradiative electron-hole recombination which leads to midgap trap states formation. By breaking the perovskite crystalline symmetry, GB creates localized states which overlap less than the conduction band and valence band. If the midgap states are absent, such localization can lower the coupling and slow down the recombination rate, which leads to efficient PSCs. The annealing of perovskite films can heal GB defects by eliminating midgap trap states. The MD simulation calculation also helps to understand the electronic structure calculations in optimized structures as well as room-temperature geometries. The electronic energy levels of the defect states can be easily altered by the external thermal energy (Li et al. 2019).

Hybrid Computational Techniques

Recent advances in computational techniques have led to the evolution of hybrid approaches that combine DFT with MD simulations. These allow for the integration of atomic-scale electronic structure calculations with large-scale dynamical simulations. This provides a more comprehensive idea of the behavior of defects such as a DFT calculation is able to determine the formation energy of a defect. An MD simulation tracks the migration of ions and defects with time. By combining these both methods, researchers are able to simulate defect formation in a real-time manner.

Hybrid Methods and Machine Learning (ML)

Machine Learning (ML) has evolved as a powerful technique to determine the defect behavior in perovskite material absorbers. These techniques are being increasingly integrated into PSCs research to increase the predictive capabilities and complex datasets analysis. By training ML models on large datasets generated through DFT or experimental studies, researchers can predict the formation energies, charge states, and electronic properties of defects more efficiently. Hybrid methods by combining DFT and MD simulations are increasingly being used to gain a more comprehensive understanding of defect behavior. Again, machine learning (ML) techniques are being employed to determine defect formation and guide the design of perovskite materials with fewer defects. Regression ML models and neural networks are used to accelerate the discovery of defect-resistant perovskite compositions and to optimize material processing conditions. These methods hold great promise for reducing the time and computational cost of defect studies (Malek et al. 2020).

ML is particularly helpful in defect prediction and classification. ML techniques can quickly predict the defect formation energies in perovskite absorbers. This helps to identify the most disadvantage types that may limit device performance. These techniques can use large datasets of known defect structures and properties for the prediction of new defect states in emerging perovskite material absorbers. The DFT computation calculations for defect generation can be expensive. This is more costly when screening many different defect configurations or materials. To make the defect analysis more efficient and scalable ML techniques, such as neural networks and regression models, can be trained to approximate DFT results. By leveraging ML tools, researchers also identify optimal material compositions for perovskite absorbers which can minimize defect formation as well as improves the long-term stability and efficiency of perovskite solar cells. This technique helps to develop novel materials by identifying correlations between defect characteristics and performance of solar cell. Researchers use ML models to know the defect-related phenomena in mixed cation perovskite materials. By training on DFT-calculated defect states in similar materials, the ML tool is able to reveal different cations (formamidinium and methylammonium). They affect the defect formation and electronic properties which accelerates the discovery of new stable and efficient perovskite absorbers.

SCAPS (Solar Cell Capacitance Simulator)

Solar cell capacitance simulator (SCAPS) technique is a widely used for the simulation of electrical characteristics of thin-film perovskite solar cells. Drift-diffusion model is used in SCAPS to simulate defect generation, charge carrier transport, recombination, and other phenomena. SCAPS allows the introduction of defects in the simulation of perovskite solar cells. Here defects lead to charge carrier recombination which affects the efficiency of charge collection, efficiency, and overall device performance. SCAPS can reveal the effects of different defect densities and types of defects (e.g., deep or shallow traps) on the current-voltage characteristics of the solar cell.

SCAPS reveals how defect-induced states impact the electronic transport properties of perovskite layers. The effect of varying defect densities in the open-circuit voltage (Voc), short-circuit current (Jsc), and fill factor (FF) in PSCs can be studied by this technique. This also confirms that the inclusion of deep-level defect states in the perovskite absorber significantly lowers the Voc due to increased recombination of carriers. SCAPS is performed with primary focus to facilitate defect passivation at the interface, with an organic passivation layer to mitigate the effect on Voc and recombination rate.

Researchers also investigated novel solar cells using $CaZrSe_3$, $BaZrSe_3$, $CaHfSe_3$, and $BaHfSe_3$ absorbers through SCAPS. The performance of the fabricated solar cells has been evaluated by varying the thickness, carrier concentration, and defect density of the electron transport layer (ETL), absorbers, and hole transport layer (HTL). By tuning the ETL carrier concentration, deep energy levels were formed at the interfaces. Light absorption is increased by optimizing the thickness of absorbers and carrier concentration. This also improved conductivity and built-in potential, which facilitates charge carrier transfer. This enhancement was due to the increased electric field at the perovskite absorber/HTL interface, as it effectively lowered the barrier for charge carriers (Srinivasan et al. 2025). Few researchers gave the lead-free perovskite solar cell (PSC) heterostructure using $CH_3NH_3SnI_3$ ($MASnI_3$) absorber layer significant attention due to their superior device performance. The power conversion efficiency was optimized by tuning the bandgap and electron affinity variations in perovskite absorber layer. Thereafter, the capacitance-voltage, Mott-Shottkey, and conductance-voltage measurements are performed through SCAPS simulation. This also confirmed a strong indication of deep defects in heterostructure perovskite solar cells (Kumar et al. 2022).

SILVACO ATLAS

SILVACO ATLAS is another strong simulation technique commonly used in device modeling. It helps researchers to model the physical behavior of materials used in perovskite solar cells. Figure 2.3 demonstrates ATLAS inputs and outputs configuration. ATLAS can simulate the carrier transport modeling, which simulates carrier transport within the perovskite absorber and other layers of the solar cell. This also demonstrates the effects of defects on mobility and recombination. ATLAS allows for the detailed modeling how defects can significantly lower the charge carrier mobility by creating trap states that impede carrier flow. ATLAS can also introduce defect states within the band structure of PSCs. By modeling the impact of specific defects (e.g., vacancies, interstitials, or grain boundaries) on carrier dynamics by simulation, ATLAS can simulate how defects can alter the performance. This can occur by reducing the open-circuit voltage (Voc) or increasing recombination losses. SILVACO ATLAS is also used to optimize perovskite solar cell architectures, taking into account the formation of defects, layer properties, and their interaction with

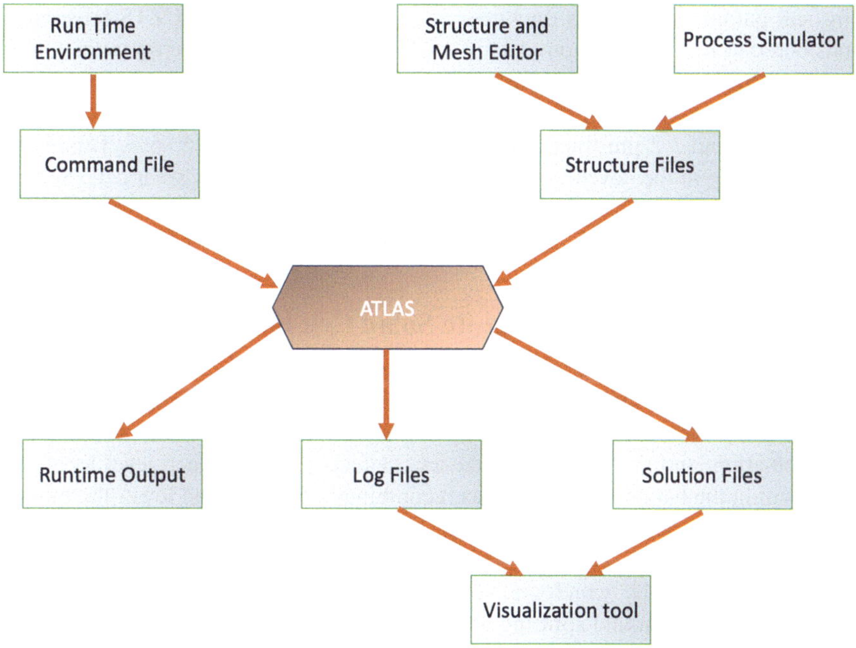

Fig. 2.3 Demonstration of ATLAS inputs and outputs configuration

among themselves. This technique helps to identify the most efficient perovskite absorber material and designs. This improves both the performance and long-term stability of the devices. ATLAS simulation of a perovskite solar cell might show that the presence of grain boundary defects in the perovskite layer leads to significant carrier recombination at the grain boundaries. This results in decrease of the fill factor (FF). The simulation could explore how lowering the defect density by surface passivation techniques and improved fabrication processes help in removing this issue.

Some researchers explained that the most prominent deficiencies of perovskite solar cells (PSCs) are their low efficiency and poor stability. In order to improve the productivity, they used a combination of silicon and perovskite. Scientists presented a 3D simulation analysis of various electrical and optical properties of PSCs using the SILVACO simulation software. Even without using silicone composites, just by using the inverted planar method with inorganic transport materials and the proper selection of anti-reflective coatings with a back contact layer, they were able to increase the efficiency to 28.064% and enhance the stability. Materials, such as CaF_2, SiO_2, and Al_2O_3, with various thicknesses have been studied to know the effect of anti-reflective coatings and to improve the efficiency of the PSC (Fakhri et al. 2021).

Other researchers studied the performance of Cu_2ZnSnS_4 (CZTS) chalcogenide-based solar cell. The FTO/ZnO:Co/CZTS heterostructure was simulated using the SILVACO-ATLAS 2D simulation. They deposited Cu_2ZnSnS_4 (CZTS), ZnO:Co, and FTO layers on glass substrates by the spray pyrolysis technique. The extracted physical parameters such as thickness and optical properties of CZTS have been considered to form a new input data of CZTS-based solar cell. The optimization of CZTS electron affinity and thickness was performed to have the best efficiency as well as performance. The use of CZTS perovskite absorber layer with 3.99 eV electron affinity and 3.2 μm thick has led to higher efficiency of 16.86%. This plays significant role in the development of new technologies and new solar cell (Seboui and Dabbabi 2023).

Impact of Defects on Perovskite Solar Cell Performance

The defects in the perovskite material absorber layer have direct effect on the performance of perovskite-based solar cells. These defects can impact charge transport, recombination, and performance of device. Defects can form deep-level localized states within the bandgap. These act as recombination centers and lower the mobility of charge carriers (Ono et al. 2020; Homes et al. 2001; Wright et al. 2016; Viswanath et al. 1998). The working mechanism of perovskite solar cells are demonstrated in four paths, which are, (1) photons absorption and excitons formation, (2) excitons diffusion and splitting, (3) charge transmission, (4) charge collection. Upon incident of sunlight on perovskite surface, this promotes the absorption of photons excitons. Due to the difference of binding energies, the exons form electrons and holes when the excitons get separated. At the interface between the hole transport layer (HTL) and charge transport layer (CTL), the exciton separates. Electrons get separated from the holes and move to the electron transport layer (ETL), mostly the anode substrate. At this time, holes are mainlined to the hole transport layer (HTL) and moved to the cathode which is metal electrode. This metal electrode collects the carriers (electrons and holes), and the electrode is connected to external circuit for current generation (Kheralla and Chetty 2021).

Charge Recombination and Defect States

Defects in perovskite materials are able to introduce deep-level states within the bandgap. These states act as recombination centers. Electrons and holes combine and reduce the number of charge carriers for current generation. Vacancies and antisite defects are particularly destructive because they tend to form deep traps. They are highly effective at recombining charge carriers. In case of PSCs, trap-assisted recombination is highly dominant recombination and limits their efficiency. The primary trap-assisted recombination channels (grain boundaries and interfaces) and their connection to defect ions are attributed in PSCs. When an electron or hole trapped in a defect recombines with a hole or electron in the valence (or conduction) band of the perovskite, nonradiative recombination takes place (Sherkar et al. 2017).

Most of the defects form at grain boundaries and surfaces in perovskite absorbers. These defects influence kinetics of charge recombination, ion migration, and degradation in the PSCs (Fig. 2.4). This also inevitably impacts their efficiency and stability. The reduction of defects in perovskite solar cells lowers the nonradiative recombination centers and improves the efficiency and long-term operational stability (Chen and Zhou 2020).

Charge Transport and Mobility: Reduced Charge Carrier Mobility

Defects can also impede charge transport by creating localized energy states that trap charge carriers. This can reduce the overall charge mobility within the material with the hinder of current flow through the solar cell. Point defects are particularly impactful in this regard.

Defects also prevent the movement of charge carriers through the material by lowering their mobility. Vacancies or interstitials point defects act as scattering centers for charge carriers, hindering their ability to flow through the perovskite layer. Figure 2.5 shows perovskite solar cell and defect states in perovskite materials.

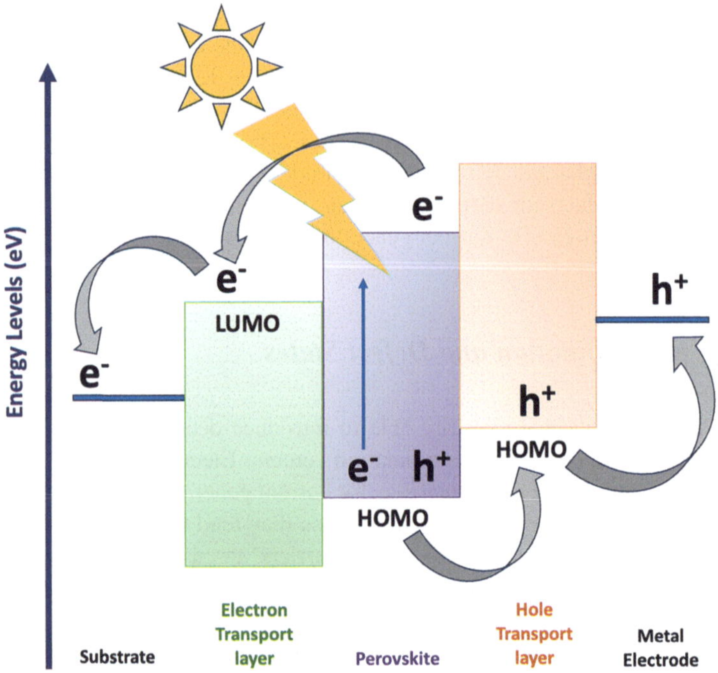

Fig. 2.4 Band diagram and working principle of perovskite solar cell

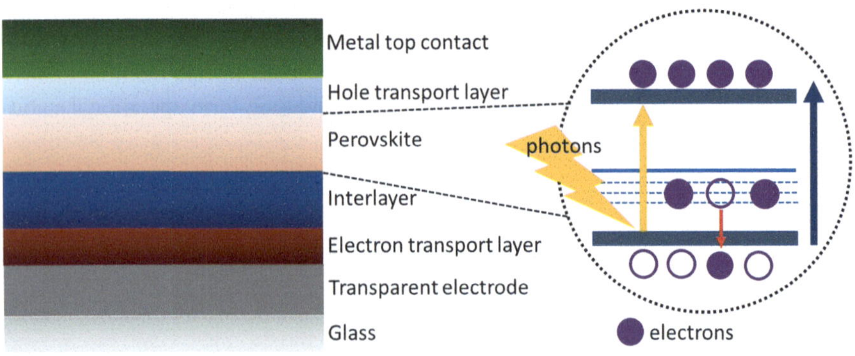

Fig. 2.5 Perovskite solar cell and defect states in perovskite materials

Device Efficiency and Lifetime: Decreased Stability

The effect of defects on device efficiency extends beyond the immediate effects on charge transport property. Defects can also reduce the critical parameters like open-circuit voltage and fill factor of perovskite-based solar cells (PSCs). Moreover, the stability of PSCs is compromised in the presence of defects. As ion migration and defect formation can lead to degradation over time. Defects also contribute to the degradation of perovskite materials under operational conditions (Guo et al. 2024).

Mitigating Defects in Perovskite Absorbers

Material Composition and Engineering

One of the most fundamental strategies to mitigate defects in perovskite absorbers is through careful composition engineering. By tuning the ratio of A, B, and X elements, researchers can lower the formation of harmful defects. The mixed halide perovskites improve stability and reduce halide vacancies. Incorporating doping elements or additives can passivate defect sites and enhance charge transport (Kojima et al. 2009; Feurer et al. 2019; Jeng et al. 2014; Toshniwal et al. 2017; Snaith 2013; Xiao et al. 2015; Tvingstedt et al. 2015; Agiorgousis et al. 2014; Homes et al. 2001; Wright et al. 2016; Viswanath et al. 1998; Guo et al. 2024; Malek et al. 2020; Yu et al. 2018). Figure 2.6 shows path toward industrialization of perovskite solar cells from lab scale to market by controlling the defects and enhancing the stability and performance of the device. The main aim is to develop highly efficient, stable, cost-effective, and large-scale PSCs. The primary challenges of the industrialization of PSCs and technological limitations, application constraints are under research for the sustainable development of large-scale devices. For this, the perovskite absorber should be reproducible and high-quality material. Combining material science, materials chemistry and characterization, and device engineering, high stability can be achieved in perovskite solar cell devices.

Passivation and Additives

Passivation is another important approach for the mitigation of effects of defects. Surface passivation techniques involve the addition of small molecules or inorganic salts that bind to defect sites and prevent charge trapping. These passivation agents

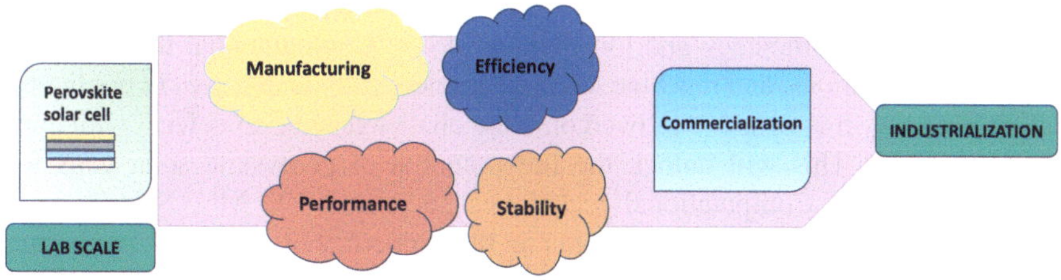

Fig. 2.6 Path toward industrialization of perovskite solar cells from lab scale to market by controlling the defects and enhancing the stability and performance of the device

help to improve the charge carrier dynamics with lowering recombination losses. Machine learning (ML) models and computational simulations are rapidly used to predict the most effective passivation techniques for specific defects. Selection of precursor, additive engineering plays a crucial role in fabrication of PSCs as they affect the kinetics of crystallization, quality, and morphology of film to control the device performance (Lee et al. 2019; Yuan et al. 2024; Zhao et al. 2023; Tan et al. 2023; Ding et al. 2024; Zhang and Park 2023; Wu et al. 2024; Jiang et al. 2023; Li et al. 2023; Chen and Park 2018). The passivation of defects bears immense significance in recombination minimization and enhancement of the photovoltaic performance of PSCs. In general, perovskite absorbers have low defect formation energy according to thermodynamics (Yin et al. 2014; Noel et al. 2014). Most defects exist within the shallow energy levels near the band edges which exhibits electrical activity (Buin et al. 2014). Again, antisite and interstitial defects exist at deeper electronic levels, functioning as nonradiative recombination centers which can damage the performance and efficiency (Buin et al. 2015). Among all the possible available methods, for passivation of defects, additive engineering has emerged as a remarkably efficient strategy. The best approach for defect passivation involves leveraging Lewis acid-base interactions for passivation of uncoordinated ions. The integration of Lewis acid and base additives into perovskite precursors can effectively lower the nonradiative recombination centers and defects near grain boundaries (Wang et al. 2020; Zhang et al. 2018; Yang et al. 2020).

Conclusion

Defects in perovskite material absorbers remain one of the biggest challenges in the development of highly efficient and stable perovskite solar cells. DFT, MD simulations, and machine learning computational modeling methods provide valuable insights into the formation, behavior, and effect of defects. Researchers develop strategies to mitigate their effects and improve the performance and stability of PSCs. Defects arise and influence material properties. The future of devices based on perovskite material absorbers depends on continued progress in defect analysis and the development of innovative defect-resistant materials. Defects in perovskite absorbers play a significant role in the efficiency and stability of perovskite-based solar cells. Understanding the sources, types, and influence of defects is necessary for improving the efficiency and longevity of PSCs. Computational tools such as DFT, MD simulations, and machine learning are necessary techniques in predicting defect behavior. It is possible to overcome the challenges due to defects with continued research. This will unlock the full potential of perovskite solar cells for source of energy. Computational techniques such as DFT, ML, SCAPS, and SILVACO ATLAS provide deep insights on the properties of defects in perovskite absorbers. But when they are used as combined technique, their true potential is unlocked. DFT gives the knowledge about defect energetics, formation, and interactions at the atomic scale. ML demonstrates the prediction and classification of defects. This helps to explore a wide range of material compositions and defect orientations. SCAPS and SILVACO ATLAS simulate the effect of defects on performance and efficiency which allows the optimization of design of perovskite solar cells. By integrating these computational tools, defect property can be predicted, perovskite absorbers can be optimized, and highly efficient and stable solar cells can be designed, with the focus on advancing the future of perovskite solar technologies. This approach will also play a significant role in impacting global energy landscape. In conclusion, the integral potency of PSCs paves the way for diverse applications which are stable under various environments. Enhancement of efficiency, performance, long-term stability, upscaling, and sustainability remains important for the successful fabrication into a wide array of PSCs. Finally, achievement of the sustainable development of PSCs is also essential. These will help in widespread adoption of clean energy, environment, and the role of renewable energy in the global energy shift.

References

Agiorgousis ML, Sun Y-Y, Zeng H et al (2014) Strong covalency-induced recombination centers in perovskite solar cell material $CH_3NH_3PbI_3$. J Am Chem Soc 136(41):14570–14575. https://doi.org/10.1021/ja5079305

Akbulatov AF, Frolova LA, Griffin MP et al (2017) Effect of electron-transport material on light-induced degradation of inverted planar junction perovskite solar cells. Adv Energy Mater 7(19):1700476. https://doi.org/10.1002/aenm.201700476

Asghar U, Qamar MA, Hakami O et al (2024) Recent advances in carbon nanotube utilization in perovskite solar cells: a review. Micromachines 15(4):529. https://doi.org/10.3390/mi15040529

Ball JM, Petrozza A (2016) Defects in perovskite-halides and their effects in solar cells. Nat Energy 1(11):1–13. https://doi.org/10.1038/nenergy.2016.149

Barbé J, Tietze ML, Neophytou M et al (2017) Amorphous tin oxide as a low-temperature-processed electron-transport layer for organic and hybrid perovskite solar cells. ACS Appl Mater Interfaces 9(13):11828–11836. https://doi.org/10.1021/acsami.6b13675

Behrouznejad F, Shahbazi S, Taghavinia N et al (2016) A study on utilizing different metals as the back contact of $CH_3NH_3PbI_3$ perovskite solar cells. J Mater Chem A Mater 4(35):13488–13498. https://doi.org/10.1039/C6TA05938D

Buin A, Pietsch P, Xu J et al (2014) Materials processing routes to trap-free halide perovskites. Nano Lett 14:6281–6286. https://doi.org/10.1021/nl502612m

Buin A, Comin R, Xu J et al (2015) Halide-dependent electronic structure of organolead perovskite materials. Chem Mater 27:4405–4412. https://doi.org/10.1021/acs.chemmater.5b01909

Chen J, Park NG (2018) Causes and solutions of recombination in perovskite solar cells. Adv Mater 31:1803019. https://doi.org/10.1002/adma.201803019

Chen Y, Zhou H (2020) Defects chemistry in high-efficiency and stable perovskite solar cells. J Appl Phys 128:060903. https://doi.org/10.1063/5.0012384

D'Innocenzo V, Grancini G, Alcocer MJP et al (2014) Excitons versus free charges in organo-lead tri-halide perovskites. Nat Commun 5(1):3586. https://doi.org/10.1038/ncomms4586

Ding B, Ding Y, Peng J et al (2024) Dopant-additive synergism enhances perovskite solar modules. Nature 628:299–305. https://doi.org/10.1038/s41586-024-07228-z

Duan L, Uddin A (2022) Defects and stability of perovskite solar cells: a critical analysis. Mater Chem Front 6(4):400–417. https://doi.org/10.1039/D1QM01250A

Elumalai NK, Mahmud MA, Wang D et al (2016) Perovskite solar cells: progress and advancements. Energies 9(11):861–881. https://doi.org/10.3390/en9110861
Fakhri N, Naderi MS, Farkoush SG et al (2021) Simulation of perovskite solar cells optimized by the inverse planar method in SILVACO: 3D electrical and optical models. Energies 14(18):5944. https://doi.org/10.3390/en14185944
Feurer T, Carron R, Sevilla GT et al (2019) Efficiency improvement of near-stoichiometric CuInSe2 solar cells for application in tandem devices. Adv Energy Mater 9(35):1901428. https://doi.org/10.1002/aenm.201901428
Galagan Y (2018) Perovskite solar cells: toward industrial-scale methods. J Phys Chem Lett 9(15):4326–4335. https://doi.org/10.1021/acs.jpclett.8b01356
Guo Z, Yuan M, Chen G et al (2024) Understanding defects in perovskite solar cells through computation: current knowledge and future challenge. Adv Sci 11:2305799. https://doi.org/10.1002/advs.202305799
Homes CC, Vogt T, Shapiro SM et al (2001) Optical response of high-dielectric-constant perovskite related oxide. Science 293(5530):673–676. https://doi.org/10.1126/science.1061655
Jeng J, Chiang YF, Lee MH et al (2014) Methylammonium lead iodide perovskite/fullerene-based hybrid solar cells. SPIE Newsroom 10(2):005033. https://doi.org/10.1117/2.1201307.005033
Jiang X, Zhang B, Yang G et al (2023) Molecular dipole engineering of carbonyl additives for efficient and stable perovskite solar cells. Angew Chem Int Ed 62:e202302462. https://doi.org/10.1002/anie.202302462
Jung EH, Jeon NJ, Park EY et al (2019) Efficient, stable and scalable perovskite solar cells using poly (3-hexylthiophene). Nature 567(7749):511–515. https://doi.org/10.1038/s41586-019-1036-3
Kheralla A, Chetty N (2021) A review of experimental and computational attempts to remedy stability issues of perovskite solar cells. Heliyon 7(2):06211. https://doi.org/10.1016/j.heliyon.2021.e06211
Kim J, Lee SH, Lee JH et al (2014) The role of intrinsic defects in methylammonium lead iodide perovskite. J Phys Chem Lett 5:1312–1317. https://doi.org/10.1021/jz500370k
Kim M, Kim G-H, Oh KS et al (2017) High-temperature-short-time annealing process for high-performance large-area perovskite solar cells. ACS Nano 11(6):6057–6064. https://doi.org/10.1021/acsnano.7b02015
Kojima A, Teshima K, Shirai Y et al (2009) Organometal halide perovskites as visible-light sensitizers for photovoltaic cells. J Am Chem Soc 131(17):6050–6051. https://doi.org/10.1021/ja809598r
Kumar M, Raj A, Kumar A et al (2022) Computational analysis of bandgap tuning, admittance and impedance spectroscopy measurements in lead-free MASnI3 perovskite solar cell device. Int J Energy Res. https://doi.org/10.1002/er.7942
Laali J, Hamedani A, Alahyarizadeh G et al (2020) Performance analysis of the perovskite solar cells by a realistic, DFT-accurate optical absorption spectrum. Superlattice Microst 143:106551. https://doi.org/10.1016/j.spmi.2020.106551
Lee TD, Ebong AU (2017) A review of thin film solar cell technologies and challenges. Renew Sust Energ Rev 70:1286–1297. https://doi.org/10.1016/j.rser.2016.12.028
Lee MM, Teuscher J, Miyasaka T et al (2012) Efficient hybrid solar cells based on meso-superstructured organometal halide perovskites. Science 338(6107):643–647. https://doi.org/10.1126/science.1228604
Lee DK, Jeong DN, Ahn TK et al (2019) Precursor engineering for a large-area perovskite solar cell with 19% efficiency. ACS Energy Lett 4:2393–2401. https://doi.org/10.1016/j.enrev.2022.100010
Li X, BI D, Decoppet J-D et al (2016) A vacuum fash–assisted solution process for high-efficiency large-area perovskite solar cells. Science 353(6294):58–62. https://doi.org/10.1126/science.aaf8060
Li W, Long R, Tang J et al (2019) Influence of defects on excited-state dynamics in Lead halide perovskites: time-domain ab initio studies. J Phys Chem Lett 10(13):3788–3804. https://doi.org/10.1021/acs.jpclett.9b00641

Li G, Su Z, Canil L et al (2023) Highly efficient pin perovskite solar cells that endure temperature variations. Science 379:399–403. https://doi.org/10.1126/science.add7331

Liu Y, Lu R, Zhang J et al (2021) Construction of a gradient type 2D/3D perovskite structure for subsurface passivation and energy level alignment of an $MAPbI_3$ film. J Mater Chem A 9(46):26086–26094. https://doi.org/10.1039/D1TA07537C

Mahjabin S, Haque M, Sobayel K et al (2020) Perceiving of defect tolerance in perovskite absorber layer for efficient perovskite solar cell. IEEE Access 8:106346–106353. https://doi.org/10.1109/ACCESS.2020.3000217

Malek NAA, Alias N, Umar AA et al (2020) Enhanced charge transfer in atom-thick 2H–WS2 nanosheets' electron transport layers of perovskite solar cells. Sol RRL 4(10):2000260. https://doi.org/10.1002/solr.202000260

Nayak PK, Mahesh S, Snaith HJ et al (2019) Photovoltaic solar cell technologies: analysing the state of the art. Nat Rev Mater 4:269–285. https://doi.org/10.1038/s41578-019-0097-0

Noel NK, Abate A, Stranks SD et al (2014) Enhanced photoluminescence and solar cell performance via Lewis base passivation of organic–inorganic lead halide perovskites. ACS Nano 8:9815–9821. https://doi.org/10.1021/nn5036476

Oga H, Saeki A, Ogomi Y et al (2014) Improved understanding of the electronic and energetic landscapes of perovskite solar cells: high local charge carrier mobility, reduced recombination, and extremely shallow traps. J Am Chem Soc 136(39):13818–13825. https://doi.org/10.1021/ja506936f

Ono LK, Liu S, Qi Y (2020) Reducing detrimental defects for high-performance metal halide perovskite solar cells. Angew Chem Int Ed 59(17):6676–6698. https://doi.org/10.1002/anie.201905521

Prasanna JL, Goel E, Kumar A (2023) Efficient modeling of double absorber layered structure in perovskite solar cells using machine learning techniques. Phys Scr 98(10):106002. https://doi.org/10.1088/1402-4896/acf535

Ran C, Xu J, Gao W et al (2018) Defects in metal triiodide perovskite materials towards high-performance solar cells: origin, impact, characterization, and engineering. Chem Soc Rev 47(12):4581–4610. https://doi.org/10.1039/c7cs00868f

Seboui Z, Dabbabi S (2023) First investigation on CZTS electron affinity and thickness optimization using SILVACO-Atlas 2D simulation. Simul Model Pract Theory 126:102758. https://doi.org/10.1016/j.simpat.2023.102758

Sherkar TS, Momblona C, Gil-Escrig L et al (2017) Recombination in perovskite solar cells: significance of grain boundaries, Interface traps, and defect ions. ACS Energy Lett 2(5):1214–1222. https://doi.org/10.1021/acsenergylett.7b00236

Simmons JG, Taylor G (1971) Nonequilibrium steady-state statistics and associated effects for insulators and semiconductors containing an arbitrary distribution of traps. Phys Rev B 4(2):502–511. https://doi.org/10.1103/PhysRevB.4.502

Snaith HJ (2013) Perovskites: the emergence of a new era for low-cost, high-efficiency solar cells. J Phys Chem Lett 4(2):3623–3630. https://doi.org/10.1021/jz4020162

Srinivasan D, Chettiar AR, Mercy ENV et al (2025) Scrutinizing the untapped potential of emerging ABSe3 (A=Ca, Ba; B=Zr, Hf) chalcogenide perovskites solar cells. Sci Rep 15:3454. https://doi.org/10.1038/s41598-024-80473-4

Srivastava S, Singh AK, Kumar P et al (2021) Comparative performance analysis of lead-free perovskites solar cells by numerical simulation. J Appl Phys 131:175001. https://doi.org/10.1063/5.0088099

Stranks SD (2017) Nonradiative losses in metal halide perovskites. ACS Energy Lett 2:1515–1525. https://doi.org/10.17863/CAM.12818

Tan Q, Li Z, Luo G et al (2023) Inverted perovskite solar cells using dimethylacridine-based dopants. Nature 620:545–551. https://doi.org/10.1038/s41586-023-06207-0

Tian J, Xue Q, Yao Q et al (2020) Inorganic halide perovskite solar cells: progress and challenges. Adv Energy Mater 10(23):2000183. https://doi.org/10.1002/aenm.202000183

Toshniwal A, Jariwala A, Kheraj V et al (2017) Numerical simulation of tin based perovskite solar cell: effects of absorber parameters and hole transport materials. J Nano Electron Phys 9(3):1–4. https://doi.org/10.21272/jnep.9(3).03038

Tress W (2017) Perovskite solar cells on the way to their radiative efficiency limit—insights into a success story of high open-circuit voltage and low recombination. Adv Energy Mater 6:1602358. https://doi.org/10.1002/aenm.201602358

Tvingstedt K, Malinkiewicz O, Baumann A et al (2015) Radiative efficiency of lead iodide based perovskite solar cells. Sci Rep 4(1):6071. https://doi.org/10.1038/srep06071

Viswanath AK, Lee JI, Kim D et al (1998) Excitonphonon interactions, exciton binding energy, and their importance in the realization of room-temperature semiconductor lasers based on GaN. Phys Rev B Condens Matter 58(24):16333. https://doi.org/10.1103/PhysRevB.58.16333

Wang F, Bai S, Tress W et al (2018) Defects engineering for high-performance perovskite solar cells. NPJ Flexible Electron 2:22. https://doi.org/10.1038/s41528-018-0035-z

Wang S, Wang A, Deng X et al (2020) Lewis acid/base approach for efficacious defect passivation in perovskite solar cells. J Mater Chem A 8:12201–12225. https://doi.org/10.1039/D0TA03957H

Wright AD, Verdi C, Milot RL et al (2016) Electron–phonon coupling in hybrid lead halide perovskites. Nat Commun 7:11755. https://doi.org/10.1038/ncomms11755

Wu Z, Sang S, Zheng J et al (2024) Crystallization kinetics of hybrid perovskite solar cells. Angew Chem Int Ed 63:e202319170. https://doi.org/10.1002/anie.202319170

Xiao Z, Yuan Y, Shao Y et al (2015) Giant switchable photovoltaic effect in organometal trihalide perovskite devices. Nat Mater 14(2):193–197. https://doi.org/10.1038/nmat4150

Xie H, Wang Z, Chen Z et al (2021) Decoupling the effects of defects on efficiency and stability through phosphonates in stable halide perovskite solar cells. Joule 5(5):1246–1266. https://doi.org/10.1016/j.joule.2021.04.003

Yang WS, Park B-W, Jung EH et al (2017) Iodide management in formamidinium-lead-halide–based perovskite layers for efficient solar cells. Science 356(6345):1376–1379. https://doi.org/10.1126/science.aan2301

Yang Z, Jinjuan D, Kou S et al (2020) Multifunctional phosphorus-containing Lewis acid and base passivation enabling efficient and moisture-stable perovskite solar cells. Adv Funct Mater 30:1910710. https://doi.org/10.1002/adfm.201910710

Yin WJ, Shi T, Yan Y (2014) Unusual defect physics in CH3NH3PbI3 perovskite solar cell absorber. Appl Phys Lett 104:063903. https://doi.org/10.1063/1.4864778

Yu M, Qu W, Xu S et al (2018) Interfacial stability, electronic property, and surface reactivity of α-MoO_3/γ-Al_2O_3 composites: DFT and DFT + U calculations. Comput Mater Sci 153:217–227. https://doi.org/10.1016/j.commatsci.2018.06.046

Yuan L, Chen X, Guo X et al (2024) Volatile perovskite precursor ink enables window printing of phase-pure FAPbI3 perovskite solar cells and modules in ambient atmosphere. Angew Chem Int Ed 63:e202316954. https://doi.org/10.1002/anie.202316954

Zhang Y, Park NG (2023) Tailoring the functionality of additives for enhancing the stability of perovskite solar cells. Adv Funct Mater 33:2308577. https://doi.org/10.1002/adfm.202308577

Zhang F, Yang B, Li Y et al (2017) Extra long electron–hole diffusion lengths in $CH_3NH_3PbI_3$–xClx perovskite single crystals. J Mater Chem C 5(33):8431–8435. https://doi.org/10.1039/C7TC02802D

Zhang F, Bi D, Pellet N et al (2018) Suppressing defects through the synergistic effect of a Lewis base and a Lewis acid for highly efficient and stable perovskite solar cells. Energy Environ Sci 11:3480–3490. https://doi.org/10.1039/C8EE02252F

Zhang X, Turiansky ME, Walle CG (2020) Correctly assessing defect tolerance in halide perovskites. J Phys Chem C 124(11):6022–6027. https://doi.org/10.1021/acs.jpcc.0c01324

Zhao J, Furer SO, McMeekin DP et al (2023) Efficient and stable formamidinium-caesium perovskite solar cells and modules from lead acetate-based precursors. Energy Environ Sci 16:138–147. https://doi.org/10.1039/D2EE01634F

Chapter 3
Interface Engineering in Perovskite Solar Cells

Vishal Singh, Vani Pawar, Ashish K. Ranjan, Manish Kumar, Ravi P. Jaiswal, and Prabhakar Singh

Abstract Organic-inorganic halide perovskite solar cells (PSCs) have achieved remarkable power conversion efficiency (PCE) of up to 26.1%, comparable to traditional silicon solar cells. However, their PCE remains below the Shockley-Queisser limit, and they suffer from poor intrinsic stability under ambient conditions. Recent studies have demonstrated that both the PCE and stability of PSCs can be significantly enhanced through careful material selection and interface engineering during fabrication. Interface engineering has emerged as a promising strategy to improve PCE and intrinsic stability by facilitating efficient carrier extraction at interfaces, reducing trap states, and mitigating undesirable degradation pathways. This chapter will discuss the following aspects like energy level alignment, interfacial defects, carrier dynamics, mitigation of ion migration, and obstacle for moisture penetration and electrode diffusion. Additionally, the discussion covered various types of materials used as interlayers at different interfaces, including electrode/ETL, ETL/perovskite layer, perovskite layer/HTL, and HTL/electrode interfaces. Furthermore, it explored interfacial engineering strategies for flexible perovskite solar cells (PSCs) and provided a future outlook for this rapidly evolving field.

V. Singh · A. K. Ranjan · P. Singh (✉)
Department of Physics, Indian Institute of Technology (Banaras Hindu University), Varanasi, India
e-mail: psingh.app@iitbhu.ac.in

V. Pawar
Centre for Nano Science and Engineering, Indian Institute of Science, Bengaluru, Karnataka, India

M. Kumar
Department of Physics, School of Sciences, IFTM University, Moradabad, Uttar Pradesh, India

R. P. Jaiswal
Department of Chemical Engineering, Indian Institute of Technology (Banaras Hindu University), Varanasi, India

M. Kumar (ed.), *Computational Analysis of Perovskite Solar Cells*, Springer Tracts in Electrical and Electronics Engineering,
https://doi.org/10.1007/978-981-95-6521-4_3

Keywords Perovskite · Interfaces · Energy level alignment · Defects · Carrier dynamics · Interface engineering

Introduction

Our society is advancing rapidly through the integration of cutting-edge technologies such as electric vehicles, robotics, mobile devices, and computers. However, the practical manufacturing and application of these technologies require energy in various forms, including heat and electricity. Currently, the majority of our energy comes from fossil fuels like coal and natural gas, which are finite resources. The combustion of these fuels not only depletes their reserves but also contributes significantly to greenhouse gas emissions and public health issues. To address these challenges, extensive research is being conducted to find sustainable substitutes for fossil fuels in the form of renewable energy sources. Among eco-friendly energy resources—such as solar, wind, biomass, and geothermal—solar energy has emerged as the best potential because of its pollution-free nature and virtually limitless availability (Celik et al. 2017; Maka and Alabid 2022). Solar energy is converted into electricity through devices (photovoltaic effect), generally recognized as solar cells (SCs). These SCs have evolved through multiple generations. The first-generation SCs, made from mono- and poly-crystalline silicon (Si), achieved power conversion efficiency (PCE) approximately 26.7% and remained stable for up to 26 years. However, the high-temperature manufacturing process, along with the silicon waste generated during cutting, drives up costs, making this technology less accessible to economically disadvantaged populations. In response, the second-generation SCs, such as copper indium gallium selenide (CIGS), cadmium telluride (CdTe), and amorphous silicon cells, were evolved. CIGS SCs have reached the highest PCE of 23.4%. Despite their improved affordability, these second-generation cells suffer from shorter lifetimes, limiting their long-term viability. To overcome these limitations, third-generation SCs have been introduced. These include organic, dye-sensitized (DSSCs), quantum dots, multi-junction, and perovskite (PSCs)-based SCs. Among these, perovskite solar cells (PSCs) have attracted a lot of scientific interest due to the low cost of perovskite materials and their remarkable optical and electrical properties such as adjustable bandgap ranging from 1.3 to 2.2 eV, long charge diffusion lengths, high carrier mobility, and a high absorption coefficient (5.7×10^4 cm^{-1} at 600 nm). The PCE of PSCs has surged from 3.8% to 26.1% in only 14 years, as illustrated in Fig. 3.1a and summarized in Table 3.1, positioning them as a strong competitor to traditional Si-based SCs. However, their efficiency is significantly less than the theoretical limit of 33%, predicted by Shockley-Queisser (SQ) for a semiconductor having a bandgap nearly equal to 1.34 eV. This efficiency gap is primarily due to defect-induced trap states, suboptimal surface contact, and misalignment of energy levels at interfaces between layers, which lead to different types of recombination (Kant and Singh 2022; Li et al. 2024; Noman et al. 2024).

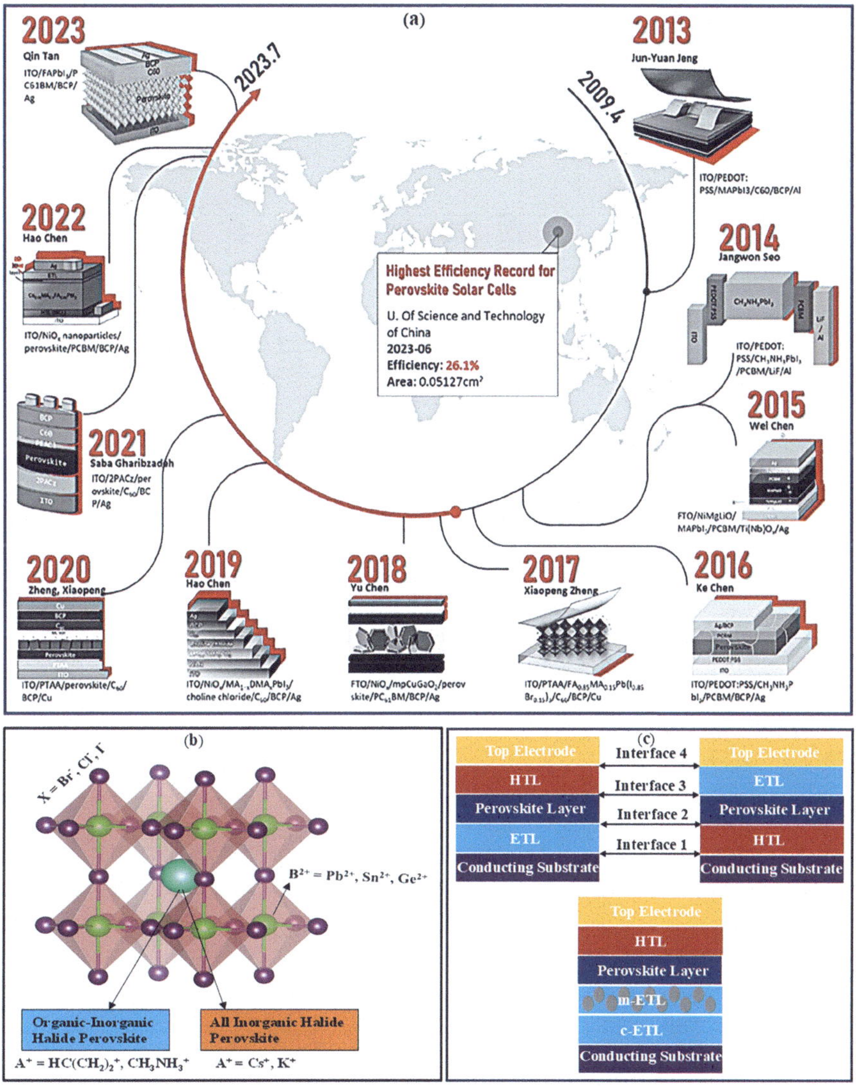

Fig. 3.1 (**a**) Representing PCE achieved by PSCs in different years (Li et al. 2024). Copyright 2024. (Reproduced with permission from American Chemical Society). (**b**) Crystallographic structure of perovskite (ABX_3) materials and (**c**) conventional n-i-p, inverted p-i-n, and conventional mesoporous n-i-p structure, respectively

The fundamental structure of perovskite materials is represented by the expression ABX_3, where A = $HC(CH_2)_2$ + (FA+), $CH_3NH_3^+$ (MA^+), Cs^+, or K^+; B = Pb2$^+$, Sn2$^+$, or Ge2$^+$; and X = Br^-, I^-, or Cl^-, shown in Fig. 3.1b. Halide perovskite's

Table 3.1 Performance data of various PSCs given in Fig. 3.1a (Li et al. 2024). Copyright 2024. Reproduced with permission from American Chemical Society)

Device configuration									
Conducting substrate	HTL	Perovskite layer	ETL	Top electrode	FF	J_{sc} (mA cm^{-2})	V_{oc} (V)	PCE (%)	Year of publication
ITO	PEDOT:PSS	$MAPbI_3$	C_{60}/BCP	Al	63	10.32	0.6	3.9	2013 (Jeng et al. 2013)
ITO	PEDOT:PSS	$CH_3NH_3PbI_3$	PCBM/LiF	Al	78.30	20.7	0.866	14.1	2014 (Seo et al. 2014)
FTO	NiMgLiO	$MAPbI_3$	PCBM/ Ti(Nb)O*x*	Ag	75	20.21	0.72	16.2 (15)	2015 (Chen n.d.)
ITO	PEDOT:PSS	$CH_3NH_3PbI_3$	PCBM/ BCP	Ag	82	22.38	1.02	18.72	2016 (Chen et al. 2016)
ITO	PTAA	$FA_{0.85}MA_{0.15}Pb(I_{0.85}Br_{0.15})_3$	C_{60}/BCP	Cu	79.1	16.46	1.13	20.59	2017 (Zheng et al. 2017)
FTO	NiO*x*/mpCu GaO_2	Perovskite	$PC_{61}BM$/ BCP	Ag	79.96	22.23	1.132	20.13	2018 (Chen et al. 2018a)
ITO	NiO*x*	$MA_{1-x}DMA_xPbI_3$/choline Chloride	C_{60}/BCP	Ag	82	23.5	1.116	21.6	2019 (Chen et al. 2019a; Zheng et al. 2020)
ITO	PTAA	$Cs_{0.05}(FA_{0.92}MA_{0.08})_{0.95}Pb(I_{0.92}Br_{0.08})_3$	C_{60}/BCP	Cu	82.0	23.9	1.14	22.34	2020 (Zheng et al. 2020)

Device configuration									
Conducting substrate	HTL	Perovskite layer	ETL	Top electrode	FF	J_{sc} (mA cm^{-2})	V_{oc} (V)	PCE (%)	Year of publication
ITO	2PACz	$Cs_{0.18}FA_{0.82}PbI_3$	C_{60}/BCP	Ag	83.2	23.4	1.161	22.3	2021 (Gharibzadeh et al. 2021; Chen et al. 2022)
ITO	NiO*x* nanoparticles	$Cs_{0.05}MA_{0.1}FA_{0.85}PbI_3$	PCBM/BCP	Ag	83.46	24.90	1.15	23.91	2022 (Chen et al. 2022)
ITO	HTL free	$FAPbI_3$	$PC_{61}BM$/BCP	Ag	84.73	25.69	1.187	25.86	2023 (Tan et al. 2023)

crystallographic structure and stability can be forecast based on the Goldschmidt tolerance factor (t) and octahedral factor (μ); the expression is given below:

$$t = \frac{R_X + R_A}{\sqrt{2}\left(R_X + R_B\right)} \tag{3.1}$$

$$\mu = \frac{R_B}{R_X} \tag{3.2}$$

where R_A, R_B, and R_X represent ionic radii of A, B, and X, respectively. If the values of t and μ will exist in the $0.81 < t < 1.11$ and $0.44 < \mu < 0.90$ range, then halide perovskite will form a 3D-crystal structure. However, according to certain reports, a perfect cubic structure will form when the value of t falls within the more constrained range of 0.89–1.0, but if the value of t falls outside of this range, a less symmetric tetragonal or orthorhombic structure may be formed (Zhou and Zhao 2019). The architecture of PSCs based on the arrangement of layers has been classified into conventional n-i-p and inverted p-i-n configurations. In both configurations, the perovskite active layer is positioned between the electron transport layer (ETL) and hole transport layer (HTL). In a conventional n-i-p configuration, ETL is deposited on a transparent conductive oxide substrate (Indium or Fluorine-doped tin oxide), through which solar energy is transmitted. This is followed by the deposition of the perovskite layer, HTL, and finally, the back electrode. The inverted p-i-n configuration follows the opposite sequence, starting with the HTL, then the perovskite layer, ETL, and back electrode. Conventional n-i-p can also be categorized as either mesoporous or planar, based on whether a mesoporous scaffold is present. The mesoporous layer promotes effective charge transport within the cell and offers structural support while the perovskite film forms. In both the n-i-p and p-i-n configurations, there are four critical interfaces: (1) electrode/ETL interface, (2) the ETL/perovskite layer interface, (3) the perovskite/HTL interface, and (4) the HTL/electrode interface, shown in Fig. 3.1c These interfaces play a vital role in charge extraction and injection when the solar cells are operated, significantly influencing the overall performance and efficiency. The efficiency of carrier extraction and injection in cells depends on quality of these interfaces. The effectiveness of charge transfer and the rate of carrier recombination at these interfaces are largely determined by the energy level alignment and defect states. Where defect states, such as ionic defects at these interfaces, can lead to substantial charge accumulation and recombination, which negatively impacts the open-circuit voltage (V_{oc}) and fill factor (FF), ultimately reducing overall device performance (Wang et al. 2018a). Moreover, these ionic defects not only affect the electrical performance but also compromise the operational durability of PSCs by accelerating device degradation (Tress 2017; Tress et al. 2018). Therefore, to achieve stable, high-performance

PSCs, it is essential to engineer interfaces with optimal energy band alignment, strong interfacial contact, and minimal defect density.

To date, extensive research has focused on reducing defect states, improving interfacial contact between layers, and optimizing energy band alignment to ensure that PSCs achieve high efficiency along with maintaining it for a longer time (Liu et al. 2021a). Solutions to these interfacial issues are necessary for the commercialization of PSCs, as their long-term viability depends on delivering high efficiency, durability, and low fabrication and maintenance costs (Bi et al. 2016). Previous research aimed at enhancing the efficiency of PSCs has primarily focused on optimizing perovskite film technology and employing passivation techniques to mitigate bulk and grain boundary defects. This includes advancements in composition tuning, additive engineering, and processing methods (Ke et al. 2016; Sherkar et al. 2017). More recently, significant advancements have been made in interface engineering to increase PSC's stability and efficiency. By reducing nonradiative recombination at the interfaces between different layers, inhibiting the undesirable movement of hetero-charge carriers, and improving adhesion between the top and bottom layers, interface engineering has emerged as an effective strategy for achieving superior device performance. This chapter will begin by examining the pivotal role of interfacial engineering in PSCs, focusing on key aspects such as interfacial charge dynamics, energy level alignment, defect passivation, suppression of ion migration, and the development of barriers against moisture ingress and electrode diffusion. Subsequently, the discussion will shift to interlayer materials employed at various PSC interfaces, emphasizing their contribution to achieving stability and enhanced performance in PSCs.

Importance of Interfacial Engineering

Mostly in functioning PSCs, multiple recombination processes can occur simultaneously, including radiative, Auger, and Shockley-Read-Hall (SRH) recombination. Among these, Auger and SRH are classified as non-radiative recombination (NRR). In PSCs, SRH recombination typically dominates NRR, which arises from carrier recombination via dangling bonds, defects traps, and other imperfections. Auger recombination, on the other hand, has a negligible impact. Deep-level defects, often located at mid-gap, are particularly detrimental as they trap carriers, enhancing the possibility of NRR. However, shallow defects located closer to valence or conduction band edges have a minimal effect on NRR. The loss in open-circuit voltage (V_{oc}) due to NRR in PSCs (under AM1.5G conditions) is captured by the following equation (Pazos-Outón et al. 2018):

$$V_{\text{oc,SHR}} = V_{\text{oc,rad}} + \frac{K_B T}{q} \ln(\text{PLQY}) \tag{3.3}$$

In this equation, KB, T, and q represent the Boltzmann constant, temperature, and elementary charge, respectively, and PLQY refers to the photoluminescence quantum yield. The V_{oc} resulting from SRH recombination ($V_{oc,SRH}$) is derived when both SRH and radiative recombination are considered, while $V_{oc,rad}$ refers to the V_{oc} when only radiative recombination is involved. PLQY is strongly connected to NRR in SCs. According to Eq. (3.3), a high PLQY indicates low NRR and, consequently, enhanced device performance. Thus, an increase in V_{oc} can be achieved through interface engineering, which mitigates deep-level defects and reduces NRR. Short-circuit current density (J_{sc}) of solar cell depends on its quantum efficiency (QE), which is determined by carrier dynamics and light absorption. Apart from defect-mediated carrier losses and optical inefficiencies within the perovskite layer, factors such as interface-driven carrier losses and energy-level misalignment at the interfaces also play a critical role in determining carrier extraction efficiency and overall photovoltaic performance. Interface engineering is a simple and effective strategy to mitigate these interfacial losses without compromising the bulk properties of the perovskite layer. Further, modifying the interfaces can enhance device stability by providing protection against degradation. In the following discussion, we will explore the specific roles and benefits of interface engineering in addressing these challenges.

Energy Level Alignment

When a system is formed by connecting two semiconductors, the law of thermodynamics and Anderson's model state that the system must have a single align fermi level (Shi et al. 2015). So, the charge carriers (electrons and holes) of both semiconductors start to diffuse into each other by leaving behind positive or negative ion cores near the interface and continue this diffusion until the fermi level of both semiconductors becomes aligned and proper band bending takes place at the interface. At equilibrium, a layer of positive and negative ion cores—known as the depletion zone or junction—will form on both sides. This junction can also be formed when two semiconductors like p-type and n-type of the same material like Ge or a different material or a semiconductor and metal come in contact. From Fig. 3.2a we can see, in the case of a homogenous p-n junction, a smooth band bending formed at the interface; however, a cliff or notch in energy becomes visible at the interface in the case of the heterogenous p-n junction. In Fig. 3.2b, c, this cliff or notch in energy arises at interface due to difference in electron affinity of both semiconductors. When the cliff is positive energy, then the transfer of electrons from n-type to p-type semiconductor encounter energy barrier, due to which start to accumulate near the interface (Pan et al. 2021). In case of PSCs, there is large number of point defects present at the interface of the ETL/Perovskite layer, which facilitates the path to recombe accumulated electrons with holes (Stolterfoht et al. 2019). Figure 3.2c illustrates when there is no barrier at the interface. Then C.B electrons and V.B holes of p-type can easily go in C.B and V.B of n-type

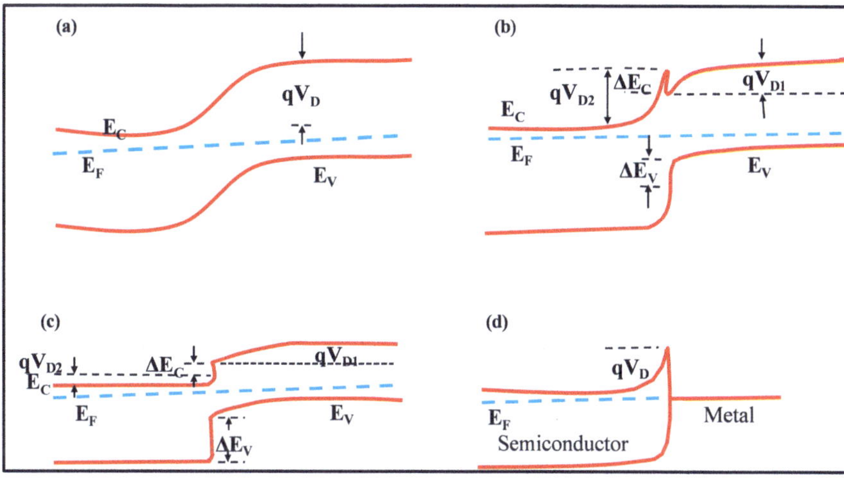

Fig. 3.2 Energy level alignment at interfaces, (**a**) p–n homojunction, (**b**) p–n heterojunction with a positive or (**c**) a negative conducting band offset (ΔE_C), and (**d**) a Schottky junction

semiconductor, respectively. But on the other side, the energy gap between the bottom of C.B n-type and the top of V.B of p-type is small that's why there can be a high probability that C.B electrons of n-type can cross the forbidden region and recombine with V.B holes of p-type semiconductor (Cho et al. 2018a). Figure 3.2d shows the setup of energy levels at Schottky contact which is formed between n-type semiconductor and metal. The energy level from the semiconductor bends upward at Schottky contact. This bending occurs, because the energy of the charge carrier near to semiconductor surface is higher than bulk's carriers. This Schottky contact will appear between indium or fluorine-doped tin oxide (ITO or FTO) and ETL in PSCs. At such type of interface, C.B electrons of ETL face an energy barrier and do not have smoothly collected by an electrode. Then short-circuit current that flows in the outer circuit will decrease and thus performance of PSCs decreases. So, the strategy of energy level alignment at interfaces in electrode/ETL/Perovskite layer/HTL/electrode device is a very important aspect for improving the PCE and stability of PSCs. By tuning energy level alignment at these interfaces, carrier extraction and injection can be boosted and, in that way, energy losses can be minimized. In recent years, interfacial engineering has become an essential technique for adjusting the energy level alignment and successfully decreases the energy obstacle, as a result, lowers the NRR at interfaces of PSCs. For example, to see the effect of this engineering, Lisha Xie et al. made a thin film of small organic molecule adenine on the NiO_x surface. They got the energy level of NiO_x reduced by 0.1 eV and increased grain size and crystallinity of the perovskite layer. Thereby charge extraction and transportation at the interface increased and then V_{oc} increased. This engineering also increased the stability against light exposure and moisture (Xie et al. 2020; Wu et al. 2023).

Defect Passivation

In a real crystal, the structure is influenced by factors like how the crystal grows and any processes it undergoes afterward. However, individual atoms acquire a fixed location within an ideal crystal. Perovskite films, which are complex poly-crystalline materials, contain multiple grain boundaries, adding to their complexity (Wu et al. 2023). During the process of preparing a film, many uncontrollable factors affect how crystals form and grow, leading to different defects both inside and on the surface of the film. These defects can be categorized based on their structural features (Min et al. 2024). For instance, in perovskite films, intrinsic defects are usually point defects, which happen when atoms are arranged in reverse, occupy incorrect spaces in the lattice, or are missing from their usual positions. Figure 3.3 shows different types of defects like vacancies, interstitials, and antisite defects. These point defects are further categorized as either shallow or deep energy level defects, depending on their position within the bandgap (Hassan et al. 2022). Shallow energy level defects, which have low formation energy, are usually caused by vacancies or interstitial defects. These defects are located near the upper edge of the valence band or the lower edge of the conduction band (Isikgor et al. 2023). Deep-level energy defects, which require significant energy to form, act as trap centers that capture holes and electrons, facilitating non-radiative carrier losses. These defects are generally situated near the mid-gap region of the band structure (Yin et al. 2014). In general, shallow energy level defects do not trap charge carriers or significantly

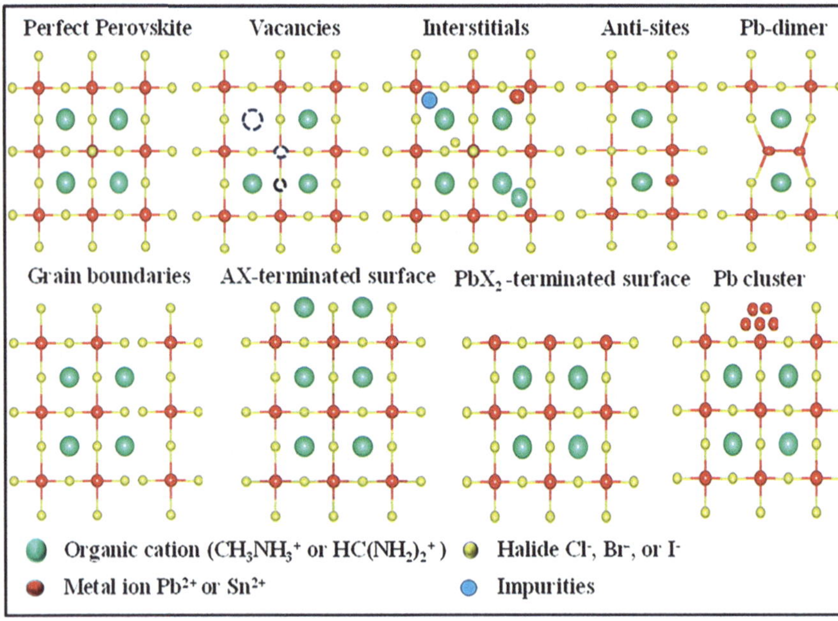

Fig. 3.3 Types of defects in perovskite

disrupt charge transport, so they have minimal impact on the device's immediate performance. However, when an electric field is applied, these defects can move, leading to phase separation and hysteresis, which can negatively affect the device's long-term durability (Zhang et al. 2022). Deep energy-level defects serve as the primary drivers of NRR, which can lead to significant voltage loss (V_{oc}), energy loss, and a decrease in power conversion efficiency (PCE). These issues ultimately reduce the overall performance of the device (Stranks 2017). Apart from the point defects, there exist high dimensional defects including 3D grain defects, 2D surface defects, and 1D dislocation defects. While exterior faults dispersed on the grain surface and at grain borders are sometimes referred to as 2D defects, localized atomic dislocations in crystals typically create 1D defects (Huang et al. 2024). Typically, 3D defects include issues like pores, contaminants, and cracks. These defects lead to high series resistance and increased surface roughness in the device, which in turn hampers performance by creating bad contact at interfaces between the perovskite layer and the charge transport layer (Akin et al. 2020). At the electrode, charge collection efficiency is directly influenced by the alignment of energy levels and the quality of the interface contact between the perovskite and the charge transport layer (CTL). Research shows that interface engineering—by managing chemical interactions at the interface, such as chemical bonding and electronic coupling—can effectively passivate defects, improve band alignment, reduce charge accumulation, and enhance film quality. This ultimately boosts the photovoltaic performance of devices (Xia et al. 2023). For instance, defects at or near the TiO_2/perovskite interface were passivated by an ultrathin layer of [6,6]-phenyl-C_{61}-butyric acid methyl ester (PCBM) and PMMA mixture, which considerably reduced interfacial recombination and improved V_{oc} by up to 80 mV (Peng et al. 2017).

Carrier Dynamics

Carrier dynamics including charge generation, separation, diffusion, recombination, and collection can be supposed as the heartbeat of PSCs during operation. These dynamics directly influence the performance of PSCs. When the perovskite layer absorbs incident photons having energy equal to or greater than its bandgap, then a charge carrier will be generated; lifetime of these generated carriers depends on the nature of the perovskite layer such as the lifetime scale for poly and single crystalline perovskite layer are 10–100 ns and >1000 ns, respectively (Shi et al. 2015, 2018). The generated carriers move through the perovskite layer, where some of them recombine. The remaining carriers reach the interface between the charge transport layer (CTL) and the perovskite layer. Here, the junction force helps extract them into the CTL, from where they are collected by the electrode, as shown in Fig. 3.4. However, some carriers recombine at the interface due to defects, reducing the short-circuit current density (Shi et al. 2015; Yang et al. 2018). Besides this, the high electrons and hole mobility of ETL and HTL are also essential for the smooth extraction of charge carriers at the interface. So, to enhance the performance of

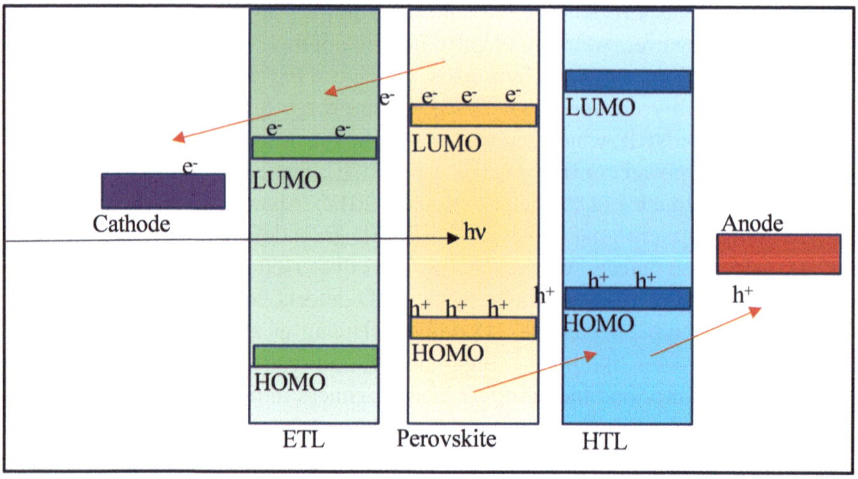

Fig. 3.4 Working mechanism of PSCs

PSCs there is a need for interfacial modification to minimize interfacial defects and grow large grain size and crystalline perovskite layer (Rajagopal et al. 2018). In these days, interfacial engineering has been emerging as a handleable route to minimize defects and boost carrier dynamics. High conductive functional groups found in interlayers can promote advantageous interfacial charge dynamics and reduce carrier losses that could result from interfacial dipoles. For instance: Zhu et al. injected graphene ink into SnO_2 and reported defect density decreases along with carrier extraction enhance compare to controlled SnO_2 at the interface between perovskite layer/SnO_2 (Zhu et al. 2019). To increase the hole's transpotation at the perovskite layer/HTL interface in inverted PSCs. Ru et al. utilized such molecules having controlled electron affinity to modify the interface between HTL/perovskite layer. That significantly enhanced the HTL conductivity and increased the V_{oc} and FF. Consequently, increased PCE of PSCs has been achieved 22.1% (Ru et al. 2020).

Mitigation of Ion Migration

The electro-ionic conducting and flexible intrinsic ionic crystal structure nature of perovskite generates the phenomenon of ion migration in whole PSC devices (Liu et al. 2021b). Basically, ion migration is originated by different types of point imperfections such as I and MA vacancies within $MAPbI_3$ perovskite. To see the movement of I_2 in device, permeation technique was first used by Joachim Maier and co-workers, in which high I_2 pressure and copper (Cu) substrate beneath the thin film of $MAPbI_3$ has been used. They attributed I^- ions move much faster toward Cu substrate from $MAPbI_3$ film uder the light shining and forms CuI, resulting in electron-ion interactions (Eames et al. 2015; Kim et al. 2018). In PSCs, ion migration frequently results in phenomena like J-V hysteresis, optical soaking, and the gradual post-illumination loss of V_{oc}. It also significantly impairs the PSCs' long-term operational stability by causing phase segregation and amorphization, which are early signs of device deterioration (Di Girolamo et al. 2020). Many attempts have been made to identify the steric hindrance of ion migration. Interfacial interlayers containing different molecules are thought to be one of the successive methods among various strategies. Yan and colleagues, for example, showed that phenylethyl ammonium iodide (PEAI) surface passivation considerably raised the activation energy needed for ion migration inside the perovskite. Such treatments improve V_{oc} and FF by reducing the buildup of charge-induce defects and suppressing ion migration (Chen et al. 2020a). These results highlight how vital it is to manage ion migration through focused interfacial engineering in order to improve PSCs stability and effectiveness.

Obstacles for Moisture Intrusion and Electrode Diffusion

The main obstacle to PSC commercialization is long-term stability because hybrid perovskites are soft natured and structurally unstable (Park and Seok 2019). Water and oxygen are two examples of external elements that can significantly impact the properties and functionality of perovskite absorber layers and other CTLs (Zhang et al. 2019). While straightforward device encapsulation can significantly reduce the harmful effects of oxygen and moisture, it cannot lessen electrode diffusion (such as that of Au and Ag) within the PSCs (Domanski et al. 2016). Furthermore, moisture and oxygen still enter the encapsulating materials at a slow pace. When the encapsulants have aged and degraded in the operating field, hydrophobic interlayers can act as secondary moisture intrusion barriers, improving PSCs' humidity stability even more. Modifying the interface is now recognized as a legitimate tactic to increase device lifespan by blocking undesirable degradation pathways. The interfacial layer that has been implanted can serve as a barrier to shield the delicate layer from harm. Polydimethylsiloxane (PDMS) chemical cross-linking at the perovskite/CuSCN interface has been shown by Park et al. to improve CuSCN-based PSC stability and lessen interfacial deterioration (Kim et al. 2019). Compared to tiny organic molecules, the PDMS interlayer significantly increased the device's resistance to heat and humidity. To improve device stability, low dimensional perovskite has also been employed (Grancini et al. 2017).

Interface

ETL/Perovskite Layer

In PSCs, the ETL and the perovskite layer interface pose significant challenges for energy level alignment, defect passivation, electron extraction and injection, and device stability. It also significantly impacts the morphology of perovskite films in PSCs with normal structures, whereas, in PSCs with inverted structures, it prevents hole transmission (Balis et al. 2018). To increase the stability and performance of the device, several interface modification techniques have been implemented. Fullerene and its derivatives, metal oxides, halides, zwitterionic compounds, and MXene have been the best options for altering the ETL/perovskite interface up to now (Zhao et al. 2018; Qin et al. 2019; Liu et al. 2019a; Palei et al. 2023). PSCs with a p-i-n structure commonly use fullerene or its derivatives, such as C_{60} and PCBM, as ETLs (Zhu et al. 2016). Moreover, fullerene and its functional group-containing derivatives have been employed as surface modifiers at the interface of ETL and perovskites (Peng et al. 2017; Wang et al. 2016; Zhong et al. 2019). These interface techniques are efficient ways to enhance device stability, diminish trap-assisted charge recombination, and promote interfacial charge transfer (Chen et al. 2019b; Cao et al. 2019). A variety of fullerene-based monomers and dimers with excellent electrical characteristics were developed by Wang et al. for the ETL/perovskite interface. With the least amount of hysteresis and remarkable stability, the chloro-terminated fullerene dimer NMBF-Cl had the greatest PCE of 22.3% among these derivatives, maintaining over 21% PCE after 1000 h under ambient environments (Wang et al. 2020a). Since metal oxides have good charge mobility, suitable band alignment, long-term stability, and low cost, they make excellent ETLs for PSCs (Shin et al. 2019). To increase PSC stability and efficiency, some metal oxides can also be added as interlayers at the ETL/perovskite interface (Ma et al. 2017). At the TiO_2/perovskite interface, metal oxides with a broad bandgap, including Al_2O_3, MgO, SnO_2, and ZrO_2, have been used to inhibit charge recombination (Ma et al. 2017; Marin-Beloqui et al. 2016). More potential for metal oxides to enhance interfacial transport in PSCs has been revealed by recent publications (Ding et al. 2018; Wang et al. 2019). Liang et al. achieved higher V_{oc} and PCE by introducing an ultra-thin Al_2O_3 interfacial layer between the mesoporous TiO_2 (mp-TiO_2) film and the perovskite layer. This was attributed to the inhibition of oxygen vacancy formation on the TiO_2 surface under UV light and the reduced surface states at TiO_2/$CH_3NH_3PbI_3$($MAPbI_3$) interfaces. The former greatly increases the device's work stability, while the latter successfully lowers the carrier recombination rate by improving the electron extraction and injection property of the TiO_2/$MAPbI_3$ interface (Liang et al. 2019). It is well known that ion defects and vacancies at the ETL/perovskite interface promote non-radiative recombination (Chen et al. 2019c). At this contact, metal halides can effectively passivate both kinds of defects due to their diverse cationic and anionic constituents (Li et al. 2016a). Furthermore, introducing metal halides has been demonstrated to improve stability over time against exposure to heat, light, and humidity. Interestingly, metal halides including KCl, CsBr, NaCl, and CsCl have been used to improve the performance of devices (Li et al. 2016a, b; Yang et al. 2019a; Wang et al. 2018b; Liu et al. 2018). At the SnO_2/perovskite interface, Wang et al. showed that a KCl passivation layer could successfully passivate both positive and negative charge defects (Wang et al. 2018b). K^+ and Cl^- diffusion into the perovskite film occurred during the thermal annealing procedure. An active passivation effect would come from the diffusion of K^+ and Cl^- stopping when they approached volume defects. PSCs' average PCE went up from 16.62% to 17.81%. Following KCl therapy, the ideal PCE was 19.44%. One type of material that has been frequently used as a surface modification in PSC applications is zwitterionic molecules, which have covalently bound cations and anions on the same molecules (Chen et al. 2019b). Because of their special capacity to create interfacial dipoles, zwitterionic compounds can effectively improve charge injection and extraction (Li et al. 2022). It creates a chemical contact with perovskite and ETL by means of its cationic and anionic groups, which passivates interface defects (Zheng et al. 2017; Hou et al. 2018). In conventional planar PSCs, tin oxide (SnO_2) has become a preferred ETL material. In contrast to mesoporous structures, SnO_2-based devices frequently suffer from hysteresis, which usually results in a lower PCE. In order to improve the SnO_2 film coverage and promote perovskite crystal formation, Park et al. chemically linked 3-(1-pyridine)1-propyl sulfonate to SnO_2 at the SnO_2/perovskite interface (Choi et al. 2018). Importantly, the redesigned ETL's interfacial dipoles provide a conductivity boost that improves charge extraction, modifies the ETL's work function, and prevents electron backflow. MXenes are known for their exceptional qualities, especially at interfaces, and have shown promise as materials for a range of applications. They are formidable rivals for closing the gap between perovskite and ETLs due to their conductive nature and versatile surface properties. MXenes promote improved charge transfer between the perovskite layer and the ETL because of their highly conductive 2D structure (Nie et al. 2023). By reducing recombination losses, this capacity could eventually increase the efficiency of solar cell systems. Functionalization enables fine adjustment of the surface characteristics of MXenes by accurately tailoring their surface. This flexibility strengthens layer adhesion, lowers energy barriers, and promotes effective charge extraction, all of which contribute to improving interfacial characteristics. Agresti et al. built a perovskite/ETL interface and adjusted the WF of the perovskite absorber in 2019 using 2D $Ti_3C_2T_x$ MXene (T_x = –O, –OH, and –F). It was found that the WF of the perovskite was raised by the O-terminated surface of MXene and decreased by the OH-terminated surface. The literature reports this nonlinear tendency (Di Vito et al. 2020). Moreover, MXene-doped perovskites showed an increase in light absorption. The WFs of halide perovskite and TiO_2 layers can be adjusted by adding $Ti_3C_2T_x$ without altering the other electronic properties. In comparison to the un-doped perovskite absorber layer (15.6%), the PSC with the MXene-doped perovskite active layer likewise shows a higher PCE (17.4%). Furthermore, the dipole that Ti_3C_2Tx induces at the perovskite/ETL interface can be used to change the band alignment between these layers. Together, WF tuning and interface engineering enable MXene-modified PSCs to achieve a 20% PCE with a V_{oc} of 1.09 V, a J_{sc} of 23.82 mA cm^{-2}, an FF of 77.6%, and a decrease in hysteresis when compared to reference cells free of MXene (Agresti et al. 2019).

HTL/Perovskite Interface

It is commonly known that cavity extraction and retarded carrier complexation occur at the interface between the HTL and perovskite layer. At this interface, efficiency, stability, and the improvement of both are the primary topics of study. The HTM's hole mobility and the interface contact between the HTL and the perovskite layer are primarily what determines efficiency. More significantly, the HTL has an impact on the perovskite film's morphology. Today, bis(4-phenyl)(2,4,6-trimethylphenyl)amine (PTAA), PEDOT:PSS, and the inorganic NiOx are the three primary HTL materials that are frequently utilized. Each of them has pros and cons, and none is ideal for the HTL, particularly since organics are expensive and not suitable for commercial preparation on a wide scale. Therefore, additional research is required to find the alternative way to improve the band alignment and remove defects at the perovskite/HTL interface to enhancing hole extraction and cell stability (Li et al. 2024). Different types of materials such as insulating polymers, conjugates polymers, 2D perovskites, organic ammonium salts, small organic molecules, metal salts, quantum dots, and carbon-based materials have been used as interfacial material at HTL/Perovskite Interface. According to reports, adding an insulating polymer to the perovskite/HTL interface strengthens device stability and inhibits defect-induced recombination. However, the polymer layer needs to stay ultrathin since it is insulating (Bi et al. 2016). The usefulness of a thin PMMA layer at the HTL/perovskite border was shown by Zhang et al. The PMMA layer was used as a powerful bridge to improve the perovskite/perovskite interface in addition to controlling the crystal development and shape of perovskite films (Zhang et al. 2018). By using this interfacial strategy, hysteresis was reduced, PCE was improved, FF was conserved, and V_{oc} was increased. Peng et al. developed a novel spatial arrangement of the passivation layer by using a nanostructured ETL to provide localized charge transport routes across the passivated interface in order to solve the inherent low electrical conductivity of such insulating polymers (Peng et al. 2021). In the quest for long-lasting and effective PSCs, this integration of a nano-ETL with a strong HTL produced an impressive certified PCE of 21.6% on a device area of 1 cm^2 with an FF of 84%, retaining 92% of its initial PCE even after 1000 h in hot, humid conditions. Conjugated polymers' higher electronic conductivity and film-forming properties give them a clear edge over insulating polymers at the perovskite/HTL interface. Conjugated polymers can promote charge tunneling, lower carrier recombination, and improve device stability, making them effective interfacial modification layers in contrast to insulating polymers that hinder carrier transfer (Li et al. 2018a, 2019). Yang et al. emphasized the effectiveness of PTQ10, a cheap conjugated polymer that was added to the perovskite/HTL interface. By significantly reducing the volatilization of volatile organic cations during thermal annealing, this PTQ10 layer promoted the directed development of perovskite crystals. This improved the hole extraction as well as the alignment of the energy levels. Notable thermal stability and an outstanding PCE of 21.2% and FF of 0.82% were attained by the resultant devices (Meng et al. 2018). Since 2D perovskites include large, hydrophobic organic cations, they have been developed to overcome the limiting moisture and heat stability of 3D perovskites (Li et al. 2018b). However, because of their poorer absorption, higher exciton binding energy, and anisotropic charge transport, 2D perovskites cannot fully attain high performance when embedded within 3D perovskites (Gao et al. 2018). The ideal solution to this challenge would be to precisely manage a conformal 2D perovskite layer on top of 3D perovskite in order to create a 2D/3D structure. The high efficiency of 3D perovskite and the stability of 2D perovskite are both retained by the 2D perovskite layer, which in fact serves as an interface modification layer (Kim et al. 2020; Liu et al. 2020). To create a 2D/3D perovskite layer, Wang et al. and Nazeeruddin et al. both used PEA2PbI4 capping layers. Although the stability of the device was greatly enhanced by the inclusion of PEA2PbI4 at the perovskite/HTL interface, the photovoltaic characteristics are still inferior to those of the 3D PSCs (Cho et al. 2018b; Chen et al. 2018b). Liu et al. added 2-(4-fluorophenyl) ethyl ammonium iodide (FPEAI), a fluorinated PEAI, to the 3D perovskite/HTL interface in order to increase efficiency. FPEAI served as a capping layer in this work to shield the 3D perovskite from ambient air moisture. It also promoted charge transfer at the perovskite/HTL interface, which decreased non-radiative recombination in the resulting film (Zhou et al. 2019). Pentafluorophenylethylammonium (FEA) was used as a fluoroarene cation in another fluorinated PEAI to develop a 2D perovskite layer on top of 3D perovskite. The interfacial layer was given an ultrahydrophobic character by the perfluorinated benzene, which inhibited ionic diffusion and shielded the perovskite from ambient moisture. The interface layer also enhanced hole injection and reduced harmful charge collecting losses. Thus, a PCE of more than 22% was attained by the 2D/3D device (Liu et al. 2019b). The study demonstrated that 2D/3D perovskite layers can be used to create operationally stable PSCs that are resistant to heat and water for future uses. Small molecules and polymers typically act as surface modifiers for perovskite/HTM, whereas organic components will make the film more hydrophobic and stop oxygen and water from diffusing over the surface (Xia et al. 2023). A certain functional group also coordinates with the perovskite, passivates the trap states, and provides the appropriate interfacial energy level, which reduces recombination and improves carrier movement at the interfaces. Additionally, when a molecule interacts with a perovskite, electrons are occasionally transferred, modifying the perovskite surface's energy level to match the HTM and enhancing the PCE of PSCs. Krishna and colleagues introduced 2,5-thiophene dicarboxylic acid at the perovskite/HTM interaction. To lessen the buildup of surface charges, the molecule's C=O group may coordinate with Pb-cluster atoms on the perovskite surface. Through the use of the interface change, the associated n-i-p PSCs' PCE increased to 23%. It is interesting to note that the molecular filling of grain boundaries also inhibits the migration of Au element to perovskite. As a result, after 1400 h of storage in N2 conditions, the PCE maintained over 90% of its initial efficiency (Krishna et al. 2021). Carbon-based materials and quantum dots (QDs) are also frequently utilized by Bakr et al. for interface engineering because they provide special chemical interactions and can change the surface energy level. The organic components will adhere to the perovskite film's surface, while the inorganic perovskite quantum dots will uniformly dope the entire film. The findings indicate that the trap state of the perovskite would be decreased by the inorganic perovskite quantum dots. Additionally, by successfully blocking the water on the perovskite's surface, the self-assembled organic layer will increase the device's stability and efficiency. In conclusion, the matching PSCs attain a 21.5% efficiency rate. Because of the passivation of the defect and the hydrophobic organic molecules on the perovskite surface, PSC stability has significantly increased, and after 500 h of aging, it retained 80% of its initial efficiency (Zheng et al. 2019).

Electrode/ETL

Energy level mismatches frequently arise at the ETL/electrode interface of inverted structure PSCs, which can impede electron extraction and lower device efficiency (Chang et al. 2016; Liu et al. 2016). ETL interface engineering techniques have been created to improve performance by fine-tuning the interfacial characteristics in order to overcome this difficulty (Wolff et al. 2017). Organic compounds and their derivatives were added by Xiong et al. as the interfacial layers in PSCs (Wang et al. 2021). Several performance improvements were brought about by the use of these interlayers, including improved charge conductivity, surface trap passivation, reduced charge transfer resistance, and higher charge mobility. These changes have been shown to promote charge transport and extraction while reducing charge recombination at the cathode by lowering the energy barrier by introducing materials with smaller work functions. In comparison to control devices that were only outfitted with a PCBM/Al cathode, devices that were altered in this way demonstrated better ambient stability. The protective layer effect, which lessens moisture-induced perovskite deterioration, is partly responsible for this improvement. Ionic liquids were used as a novel modification layer by Liao et al. to improve environmental stability. By passivating surface defects, the liquids were found to improve electron transport and extraction and refine the interface contact (Xiong et al. 2019). In order to increase stability, ionic liquids can also be used as protective capping layers because of their remarkable thermal stability. This will shield the device from deterioration caused by oxygen and water. These developments in ETL/electrode interface modification help to improve the stability and environmental resilience of perovskite devices while also advancing the field toward more effective PSCs.

HTL/Electrode

Even in a dry environment or with encapsulation, the performance of PSC devices exhibits a noticeable decline due to their intrinsic instability (Han et al. 2015). Unwanted deterioration of the Ag electrode may result from the unfavorable reaction between the I-ion and Ag at the HTL/electrode interface. In addition, I-ion migration and Au/Ag diffusion will be worsened by the constant light irradiation and positive bias, which would further lower the PCE of PSCs (Domanski et al. 2016; Chen et al. 2017). Researchers have created a variety of interfacial barriers between the metal electrode and HTL to improve device stability in order to overcome the aforementioned problems (Lee et al. 2020; Chen et al. 2020b). Most notably, Arora et al. used copper(I) thiocyanate (CuSCN) as the hole extraction layer to produce extremely effective PSCs. A conductive reduced graphene oxide (rGO) barrier layer was positioned between CuSCN and gold to address the instability brought on by the potential-induced deterioration of the CuSCN/Au contact. After aging at a maximum power point for 1000 h at 60 °C, the produced PSCs with the changed layer showed outstanding stability, retaining above 95% of their initial PCE (Arora et al. n.d.). Another application for metallic oxides is as an interlayer between the electrode and HTL. Yang et al. presented MoO_3-layered sandwiched gold nanomesh (Wang et al. 2020b). The wettability of gold was clearly enhanced by the high surface tension of MoO_3, which led to Frank-van der Merwe growth and the creation of an ultrathin layer of gold nanomesh that ensured both conductivity and optical transparency. Better performance could result from the top MoO_3 layer reducing reflection at the gold layer and increasing light transmission. Despite the implementation of various methodologies, there are still few pertinent studies regarding the HTL/electrode interface, which require further optimization (Zai et al. 2021).

Interface Engineering for the Flexible PSCs

One of the most crucial directions for PSC applications is flexible PSCs, which are well-liked for flexible wearable electronics and aerospace due to their bending and lightweight characteristics. Additionally, interface engineering has been used extensively to enhance the blending stability and photovoltaic characteristics of flexible PSCs (Yang et al. 2019b). For extremely effective flexible PSCs, inorganic interfacial materials have been used. In order to alleviate the interface stress caused by lattice mismatch, Zhang et al. added an inorganic CsBr buffer layer between the NiOx HTL and the perovskite layer. The findings demonstrate that the CsBr buffer layer improved the interface between the NiO_x HTL and the perovskite absorber, decreased interface trap states, and encouraged the injection of hole extraction. Lastly, the PCE of ideal PSCs can reach 19.2%. In contrast, the device changed by CsBr retains 82% of the initial efficiency under the same conditions (RH = 30%), while the PCE of the control device drops to 65% of the initial efficiency after 900 h (Zhang et al. 2020). In order to make the energy levels of ETL and perovskite well-matched, Ren et al. used a cross-linking agent of aluminum acetylacetonate ($Al(acac)_3$) as an interface layer between the two layers. This improved the perovskite's grain size and crystallinity at the same time. The flexible PSCs finally attain a high PCE of 20.87%. When the device is aged for 1000 h without encapsulation (RH > 50%), it maintains about 80% of its original performance. Furthermore, the devices maintained more than 91% of the original PCE even after 1500 cycles of mixing (Ren et al. 2021).

Conclusions and Outlook

Interface engineering has become the best strategy for improving PSC's stability and PCE. PSCs consist of several functional layers including ETL, perovskite layer, HTL, and electrodes, where the perovskite layer absorbs incident photons with energy more than its band gap and generates electron-hole pairs. These generated carriers cross the interfaces and are collected by electrodes. Band alignments at these interfaces is crucial for enhancing the carrier collection probability, which in turn determines the PCE of PSCs. Basically, high PSCs efficiency depends on the consistency of interfacial energy levels in addition to the inherent material qualities. PSC performance can be improved by introducing interface modifications that increase energy level alignment between the perovskite and carrier transport layers. These changes can increase the internal electric field in the perovskite absorber layer, strengthen interfacial charge transfer, and reduce charge recombination. This increases the effectiveness of charge extraction and significantly reduces non-radiative recombination inside the perovskite device. Additionally, interface engineering passivates defects at interfaces.

In this chapter, we discussed the significance of interfacial engineering and its critical role in energy level alignment, defect passivation, carrier dynamics, mitigation of ion migration and preventing moisture penetration and electrode diffusion in perovskite solar cells (PSCs). Furthermore, we explored the various materials employed as interfacial layers at different interfaces—electrode/ETL/perovskite layer/HTL/electrode—aimed at improving band alignment, reducing defect states, enhancing carrier dynamics, suppressing ion migration, and minimizing moisture penetration and electrode diffusion. These improvements collectively contribute to higher PCE and enhanced operational stability of PSCs. Finally, we discussed interfacial engineering strategies tailored for flexible PSCs.

Even though interface engineering has been shown to improve PSC stability and efficiency, more empirical research is still required to make commercialization easier. A careful choice of materials for interface engineering is necessary in the quest for devices that are sturdy and highly efficient. In addition to mitigating surface imperfections and interface contact problems, optimizing energy mismatch at interfaces by structural alterations and low-temperature treatments is essential. Interface engineering plays a crucial role in the development of PSC technology since it improves photovoltaic performance and device stability regardless of the device architecture.

References

Agresti A et al (2019) Titanium-carbide MXenes for work function and interface engineering in perovskite solar cells. Nat Mater 18(11):1228–1234. https://doi.org/10.1038/s41563-019-0478-1

Akin S, Arora N, Zakeeruddin SM, Grätzel M, Friend RH, Dar MI (2020) New strategies for defect passivation in high-efficiency perovskite solar cells. Wiley, Hoboken. https://doi.org/10.1002/aenm.201903090

Arora N et al (n.d.) "Perovskite solar cells with CuSCN hole extraction layers yield stabilized efficiencies greater than 20%. https://www.science.org

Balis N et al (2018) Dye sensitization of titania compact layer for efficient and stable perovskite solar cells. ACS Appl Energy Mater 1(11):6161–6171. https://doi.org/10.1021/acsaem.8b01221

Bi D et al (2016) Polymer-templated nucleation and crystal growth of perovskite films for solar cells with efficiency greater than 21%. Nat Energy 1(10). https://doi.org/10.1038/nenergy.2016.142

Cao T, Chen K, Chen Q, Zhou Y, Chen N, Li Y (2019) Fullerene derivative-modified SnO2 electron transport layer for highly efficient perovskite solar cells with efficiency over 21%. ACS Appl Mater Interfaces 11(37):33825–33834. https://doi.org/10.1021/acsami.9b09238

Celik et al (2017) Environmental analysis of perovskites and other relevant solar cell technologies in a tandem configuration. Energy Environ Sci 10(9):1874–1884. https://doi.org/10.1039/c7ee01650f

Chang CY, Huang WK, Chang YC (2016) Highly-efficient and long-term stable perovskite solar cells enabled by a cross-linkable n-doped hybrid cathode interfacial layer. Chem Mater 28(17):6305–6312. https://doi.org/10.1021/acs.chemmater.6b02583

Chen W et al (n.d.) Efficient and stable large-area perovskite solar cells with inorganic charge extraction layers. https://www.science.org

Chen K et al (2016) Charge-carrier balance for highly efficient inverted planar heterojunction perovskite solar cells. Adv Mater 28(48):10718–10724. https://doi.org/10.1002/adma.201604048

Chen J, Lee D, Park NG (2017) Stabilizing the ag electrode and reducing J-V hysteresis through suppression of iodide migration in perovskite solar cells. ACS Appl Mater Interfaces 9(41):36338–36349. https://doi.org/10.1021/acsami.7b07595

Chen Y et al (2018a) Design of an inorganic mesoporous hole-transporting layer for highly efficient and stable inverted perovskite solar cells. Adv Mater 30:52. https://doi.org/10.1002/adma.201805660

Chen P, Bai Y, Wang S, Lyu M, Yun JH, Wang L (2018b) In situ growth of 2D perovskite capping layer for stable and efficient perovskite solar cells. Adv Funct Mater 28(17). https://doi.org/10.1002/adfm.201706923

Chen H et al (2019a) Efficient and stable inverted perovskite solar cells incorporating secondary amines. Adv Mater 31:46. https://doi.org/10.1002/adma.201903559

Chen Q et al (2019b) Zwitterionic polymer: a facile interfacial material works at both anode and cathode in p-i-n perovskite solar cells. Sol RRL 3(9). https://doi.org/10.1002/solr.201900118

Chen B, Rudd PN, Yang S, Yuan Y, Huang J (2019c) Imperfections and their passivation in halide perovskite solar cells. Royal Society of Chemistry, London. https://doi.org/10.1039/c8cs00853a

Chen C et al (2020a) Arylammonium-assisted reduction of the open-circuit voltage deficit in wide-bandgap perovskite solar cells: the role of suppressed ion migration. ACS Energy Lett 5(8):2560–2568. https://doi.org/10.1021/acsenergylett.0c01350

Chen C et al (2020b) Chemical inhibition of reversible decomposition for efficient and superstable perovskite solar cells. Nano Energy 68. https://doi.org/10.1016/j.nanoen.2019.104315

Chen H et al (2022) Quantum-size-tuned heterostructures enable efficient and stable inverted perovskite solar cells. Nat Photonics 16(5):352–358. https://doi.org/10.1038/s41566-022-00985-1

Cho Y et al (2018a) Mixed 3D–2D passivation treatment for mixed-cation lead mixed-halide perovskite solar cells for higher efficiency and better stability. Adv Energy Mater 8(20). https://doi.org/10.1002/aenm.201703392

Cho KT et al (2018b) Selective growth of layered perovskites for stable and efficient photovoltaics. Energy Environ Sci 11(4):952–959. https://doi.org/10.1039/c7ee03513f

Choi K et al (2018) Thermally stable, planar hybrid perovskite solar cells with high efficiency. Energy Environ Sci 11(11):3238–3247. https://doi.org/10.1039/c8ee02242a

Di Girolamo D et al (2020) Ion migration-induced amorphization and phase segregation as a degradation mechanism in planar perovskite solar cells. Adv Energy Mater 10(25). https://doi.org/10.1002/aenm.202000310

Di Vito A, Pecchia A, Auf der Maur M, Di Carlo A (2020) Nonlinear work function tuning of Lead-halide perovskites by MXenes with mixed terminations. Adv Funct Mater 30(47). https://doi.org/10.1002/adfm.201909028

Ding B et al (2018) Low-temperature SnO2-modified TiO2 yields record efficiency for normal planar perovskite solar modules. J Mater Chem A Mater 6(22):10233–10242. https://doi.org/10.1039/c8ta01192c

Domanski K et al (2016) Not all that glitters is gold: metal-migration-induced degradation in perovskite solar cells. ACS Nano 10(6):6306–6314. https://doi.org/10.1021/acsnano.6b02613

Eames C, Frost JM, Barnes PRF, O'Regan BC, Walsh A, Islam MS (2015) Ionic transport in hybrid lead iodide perovskite solar cells. Nat Commun 6:8497. https://doi.org/10.1038/ncomms8497

Gao P, Yusoff ARBM, Nazeeruddin MK (2018) Dimensionality engineering of hybrid halide perovskite light absorbers. Nature Publishing Group, London. https://doi.org/10.1038/s41467-018-07382-9

Gharibzadeh S et al (2021) Two birds with one stone: dual grain-boundary and interface passivation enables >22% efficient inverted methylammonium-free perovskite solar cells. Energy Environ Sci 14(11):5875–5893. https://doi.org/10.1039/d1ee01508g

Grancini G et al (2017) One-year stable perovskite solar cells by 2D/3D interface engineering. Nat Commun 8. https://doi.org/10.1038/ncomms15684

Han Y et al (2015) Degradation observations of encapsulated planar CH3NH3PbI3 perovskite solar cells at high temperatures and humidity. J Mater Chem A Mater 3(15):8139–8147. https://doi.org/10.1039/c5ta00358j

Hassan A et al (2022) Recent defect passivation drifts and role of additive engineering in perovskite photovoltaics. Elsevier, Amsterdam. https://doi.org/10.1016/j.nanoen.2022.107579

Hou M, Zhang H, Wang Z, Xia Y, Chen Y, Huang W (2018) Enhancing efficiency and stability of perovskite solar cells via a self-assembled dopamine interfacial layer. ACS Appl Mater Interfaces 10(36):30607–30613. https://doi.org/10.1021/acsami.8b10332

Huang J et al (2024) Advances in crystallization regulation and defect suppression strategies for all-inorganic CsPbX3 perovskite solar cells. Elsevier, Amsterdam. https://doi.org/10.1016/j.pmatsci.2023.101223

Isikgor FH, Zhumagali S, Luis LV, De Bastiani M, McCulloch I, De Wolf S (2023) Molecular engineering of contact interfaces for high-performance perovskite solar cells. Nat Res. https://doi.org/10.1038/s41578-022-00503-3

Jeng JY et al (2013) CH3NH3PbI3 perovskite/fullerene planar-heterojunction hybrid solar cells. Adv Mater 25(27):3727–3732. https://doi.org/10.1002/adma.201301327

Kant N, Singh P (2022) Review of next generation photovoltaic solar cell technology and comparative materialistic development. Mater Today Proc 56:3460–3470. https://doi.org/10.1016/j.matpr.2021.11.116

Ke W et al (2016) Employing lead thiocyanate additive to reduce the hysteresis and boost the fill factor of planar perovskite solar cells. Adv Mater 28(26):5214–5221. https://doi.org/10.1002/adma.201600594

Kim GY, Senocrate A, Yang TY, Gregori G, Grätzel M, Maier J (2018) Large tunable photoeffect on ion conduction in halide perovskites and implications for photodecomposition. Nat Mater 17(5):445–449. https://doi.org/10.1038/s41563-018-0038-0

Kim J, Lee Y, Yun AJ, Gil B, Park B (2019) Interfacial modification and defect passivation by the cross-linking interlayer for efficient and stable CuSCN-based perovskite solar cells. ACS Appl Mater Interfaces 11(50):46818–46824. https://doi.org/10.1021/acsami.9b16194
Kim H et al (2020) Self-crystallized multifunctional 2D perovskite for efficient and stable perovskite solar cells. Adv Funct Mater 30(19). https://doi.org/10.1002/adfm.201910620
Krishna A et al (2021) Nanoscale interfacial engineering enables highly stable and efficient perovskite photovoltaics. Energy Environ Sci 14(10):5552–5562. https://doi.org/10.1039/d1ee02454j
Lee J, Singh S, Kim S, Baik S (2020) Graphene interfacial diffusion barrier between CuSCN and au layers for stable perovskite solar cells. Carbon 157:731–740. https://doi.org/10.1016/j.carbon.2019.10.101
Li W et al (2016a) Enhanced UV-light stability of planar heterojunction perovskite solar cells with caesium bromide interface modification. Energy Environ Sci 9(2):490–498. https://doi.org/10.1039/c5ee03522h
Li W, Li J, Niu G, Wang L (2016b) Effect of cesium chloride modification on the film morphology and UV-induced stability of planar perovskite solar cells. J Mater Chem A Mater 4(30):11688–11695. https://doi.org/10.1039/c5ta09165a
Li F et al (2018a) A universal strategy to utilize polymeric semiconductors for perovskite solar cells with enhanced efficiency and longevity. Adv Funct Mater 28(15). https://doi.org/10.1002/adfm.201706377
Li P et al (2018b) Phase pure 2D perovskite for high-performance 2D–3D heterostructured perovskite solar cells. Adv Mater 30(52). https://doi.org/10.1002/adma.201805323
Li X et al (2019) Defects passivation with dithienobenzodithiophene-based π-conjugated polymer for enhanced performance of perovskite solar cells. Sol RRL 3(6). https://doi.org/10.1002/solr.201900029
Li Y, Xie H, Lim EL, Hagfeldt A, Bi D (2022) Recent progress of critical interface engineering for highly efficient and stable perovskite solar cells. Wiley, Hoboken. https://doi.org/10.1002/aenm.202102730
Li Y, Wang Y, Xu Z, Peng B, Li X (2024) Key roles of interfaces in inverted metal-halide perovskite solar cells. American Chemical Society, Washington. https://doi.org/10.1021/acsnano.3c11642
Liang Z et al (2019) Interface modification via Al2O3 with retarded charge recombinations for mesoscopic perovskite solar cells fabricated with spray deposition process in the air. Appl Surf Sci 463:939–946. https://doi.org/10.1016/j.apsusc.2018.08.077
Liu Y et al (2016) Understanding interface engineering for high-performance fullerene/perovskite planar heterojunction solar cells. Adv Energy Mater 6(2). https://doi.org/10.1002/aenm.201501606
Liu X et al (2018) Exploring inorganic binary alkaline halide to passivate defects in low-temperature-processed planar-structure hybrid perovskite solar cells. Adv Energy Mater 8(20). https://doi.org/10.1002/aenm.201800138
Liu D et al (2019a) Improved crystallinity of perovskite via molecularly tailored surface modification of SnO2. J Power Sources 441. https://doi.org/10.1016/j.jpowsour.2019.227161
Liu Y et al (2019b) Ultrahydrophobic 3D/2D fluoroarene bilayer-based water-resistant perovskite solar cells with efficiencies exceeding 22%. https://www.science.org
Liu C et al (2020) Alkali cation doping for improving the structural stability of 2D perovskite in 3D/2D PSCs. Nano Lett 20(2):1240–1251. https://doi.org/10.1021/acs.nanolett.9b04661
Liu C et al (2021a) Interphases, interfaces, and surfaces of active materials in rechargeable batteries and perovskite solar cells. Wiley, Hoboken. https://doi.org/10.1002/adma.201905245
Liu J, Hu M, Dai Z, Que W, Padture NP, Zhou Y (2021b) Correlations between electrochemical ion migration and anomalous device Behaviors in perovskite solar cells. ACS Energy Lett 6(3):1003–1014. https://doi.org/10.1021/acsenergylett.0c02662
Ma J et al (2017) MgO nanoparticle modified anode for highly efficient SnO2-based planar perovskite solar cells. Adv Sci 4(9). https://doi.org/10.1002/advs.201700031

Maka AOM, Alabid JM (2022) Solar energy technology and its roles in sustainable development. Clean Energy 6(3):476–483. https://doi.org/10.1093/ce/zkac023
Marin-Beloqui JM, Lanzetta L, Palomares E (2016) Decreasing charge losses in perovskite solar cells through mp-TiO2/MAPI interface engineering. Chem Mater 28(1):207–213. https://doi.org/10.1021/acs.chemmater.5b03902
Meng L et al (2018) Tailored phase conversion under conjugated polymer enables thermally stable perovskite solar cells with efficiency exceeding 21%. J Am Chem Soc 140(49):17255–17262. https://doi.org/10.1021/jacs.8b10520
Min J, Choi Y, Kim D, Park T (2024) Beyond imperfections: exploring defects for breakthroughs in perovskite solar cell research. Wiley, Hoboken. https://doi.org/10.1002/aenm.202302659
Nie J et al (2023) Multi-functional MXene quantum dots enhance the quality of perovskite polycrystalline films and charge transport for solar cells. J Colloid Interface Sci 646:517–528. https://doi.org/10.1016/j.jcis.2023.05.058
Noman M, Khan Z, Jan ST (2024) A comprehensive review on the advancements and challenges in perovskite solar cell technology. Royal Society of Chemistry, London. https://doi.org/10.1039/d3ra07518d
Palei S, Murali G, Kim CH, In I, Lee SY, Park SJ (2023) A review on interface engineering of MXenes for perovskite solar cells. Springer, Cham. https://doi.org/10.1007/s40820-023-01083-9
Pan H, Shao H, Zhang XL, Shen Y, Wang M (2021) Interface engineering for high-efficiency perovskite solar cells. American Institute of Physics, College Park. https://doi.org/10.1063/5.0038073
Park BW, Seok S (2019) Intrinsic instability of inorganic–organic hybrid halide perovskite materials. Wiley, Hoboken. https://doi.org/10.1002/adma.201805337
Pazos-Outón LM, Xiao TP, Yablonovitch E (2018) Fundamental efficiency limit of lead iodide perovskite solar cells. American Chemical Society, Washington. https://doi.org/10.1021/acs.jpclett.7b03054
Peng J et al (2017) Interface passivation using ultrathin polymer-fullerene films for high-efficiency perovskite solar cells with negligible hysteresis. Energy Environ Sci 10(8):1792–1800. https://doi.org/10.1039/c7ee01096f
Peng J et al (2021) Nanoscale localized contacts for high fill factors in polymer-passivated perovskite solar cells. https://www.science.org
Qin P et al (2019) Vitrification transformation of poly(ethylene oxide) activating interface passivation for high-efficiency perovskite solar cells. Sol RRL 3(10). https://doi.org/10.1002/solr.201900134
Rajagopal A, Yao K, Jen AKY (2018) Toward perovskite solar cell commercialization: a perspective and research roadmap based on interfacial engineering. Wiley, Hoboken. https://doi.org/10.1002/adma.201800455
Ren N et al (2021) Humidity-resistant flexible perovskite solar cells with over 20% efficiency. Solar RRL 5(4). https://doi.org/10.1002/solr.202000795
Ru P et al (2020) High electron affinity enables fast hole extraction for efficient flexible inverted perovskite solar cells. Adv Energy Mater 10(12). https://doi.org/10.1002/aenm.201903487
Seo J et al (2014) Benefits of very thin PCBM and LiF layers for solution-processed p-i-n perovskite solar cells. Energy Environ Sci 7(8):2642–2646. https://doi.org/10.1039/c4ee01216j
Sherkar TS et al (2017) Recombination in perovskite solar cells: significance of grain boundaries, interface traps, and defect ions. ACS Energy Lett 2(5):1214–1222. https://doi.org/10.1021/acsenergylett.7b00236
Shi J, Xu X, Li D, Meng Q (2015) Interfaces in perovskite solar cells. Small 11(21):2472–2486. https://doi.org/10.1002/smll.201403534
Shi J et al (2018) From ultrafast to ultraslow: charge-carrier dynamics of perovskite solar cells. Cell Press, Cambridge. https://doi.org/10.1016/j.joule.2018.04.010
Shin SS, Lee SJ, Il Seok S (2019) Metal oxide charge transport layers for efficient and stable perovskite solar cells. Wiley, Hoboken. https://doi.org/10.1002/adfm.201900455

Stolterfoht M et al (2019) The impact of energy alignment and interfacial recombination on the internal and external open-circuit voltage of perovskite solar cells. Energy Environ Sci 12(9):2778–2788. https://doi.org/10.1039/c9ee02020a

Stranks SD (2017) Nonradiative losses in metal halide perovskites. American Chemical Society, Washington. https://doi.org/10.1021/acsenergylett.7b00239

Tan Q et al (2023) Inverted perovskite solar cells using dimethylacridine-based dopants. Nature 620(7974):545–551. https://doi.org/10.1038/s41586-023-06207-0

Tress W (2017) Perovskite solar cells on the way to their radiative efficiency limit—insights into a success story of high open-circuit voltage and low recombination. Adv Energy Mater 7(14):1602358. https://doi.org/10.1002/aenm.201602358

Tress W et al (2018) Interpretation and evolution of open-circuit voltage, recombination, ideality factor and subgap defect states during reversible light-soaking and irreversible degradation of perovskite solar cells. Energy Environ Sci 11(1):151–165. https://doi.org/10.1039/c7ee02415k

Wang C et al (2016) Low-temperature plasma-enhanced atomic layer deposition of tin oxide electron selective layers for highly efficient planar perovskite solar cells. J Mater Chem A Mater 4(31):12080–12087. https://doi.org/10.1039/c6ta04503k

Wang S, Sakurai T, Wen W, Qi Y (2018a) Energy level alignment at interfaces in metal halide perovskite solar cells. Wiley, Hoboken. https://doi.org/10.1002/admi.201800260

Wang P et al (2018b) Boosting the performance of perovskite solar cells through a novel active passivation method. J Mater Chem A Mater 6(32):15853–15858. https://doi.org/10.1039/c8ta05593a

Wang F et al (2019) Exploring low-temperature processed a-WOx/SnO2 hybrid electron transporting layer for perovskite solar cells with efficiency >20.5%. Nano Energy 63. https://doi.org/10.1016/j.nanoen.2019.06.021

Wang H et al (2020a) Chlorinated fullerene dimers for interfacial engineering toward stable planar perovskite solar cells with 22.3% efficiency. Adv Energy Mater 10(21). https://doi.org/10.1002/aenm.202000615

Wang Z et al (2020b) 27%-efficiency four-terminal perovskite/silicon tandem solar cells by sandwiched gold nanomesh. Adv Funct Mater 30(4). https://doi.org/10.1002/adfm.201908298

Wang J et al (2021) Tuning an electrode work function using organometallic complexes in inverted perovskite solar cells. J Am Chem Soc 143(20):7759–7768. https://doi.org/10.1021/jacs.1c02118

Wolff CM, Zu F, Paulke A, Toro LP, Koch N, Neher D (2017) Reduced interface-mediated recombination for high open-circuit voltages in CH3NH3PbI3 solar cells. Adv Mater 29(28). https://doi.org/10.1002/adma.201700159

Wu Z et al (2023) Passivation strategies for enhancing device performance of perovskite solar cells. Elsevier, Amsterdam. https://doi.org/10.1016/j.nanoen.2023.108731

Xia J, Sohail M, Nazeeruddin MK (2023) Efficient and stable perovskite solar cells by tailoring of interfaces. Wiley, Hoboken. https://doi.org/10.1002/adma.202211324

Xie L et al (2020) Improving energy level alignment by adenine for efficient and stable perovskite solar cells. Nano Energy 74. https://doi.org/10.1016/j.nanoen.2020.104846

Xiong S et al (2019) Engineering of the back contact between PCBM and metal electrode for planar perovskite solar cells with enhanced efficiency and stability. Adv Opt Mater 7(19). https://doi.org/10.1002/adom.201900542

Yang Z, Dou J, Wang M (2018) Interface engineering in n-i-p metal halide perovskite solar cells. Wiley, Hoboken. https://doi.org/10.1002/solr.201800177

Yang Y et al (2019a) Beneficial effects of potassium iodide incorporation on grain boundaries and interfaces of perovskite solar cells. RSC Adv 9(49):28561–28568. https://doi.org/10.1039/c9ra05371a

Yang D, Yang R, Priya S, Liu SF (2019b) Flexible perowskit-solarzellen: herstellung und anwendungen. Angew Chem 131(14):4512–4530. https://doi.org/10.1002/ange.201809781

Yin WJ, Shi T, Yan Y (2014) Unique properties of halide perovskites as possible origins of the superior solar cell performance. Adv Mater 26(27):4653–4658. https://doi.org/10.1002/adma.201306281

Zai H et al (2021) Sandwiched electrode buffer for efficient and stable perovskite solar cells with dual back surface fields. Joule 5(8):2148–2163. https://doi.org/10.1016/j.joule.2021.06.001

Zhang F et al (2018) Interfacial passivation of the p-doped hole-transporting layer using general insulating polymers for high-performance inverted perovskite solar cells. Small 14(19). https://doi.org/10.1002/smll.201704007

Zhang H, Nazeeruddin MK, Choy WCH (2019) Perovskite photovoltaics: the significant role of ligands in film formation, passivation, and stability. Wiley, Hoboken. https://doi.org/10.1002/adma.201805702

Zhang B et al (2020) NiO/perovskite heterojunction contact engineering for highly efficient and stable perovskite solar cells. Adv Sci 7(11). https://doi.org/10.1002/advs.201903044

Zhang J et al (2022) Sulfonic zwitterion for passivating deep and shallow level defects in perovskite light-emitting diodes. Adv Funct Mater 32:22. https://doi.org/10.1002/adfm.202111578

Zhao X et al (2018) Efficient planar perovskite solar cells with improved fill factor via interface engineering with graphene. Nano Lett 18(4):2442–2449. https://doi.org/10.1021/acs.nanolett.8b00025

Zheng X et al (2017) Defect passivation in hybrid perovskite solar cells using quaternary ammonium halide anions and cations. Nat Energy 2(7). https://doi.org/10.1038/NENERGY.2017.102

Zheng X et al (2019) Quantum dots supply bulk- and surface-passivation agents for efficient and stable perovskite solar cells. Joule 3(8):1963–1976. https://doi.org/10.1016/j.joule.2019.05.005

Zheng X et al (2020) Managing grains and interfaces via ligand anchoring enables 22.3%-efficiency inverted perovskite solar cells. Nat Energy 5(2):131–140. https://doi.org/10.1038/s41560-019-0538-4

Zhong M, Liang Y, Zhang J, Wei Z, Li Q, Xu D (2019) Highly efficient flexible MAPbI3 solar cells with a fullerene derivative-modified SnO2 layer as the electron transport layer. J Mater Chem A Mater 7(12):6659–6664. https://doi.org/10.1039/c9ta00398c

Zhou Y, Zhao Y (2019) Chemical stability and instability of inorganic halide perovskites. Royal Society of Chemistry, London. https://doi.org/10.1039/c8ee03559h

Zhou Q et al (2019) High-performance perovskite solar cells with enhanced environmental stability based on a (p-FC6H4C2H4NH3)2 [PbI 4] capping layer. Adv Energy Mater 9(12). https://doi.org/10.1002/aenm.201802595

Zhu Z, Bai Y, Liu X, Chueh CC, Yang S, Jen AKY (2016) Enhanced efficiency and stability of inverted perovskite solar cells using highly crystalline SnO2 nanocrystals as the robust electron-transporting layer. Adv Mater 28(30):6478–6484. https://doi.org/10.1002/adma.201600619

Zhu M, Liu W, Ke W, Xie L, Dong P, Hao F (2019) Graphene-modified tin dioxide for efficient planar perovskite solar cells with enhanced electron extraction and reduced hysteresis. ACS Appl Mater Interfaces 11(1):666–673. https://doi.org/10.1021/acsami.8b15665

Chapter 4
Physical Impedance Models in Perovskite Solar Cells

Rajat Kumar Goyal, Nidhi Bharadwaj, and M. Chandrasekhar

Abstract Recent progress in perovskite solar cell (PSC) research has hope for the adoption of solar energy. To gain a deeper understanding of the underlying mechanisms is crucial to advancing current solar technologies. Impedance spectroscopy (IS) is a key method for exploring the internal dynamics of solar cells. Ongoing analysis and optimization of electrochemical spectroscopy data have revealed complex operational insights. This section collects the physical principles driving PSC expansion, supplying a complete resource on resistance plots, corresponding circuits, and allied procedures.

Keywords Impedance spectroscopy · Solar energy · Perovskite solar cell

Introduction

Perovskite solar cells (PSC) have appeared as a capable frontier presenting unmatched competences and flexibility in the fast-developing photovoltaic study. They show exceptional light-absorption properties and control transport features. The tuneable properties of these materials support them as promising candidates for next-generation solar cell technologies. Perovskites' exceptional properties and

R. K. Goyal
Interdisciplinary Research Division, Indian Institute of Technology, Jodhpur, Rajasthan, India

Department of Physics and Astronomy, National Institute of Technology Rourkela, Rourkela, Odisha, India

N. Bharadwaj
Interdisciplinary Research Division, Indian Institute of Technology, Jodhpur, Rajasthan, India

M. Chandrasekhar (✉)
Department of Physics and Astronomy, National Institute of Technology Rourkela, Rourkela, Odisha, India

Department of Physics, CVR College of Engineering, Hyderabad, Telangana, India

M. Kumar (ed.), *Computational Analysis of Perovskite Solar Cells*, Springer Tracts in Electrical and Electronics Engineering,
https://doi.org/10.1007/978-981-95-6521-4_4

innovative architectural designs have increased solar cell efficiency from 3.8% (Kojima et al. 2009) to 26.7% within a decade (Mahjabin et al. 2020). A better understanding of operating and degradation mechanisms can lead to future efficiency and stability enhancements. Solar cells operate through various physical processes, including recombination rates, carrier mobility, and the effectiveness of contact points in extracting charge (Herz 2017; Johnston and Herz 2016). Understanding these mechanisms is crucial for optimizing their performance. The primary focus is analyzing their steady-state behavior (current density-voltage J-V curve), representing the power conversion efficiency (PCE). However, solely examining the J-V characteristics does not provide insights into the dynamic phenomena governing their performance. To unravel these dynamics, it is essential to introduce external signals and observe the system's response.

The orthodox technique used for the electrical properties of PSCs comprises gauging the current-voltage (J–V) curves in forward and reverse bias scans (Zimmermann et al. 2016; Juarez-Perez et al. 2014a; Chen et al. 2015). The precision of factors like short circuit current (J_{SC}), open-circuit voltage (V_{OC}), and fill factor (FF) are vital for solar cells. The presence of anomalous hysteresis in these measurements can compromise the parameters. Hysteresis and the scan rate play a significant role in determining these parameters (Akriti et al. 2022; Fang et al. 2017; Herz 2017; Mahapatra et al. 2024). Hysteresis in PSCs is governed by ionic conduction, contact properties, and surface polarization effect (Gonzalez-Pedro et al. 2014; Ravishankar et al. 2017; Wu et al. 2016; Shao et al. 2014; Meloni et al. 2016; Sanchez et al. 2014). These complexities suggest a need for advanced techniques to enhance the characterization of PSCs.

The most considered perovskite (ABX_3) structure in literature is $MAPbI_3$ (Kojima et al. 2009). It has a little bandgap (1.55 eV) (Grätzel 2014) with a high light extermination coefficient, extended charge carrier dispersion length and lifetime, also high charge carrier movement (Tavakoli et al. 2015; Prochowicz et al. 2019). There are lots of aspects for refining photovoltaic technology like a high optical absorption coefficient, suppressed recombination, an elongated carrier diffusion length, with a well-proportioned charge transference. However, $MAPbI_3$-based PSCs exhibit low stability under environmental conditions (light, heat, and oxygen), ionic migration, and interface/contact issues (Kim et al. 2022). Various strategies, such as approaching lower dimensions, chemical composition modifications, and surface passivation, are employed to overcome these (Wei et al. 2022). PSCs exhibit unique properties that are uncommon in semiconductor devices, including anomalous hysteresis during J–V measurements (Wu et al. 2016) and huge switchable photocurrents (Juarez-Perez et al. 2014b).

Electrochemical impedance spectroscopy (EIS) introduces a small perturbation with an electrical signal and then analyzes its response. Two main approaches are used for perturbation: current and voltage. In potentiodynamic EIS (PEIS), perturbation is due to voltage, and galvano dynamic EIS (GEIS) is due to current. The system's resistance is measured using alternating-current (AC) impedance. Impedance consists of a contribution from all Resistance (R), Capacitance (C), and Inductance (L) and varies with signal frequency. Macdonald outlines several benefits of EIS: (1) its linearity allows for analysis using Linear Systems Theory, (2) frequency impedance spectrum provides the same insights as linear electrical techniques, (3) EIS efficiently produces a large volume of information in comparison to the collected data, and (4) mathematical methods can verify data accuracy independently of the experimental setup (MacDonald 2006). EIS makes a thorough understanding of device behavior possible, which also assists in discovering critical insights into the performance of devices (Dualeh et al. 2014; Gonzalez-Pedro et al. 2014; Pascoe et al. 2015; Hishikawa et al. 2016; Fabregat-Santiago et al. 2018). EIS can offer crucial details such as charge recombination, dielectric variables, and interfacial effects, in addition to the position and density of traps (Shao et al. 2014; Juarez-Perez et al. 2014b; Kim et al. 2013; Bisquert et al. 2014a). This through examination not only helps to understand phenomena such as degradation, stability, and hysteresis but also enables to gain deeper insight into the underlying (Bisquert et al. 2016). Despite the fact that existing simulation technology makes the course of gathering and fitting impedance arcs with related circuits straightforward, complications arise when trying to associate and attribute specific components of the impedance spectrum to implied mechanisms that exist inside the PSC. Getting rid of this uncertainty continues to be of interest in this subject. The graphical abstract figure represents the systemic overview of IS (Fabregat-Santiago et al. 2011; Impedance Spectroscopy 2018).

This chapter focuses on the significance of utilizing equivalent circuits in modeling impedance spectra to better understand the physical processes inherent in PSCs. The main goal is to elucidate the role of impedance spectroscopy (IS) measurements in unraveling the operational principles of PSCs. Initially, a concise overview of IS and fundamental Equivalent Circuit Models (ECMs) is provided. After that, various phenomena for PSCs through ECMs will be explored. The chapter delves into how IS has evolved into a pivotal tool for furnishing detailed device-level insights to enhance the efficiency and reliability of PSCs. Furthermore, it delves into prospective research directions, targeting research areas for further exploration.

Understanding of IS and Basic Models

IS involves applying a slight amplitude stimulus (voltage or current) to a system and measuring the succeeding response (current or voltage) over numerous frequencies. A small amplitude perturbation is necessary for accurate measurements to establish a linear relationship between the signal and system response. This can be done by either a two-electrode setup or a three-electrode configuration. In two-electrode setups, the reference and counter electrodes of the inspected cell are in electrical contact; it aids in analyzing attributes where it is needless to differentiate between the discrete electrode procedures, such as in symmetric half-cells. The electrode setup configuration gives the advantage of distinguishing impedances from counter or working electrodes; it is preferable because it enables accurate differentiation of the interface properties related to each electrode (cathode or anode). In this section, we will cover the mathematical foundation of EIS, the general (Basic/foundation) models used for IS, validation of EIS data, and different timescales.

Mathematical Formalism for Impedance Data

The system can be evaluated by means of IS, based on the frequency-dependent potential or current response. A potential-controlled excitation Impedance Spectroscopy (PEIS) and corresponding current response can be represented as:

$$E(t) = E_0 \sin(\omega t) \tag{4.1}$$

$$I(t) = I_0 \sin(\omega t + \phi) \tag{4.2}$$

where E_0 is low voltage amplitude, I_0 is current amplitude, and ϕ represents the phase shift. The impedance of the circuit can be defined as:

$$Z = \frac{E(t)}{I(t)} = \frac{V_0 \sin(\omega t)}{I_0 \sin(\omega t + \phi)} = Z_0 (\cos(\phi) + jsin(\phi)) = Z_0 \exp(j\phi) = Z' + jZ'' \tag{4.3}$$

where phase $\phi = (\omega t)$ and Z' and Z'' is the real and imaginary parts of the impedance, denoted in the complex plane on the X-axis and Y-axis, respectively. As revealed in the graphical abstract figure. Impedance can be defined either by its amplitude and phase ($|Z|$, ϕ) or by its real and imaginary parts (Re(Z), Im(Z)), providing insights into both the resistive and reactive elements within the circuit. This representation is valid for each frequency in the spectroscopic range, creating a three-dimensional array with frequency as a discrete variable. Typical representations of this data include Nyquist (Im(Z) vs. Re(Z)) or Bode ($|Z|$ vs. ω) plots, which proposes projections along one of the three axes, permitting for visualization of impedance behavior at diverse frequencies.

Overview of EEC Elements for PSCs

IS data interpretation is based on Equivalent Electrical Circuit (EEC) models. It combines circuit elements like a resistor, capacitor, constant phase element (CPE), and inductor arranged in series or parallel configurations. This approach may not adequately distinguish between proposed local processes, such as reaction mechanisms. To efficiently illustrate PSCs using IS, it is crucial to comprehend capacitance, inductance, resistance, and their association with diverse physical approaches (Barsoukov and Macdonald 2018). Figure 4.1 illustrates the diverse impedance (R,

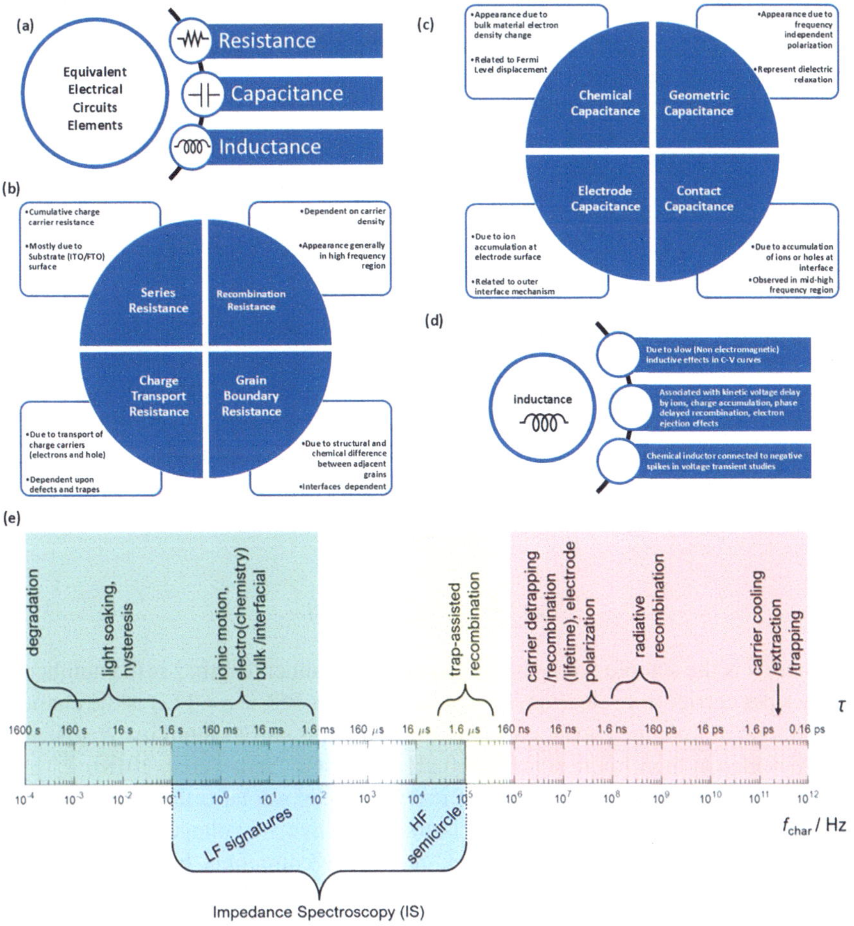

Fig. 4.1 (**a**) Illustration of various types of EEC elements, (**b**) Various types of resistance, (**c**) various types of capacitances, and (**d**) inductors used in PSCs (**e**) various extractable underlying phenomena in impedance spectra across different frequency and time range (Von Hauff and Klotz 2022)

L, C) relevant to PSC studies, extensively discussed in existing IS literature. This section discussed EEC elements (R, L, C).

Overview of Resistance

Resistance plays a crucial role in understanding ionic and electronic conductivity. IS technique offers a detailed understanding of frequency-dependent and frequency-independent resistance. This analysis helps identify and quantify different types of resistances present within the system. Many types of resistance are observed in physical systems: charge transfer, recombination, and ohmic. Figure 4.1b illustrates the various resistances typically observed in PSCs. Mathematically, resistance can be expressed as follows:

$$R_{Tot} = \left(\frac{\partial J}{\partial V}\right)^{-1} \tag{4.4}$$

where the total resistance of the system (R_{Tot}), the current density (J), and the potential difference (V).

Charge Transfer Resistance (R_{CT}) or Transport Resistance (R_{TR}) It is generally studied in electrochemistry, which replicates the R_{CT} at interfaces and is influenced by band bending, voltage, and temperature. It plays a central role in charge collection and extraction in PSCs, mainly at the interfaces between the perovskite layer and transport layers (ETL and HTL). R_{CT} estimates the energy necessary for mobile charge transfer and depends on material properties. It is in reverse proportional to substance conductivity and mathematically stated as (Von Hauff and Klotz 2022):

$$R_{TR} = d.\sigma^{-1} = d.(z.q.n.\mu)^{-1} \tag{4.5}$$

where d is the sample thickness or conduction channel length, z is the number of charges per carrier (e.g. electron, hole, or ion, and for PSCs, $z = 1$), q is the elementary charge, n is charge carrier concentration, and m is carrier mobility. R_{TR} is typically detected at high frequencies (~100 kHz) in PSCs (Yadav et al. 2018).

Dynamic resistance (R_d): The dynamic resistance of a solar cell changes with the change in voltage and current. This is the derivative of voltage versus current. Dynamic models, parasitic capacitance, and R_d are commonly linked in the solar cell. Existing literature has not extensively explored the R_d in PSCs. Previous studies have ignored this element despite its potential importance in explaining PSC behavior (Touré 2022).

Recombination resistance (R_{rec}): Resistance happens when an electron/hole travels from the band to a bandgap defect/surface state. Conventionally, R_{rec} is measured at low frequencies (<100 Hz), but practically, it is measured in EIS circuits at frequencies above 100 kHz (Castro-Chong et al. 2021; Fabregat-Santiago et al. 2004). Perovskite Rrec is affected by carrier density, which is altered by voltage, light, and temperature. Therefore, R_{rec} usually depends on carrier density (n) and temperature (T). R_{TR} increases as carrier density falls due to recombination. This links R_{rec} with transport resistance (Von Hauff and Klotz 2022). Quantifying such linked processes with IS is difficult. A mathematical expression for R_{rec} is:

$$R_{\mathrm{rec}} = \frac{mk_bT}{qj_o}\frac{e^{-qV}}{nk_bT}. \tag{4.6}$$

Series resistance (R_s): It is recommended to be kept low because it frequently originates from device contacts. ITO/FTO, which stands for transparent conductive oxide windows, causes this phenomenon in thin film PSCs (Von Hauff and Klotz 2022). It corresponds to a shift along the real axis away from the origin; determining R_s in the Nyquist plot is straightforward (Suresh 1996). The presence of sheet resistance of contact materials can also be verified by using the Nyquist plot, a measurement technique.

Overview of Capacitance

Capacitance is essential in impedance spectroscopy, enabling the characterization of material electrical properties. It provides insights into charge transfer kinetics, diffusion processes, and double-layer formation at interfaces. Figure 4.1c illustrates the various capacitance typically observed in PSCs.

Chemical capacitance (**$C\mu$**) or diffusion capacitance plays a key role in electrochemistry and semiconductor physics. It represents charge storage linked to chemical reactions at interfaces or carrier transport in devices like P-N junction diodes. Defined as $C_{\mu} = q^2 \frac{\partial n}{\partial E_{Fn}}$, where n is carrier density, it highlights the relationship between charge density and Fermi energy. Typically, it is observed in low-frequency ranges (0.1–10 Hz). In forward-biased diodes, diffusion capacitance arises from minority carrier accumulation near the depletion region, $C_{\mathrm{diff}} \sim \frac{\mathrm{qnd}}{\mathrm{kT}} \exp\left(\frac{qV}{kT}\right)$. And is significant in mid-high frequency ranges (>1 kHz). Additionally, it complements drift transport in solar cells, contributing to charge dynamics. (Fabregat-Santiago et al. 2004)

Geometric capacitance: Dielectric relaxation is key phenomenon for polarization and charge separation in an electrical field. Geometrical capacity rises from the dislocation of charge carriers from their equilibrium. This phenomenon is determined by the dielectric material's response to an applied field that delivers valued insights into the performance of electrical systems, mostly at high frequencies (<10^5 Hz). Geometrical capacitance is defined as:

$$C_g = \frac{\varepsilon_0 \varepsilon_\infty A}{d} \tag{4.7}$$

where ε_0 dielectric constant, ε_∞ dielectric constant at high frequency, A is an area, and d is the thickness of the material.

Electrode capacitance: Electrode polarization is a phenomenon that occurs in ionic conducting systems for the reason that ions are incapable of traveling through the metallic accumulator interaction, results in reduction of conductivity in low-frequency window. Within the contact interface, the concentration of ions results in the production of a capacitance that is linked to the surface changes. This can be close to the contact if the film is adequately heavy and can remain to exist irrespective of the thickness of the film (*d*). The polarization of the electrode results in a significant rise in the values of apparent permittivity. This is a characteristic that is shared by a multitude of ionic systems. It is described as the capacitance related to electrode polarization, which happens within a particular frequency range.

$$C_{sc} = \frac{\varepsilon\varepsilon_0}{\gg_D} \tag{4.8}$$

Here, $\boldsymbol{\lambda}_D$ represents the Debye length, defined as

$$\lambda_D = \sqrt{\frac{\varepsilon\varepsilon_0 K_b T}{q^2 N}} \tag{4.9}$$

Contact capacitance: A work function offset at a metal-semiconductor interface forms a Schottky barrier that creates depletion zones that generate capacitive responses. Its magnitude depends on the depletion layer width (*w*), which can be adjusted to influence the system's capacitive characteristics, which is crucial for optimizing devices like diodes and transistors. Contact capacitance is typically detected at frequencies above 100 kHz and can be characterized mathematically as:

$$C_{dl} = \frac{\varepsilon_0 \varepsilon_\infty}{w} \tag{4.10}$$

Overview of the Constant Phase Element

Constant phase element (CPE) when the impedance does not follow ideal capacitor characteristics. It represents various phenomena, such as imperfect capacitors, double-layer capacitance in electrochemical systems, and other systems with complex impedance responses. The CPE is used to represent a depressed semicircle in Nyquist plots. Mathematically, the impedance of a CPE is expressed as

$$Z_{CPE} = \frac{1}{Q_0 (j\omega)^n} \tag{4.11}$$

where Q_0 is admittance at $\omega = 1$ rad/s, and n is a constant phase angle between 0 and 1. The CPE can represent systems with varying degrees of imperfection, from ideal capacitors ($n = 1$) to pure resistors ($n = 0$), with values in between indicating different deviation levels from ideal behavior. The CPE is commonly used in electrochemical systems to model phenomena like double-layer capacitance, rough surfaces, and heterogeneous reactions, providing a more accurate representation of the system's electrical properties (Gateman et al. 2022).

Overview of Inductance

The inductive behavior in PSCs has gained significant attention due to its association with dynamic hysteresis in current density-voltage characteristics (Hernández-Balaguera et al. 2023). It is primarily caused by ionic migration (Guerrero et al. 2016; Alvarez et al. 2020), phase-delayed recombination (Jacobs et al. 2018), and other kinetic effects (Khan et al. 2021; Ebadi et al. 2019). Recent studies describe a "chemical inductor" concept that explains negative voltage transients and hysteresis using a neuron-style dynamic model derived from surface polarization (Ravishankar et al. 2017; Gonzales et al. 2022). This model employs kinetic equations to relate slow ionic relaxation to fast external voltage changes, revealing essential parameters like capacitance, resistance, and inductance (Alvarez et al. 2020; Bou et al. 2020). These insights, supported by equivalent circuit modeling, advance the understanding of electrochemical processes and their influence on PSC performance.

Understanding of Simulated Basic EEC Models

Understanding basic equivalent circuit (EEC) models is crucial for analyzing EIS data. EEC models represent the internal behavior of electrochemical systems, like batteries, through external characteristics using elements such as resistors, capacitors, and Warburg diffusion components. These models help researchers interpret complex impedance spectra, predict system behavior, and gain insights into electrochemical processes. Using tools like Zview, Zplot, Zsimpwin, and EIS analyzer, EEC models are constructed to simulate system behavior, which serves as a foundation for decision-making in electrochemical research. This section explores the building blocks of EEC models.

When resistance and capacitance are present in the circuit, as shown in Fig. 4.2a. equation of impedance

$$Z = R_0 + \frac{1}{j\omega C} = R_0 - j\left(\frac{1}{\omega C_1}\right) = Z' + jZ'' \Rightarrow Z' = R_0, Z'' = -\frac{1}{\omega C_1} \tag{4.12}$$

If we look at the pattern in the Nyquist plot (Fig. 4.2a), it is a combination of the Nyquist plot for only capacitance and resistance. Suppose we observe the real and imaginary parts at higher and lower frequency regions. In that case, the imaginary part is neglected in the high-frequency areas, and the fundamental part dominates similarly in the lower frequency imaginary part. So, we observe (Fig. 4.2a) a straight line (independent of frequency) in higher frequency regions. In the low-frequency window, the imaginary part surges at a precise frequency, $\omega = 1/R_0C$, when the genuine part is identical to the imaginary part and after the imaginary part controls this response. An S-shaped curve in the Bode phase plot can also be detected in this conversion.

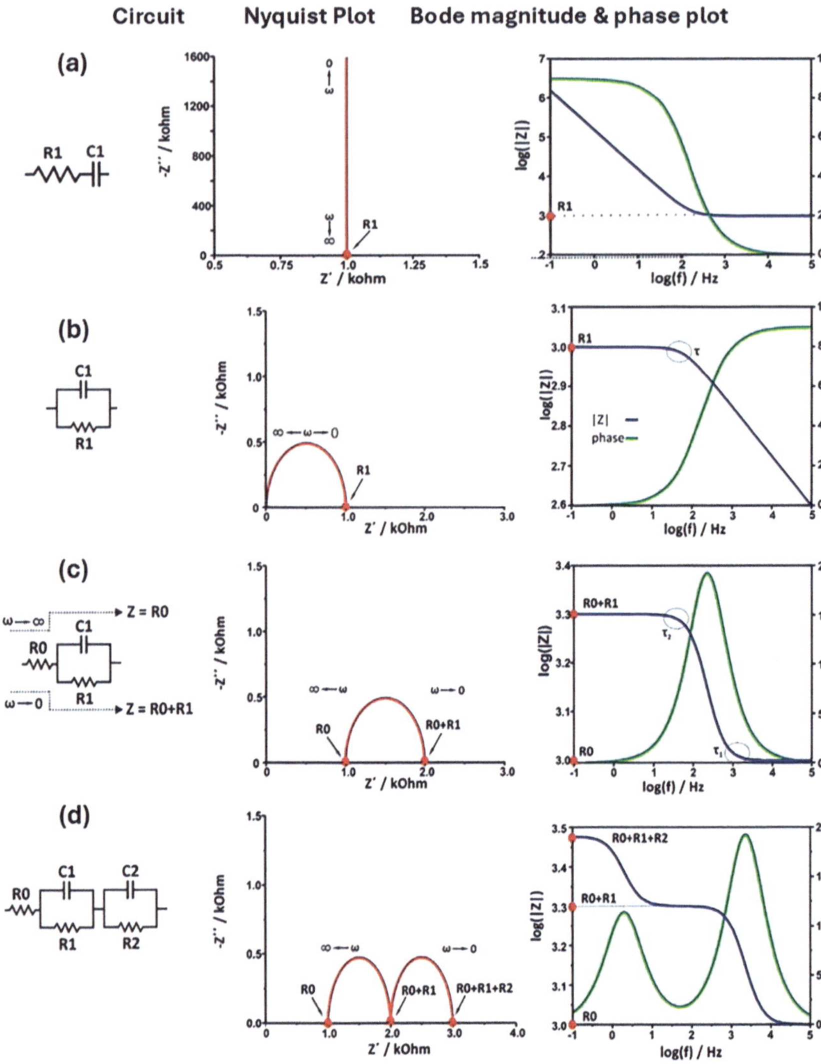

Fig. 4.2 Nyquist, Bode magnitude, and phase angle plots of some basic model circuits simulated by Z-view for understanding. (Lazanas and Prodromidis n.d.)

When the circuit contains a capacitance and resistance with parallel connection, Fig. 4.4b, the equation of impedance is as follows:

$$Z(\omega) = \frac{1}{\frac{1}{R_1} + jwC_1} = \frac{R_1}{1 + j\omega R_1 C_1} = \frac{R_1}{1 + (\omega R_1 C_1)^2} - j\frac{\omega R_1^2 C_1}{1 + (\omega R_1 C_1)^2} \quad (4.13)$$

It is observed as a semicircle in the Nyquist plot. In the high-frequency case, capacitive reactance tends to zero, so it works as a short circuit, and all current passes through it. In the lower frequency region, capacitive reactance inclines to endlessness, so all current passes through the resistor, so impedance contains only the real part (R_1). From the High to the low-frequency change, a single characteristic frequency (impedance part is maximum) corresponds to an equal value of reactance and resistance.

$$\frac{1}{\omega C_1} = R_1 \Rightarrow \omega_{Z''_{max}} = \frac{1}{\tau} = \frac{1}{R_1 C_1} \tag{4.14}$$

Here $\tau = R_1C_1$ is the characteristic time constant of the system. At this frequency, the real and imaginary part becomes equal. Bode mode plot 4(a), according to the above equations, $|Z| = R_1$ (at very low frequencies) and $|Z| = 1/\omega C_1$ (while at very high frequencies). The slant of the bend is varying, represented in Fig. 4.4b by a spotted line circle about the frequency matching to the system's time constant. If we link a R_0 resistance shown in Fig. 4.4b, then circuit seems alike Fig. 4.4c and impedance is denoted by:

$$Z(\omega) = R_0 + \frac{R_1}{1+(\omega R_1 C_1)^2} - j\frac{\omega R_1^2 C_1}{1+(\omega R_1 C_1)^2} \tag{4.15}$$

The circuit responds resistively at both very high and very low frequencies, which causes a shift in the Nyquist plot's semicircle to the real axis at the circuit's ohmic resistance (R_0). At high frequencies ($\omega \rightarrow \infty$), the impedance simplifies to (R_0) while at low frequencies ($\omega \rightarrow 0$), it becomes $R_0 + R_1$. This circuit exhibits two-time constants: τ_1 that is influenced by R_0 and $\tau_1 = [(R_0R_1)/(R_0 + R_1)]C_1$ represents the slower response that can be derived on Nyquist plane similar to a simple RC parallel circuit, where $\omega_{Z''_{max}} = \frac{1}{\tau} = \frac{1}{R_1 C_1}$. The Bode plot reflects these dynamics, showing an S-shaped variation in $|Z|$ and a bell-shaped phase curve across frequencies, highlighting the impedance and phase transitions.

If we add a parallel RC circuit into Fig. 4.4c, the circuit is represented in Fig. 4.4d. The equation of impedance

$$Z(\omega) = R_0 + \left[\frac{R_1}{1+(\omega R_1 C_2)^2} + \frac{R_2}{1+(\omega R_2 C_1)^2}\right] - j\left[\frac{\omega R_1^2 C_2}{1+(\omega R_1 C_2)^2} + \frac{\omega R_2^2 C_1}{1+(\omega R_2 C_1)^2}\right] \tag{4.16}$$

The circuit in Fig. 4.4d, which is analogous to the examination offered before in Fig. 4.4c, reveals two distinct time constants τ_1 and τ_2, those relate to the highest value of $\omega_{Z''_{max}}$ at each semicircle. Depending on the ratio of the values of τ_1 and τ_2, the Nyquist plot exhibits either two semicircles that are well-resolved (where τ_1 is more significant than 100 τ_2) or poorly resolved (where τ_1 is less than 100 τ_2). Nevertheless, when τ_1 equals τ_2, a single semicircle is observed. In addition, the ratio between the $\tau 1$ and $\tau 2$ values impacts the Bode magnitude and phase chart design, including a second step and peak in the plots formed since the second time constant arises when the value of τ_1 is equivalent to or greater than 100 times τ_2.

Classification of Relaxation Processes in the Impedance Spectra

Relaxation processes observed in impedance spectra can be systematically classified based on their characteristic frequency response and behavior, aiding in the interpretation of electrochemical systems. Figure 4.1e illustrates the different dynamic processes associated with distinct relaxation times. The relaxation processes in impedance spectra can be classified into resistive-capacitive and resistive-inductive distribution of relaxation times functions, depending on the nature of the system under study. The relaxation procedures attribute numerous physical phenomena like charge transmission, diffusion, and ionic relocation, which can disturb the performance and steadiness of electrochemical systems.

Classifying impedance spectra involves identifying relaxation processes through Nyquist plots and equivalent circuit models. Debye processes appear as semicircles, indicating ideal capacitive behavior, while Warburg impedance manifests as 45° linear trends, reflecting diffusion-controlled processes. Complex systems often show multiple arcs or peaks due to diverse electrode reactions or charge transfer mechanisms. In PSCs, low-frequency IS spectra (<100 Hz) reveal ionic mobility influenced by bulk and interfacial effects. In contrast, high-frequency responses (>10^4 Hz) highlight trap-assisted recombination and defect-related phenomena. Analyzing IS spectra uncovers parameters like trap densities, energy levels, and relaxation time constants ($\tau = RC$), offering insights into resistive, capacitive, and dispersive behaviors that shape device performance.

Physical EIS Modeling for PSCs

Equivalent Circuit Modeling (ECM) is widely used in IS analysis. Still, its success relies on accurately assigning physical components to electrical elements, a process complicated by challenges like evaluating time constants and interpreting IS spectra. The Distribution of Relaxation Times (DRT) technique aids by converting EIS data into the time domain to identify time constants but is limited by statistical deviations. Selecting a suitable EEC model involves either empirical fitting or deriving physical models, with the latter offering more profound insights into electrochemistry and charge transfer. Care must be taken to avoid overfitting by only adding parameters that meaningfully enhance the fit quality.

In real life, things are more complicated than in an ideal EEC. So, to make a better model for EECs, we need to consider how physical things work. This means looking at real data and the processes in the system we are interested in based on what we see in experiments. In this part, we will talk about ideas and different types of models for interpreting the spectra of PSCs. Each type of model we discuss will explain specific parts of a circuit, how they are connected, and what they mean in terms of what is happening in PSCs. We will also look at how different circuit parts affect frequencies like low, medium, and high frequencies.

Type 1: Voigt Circuit

The Voigt Circuit is one of the circuits used in literature the most frequently over the years. As shown in Fig. 4.3b, it has become the most common EC utilized to fit the IS spectra of PSC measurements. Among the important components that are encompassed in this circuit are R_s, R_{ct}/R_{tr}, and R_{rec}. These components are associated in parallel with their corresponding constant phase elements (CPE1 and CPE2). It provides data about these elements. A lot of studies are available in the body of previous literature. We took a study for analyzing the IS by Bhaskar Parida, which is illustrated in Fig. 4.3a, and corresponding their circuit and impedance of circuit (Parida et al. 2019). Bhaskar Parida explored charge transport and recombination in PSCs by conducting IS tests on devices with mp-TiO_2 and c-TiO_2/m-TiO_2 layers. The results of the study confirmed that the c-TiO_2/m-TiO_2 layer exhibited lower values for R_s and R_{tr} in contrast to the mp-TiO_2 layer under examination. This reflection proves that the c-TiO_2/m-TiO_2 layer is an effective charge transport layer, and it also underscores the effect of this layer on recombination rates. It was determined that the interface between the c-TiO_2/m-TiO_2 layer and the ETL and the Perovskite layer was responsible for the increased charge transport capabilities of the c-TiO_2/m-TiO_2 layer. The rate of charge recombination has been successfully lowered as a result of their successful efforts. As a consequence of this, the c-TiO_2/m-TiO_2 layer makes a significant contribution to the optimization of charge transport and the reduction of recombination, which ultimately results in an improvement in the overall efficiency and functionality of the PSCs.

With the use of the Voigt circuit, the researchers were able to determine the charge collecting efficiency (CCE) of PSCs by applying the equation that is presented below: (Obrero-Perez et al. 2022):

$$CCE \approx \left(1 - \frac{R_{rec}(OC)}{R_{rec}(NOC)} \times 100\right) \tag{4.17}$$

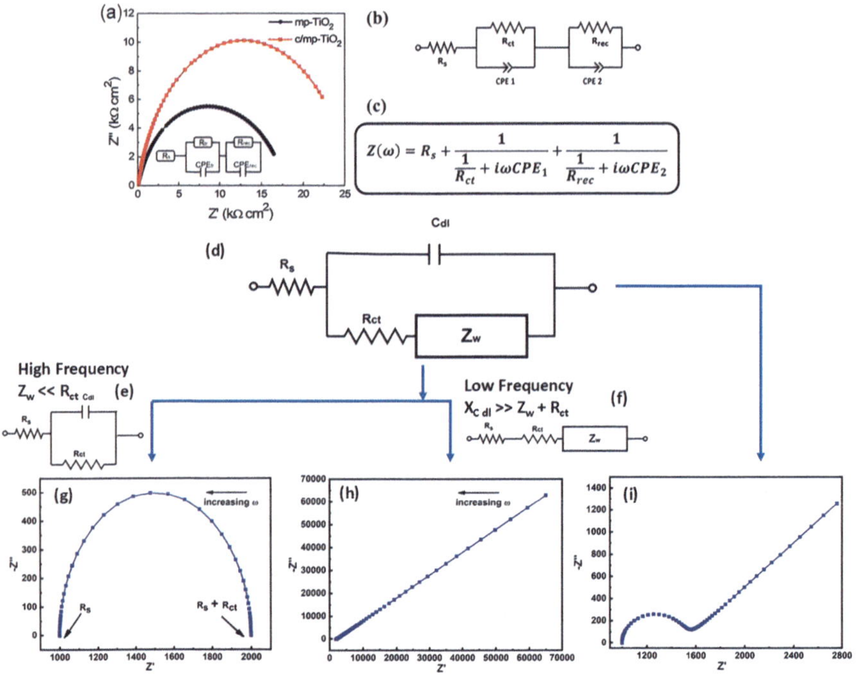

Fig. 4.3 (a) IS on devices with mp-TiO_2 and c-TiO_2/m-TiO_2, (b) Voigt Circuit (widely used circuit), utilized for circuit fitting (c) corresponding to their total impedance (Parida et al. 2019), (d) Randles circuit, (e) conversion at high frequency, and (f) at low frequency (g–i) simulated IS by EIS analyzer according to the circuit

The symbol R_{rec} denotes the R_{rec} in high-frequency region, whereas OC and NOC stand for open circuit and non-open circuit conditions, respectively.

It is well reported like (Almora et al. 2020; Xu et al. 2023) that the extraction of extracting high-frequency resistance in PSCs using Voigt circuit technique. In order to contribute to the development of PSCs technology, the purpose of the study was to obtain insights into the recombination behavior and CCE of PSCs. This was accomplished by the utilization of the Voigt circuit and the equations that were described earlier.

The Voigt circuit struggles to fully interpret electrochemical systems (particularly at interfaces) where it fails to account for inductive effects and frequency-dependent impedance behaviors observed in many PSCs. Its reliance on resistors and capacitors/CPEs limits its ability to analyze complex phenomena like charge transfer kinetics. To overcome these limitations, the modified Randles circuit is often employed. It offers enhanced features to better capture interfacial effects and provide a more accurate interpretation of impedance data across diverse PSCs.

Type 2: Randles Circuit

The Randles circuit is a widely used in the field of electrochemistry and electrochemical impedance spectroscopy (EIS). It is a model that is frequently utilized for the purpose of researching ion migration as well as oxidation and reduction processes. Planar electrode-faradaic reaction (controlled by a semi-infinite diffusion) is modeled by the Randles circuit. It comprises the ionic resistance (R_i), double-layer capacitance (C_{dl}), charge transfer resistance (R_{ct}), and Warburg impedance (Z_W) elements. The Randles circuit is arranged in a specific way to represent the physical processes occurring at the electrode-electrolyte interface, and it produces a distinctive Nyquist plot with a semicircle followed by a 45° line.

Figure 4.3d illustrates the Randles circuit. At low frequency, the reactance of C_{dl} is very high so all the current passes through the R_{ct}, R_s, and Z_W corresponding circuit is represented in Fig. 4.3f and IS in Fig. 4.3h. At higher frequencies, R_{ct} is much bigger than Z_W so current flow is implemented by R_s, R_{ct}, C_{dl}. Corresponding to this EEC is represented in Fig. 4.3e and IS in Fig. 4.3g. In real system, we observe both pattern (semicircle and straight line) of the Nyquist plot over a wide range of frequencies. And whole system is controlled by all parameters corresponding to this EEC is represented in Fig. 4.3d and IS in Fig. 4.3i. If we fix all the parameters except C_{dl}, by simulation we can see that distortion increase with along with C_{dl}. Clear explanation of Randles circuit is given in the study effects of ionic functional groups on ion transport at perovskite interfaces (Liu et al. 2017). Liu et al. studied the effects of ionic functional groups on ion transport at perovskite interfaces (Liu et al. 2017), high-frequency semicircle (closest to the origin) reflects transportation of electrons, while the low frequency semicircle is from transportation of MA^+ ions. IS response of each device was symbolized by an initial high-frequency semicircle, which was displayed by all of the devices. At low frequency, for devices with cationic PVBT-TMA and anionic PVBT-SO_3 HTLs, impedance is represented by linear semi-infinite ion diffusion. EEC was utilized in order to model IS, this system combines high-frequency charge transfer with low-frequency ionic transport. Here R_s is used to represent the contact resistance between layers as well as in the electrode connections. The impedance at high frequencies is embodied by bulk ionic and electronic charge transport/recombination resistance ($R_{tr/rec}$). Moreover, the charge stored in the perovskite is portrayed, in parallel, with a chemical capacitance element (C_μ). The electronic transport that is not linked with the ionic transport is denoted by a parallel resistor (R_{elec}). The low-frequency feature observed is attributed to ion diffusion, which is dependent on ion size and temperature. To model this, W_s was used in series with R_{CT} that is similar to a Randles circuit and with double-layer capacitance (C_{DL}) representing charge accumulation. In zwitterionic PVBT-SB HTL devices, the EIS spectrum shows a linear component transitioning to a semicircle, indicative of two ion transport scenarios: reactive barriers with low resistance or diffusion to an adsorbing boundary. The PVBT-SB acts as an adsorbing interface for MA^+ ions in $MAPbI_3$, confirmed by separating the interface with PVBT-SO_3, yielding a Warburg-like spectrum indicative of semi-infinite diffusion.

To conclude, the Randles circuit serves as an important basic model in EIS for interpreting IS. This circuit operates under the assumption that the rate of the faradaic reaction is governed by the diffusion of reactants to the electrode surface. While the model can be extended to represent more intricate electrochemical systems, it has limitations in accurately capturing the complexity of real systems due to its simplicity. This circuit functions under the supposition that the rate of the faradaic reaction is administered by the diffusion of reactants to the electrode surface. Besides, the supposition that the frequency of the faradaic response is exclusively dictated by diffusion might overgeneralize real electrochemical processes, leading to potential inaccuracies in data interpretation, particularly in systems with more intricate behaviors. Furthermore, the Randles circuit may not be suitable for systems with multiple overlapping processes, as it is primarily tailored for semi-infinite diffusion-controlled faradaic reactions to planar electrodes, thus restricting its applicability to more diverse electrochemical systems.

Type 3: Transmission Line Model

It has been demonstrated that the transmission line model can be effectively utilized for the purpose of analyzing Dye Sensitized Solar Cells (DSSCs) and their variations, particularly those that make use of changeable colors like chlorophyll, anthocyanin, and beta-carotene. On the other hand, this model is most commonly seen in mesoporous TiO_2 arrangements for solid-state capacitors. This particular model's transmission design is characterized by a straight line (signifies carrier transport), followed by an arc at lower frequencies (linking of capacitance with recombination resistance). This pattern is characterized by the fact that the arc occurs at lower frequencies. (Fabregat-Santiago et al. 2011; Cari et al. 2018)

The Transmission Line Model (TLM) is a widely used equivalent circuit for PSCs, particularly for analyzing the electrical behavior of the device. The TLM is based on the concept of a series of transmission line segments, each representing a small section of the PSCs, where the electrical properties of the material are assumed to be constant (Yoo et al. 2019). The TLM is used to extract important parameters of PSC, such as the R_{TR} for both electrons and holes, which can provide insights into the charge transport behavior of the device (Yoo et al. 2019). The TLM is also used to model the electrical behavior of the device under varying illumination intensity, which is important for understanding the mechanisms limiting energy harvesting in medium- and long-range wireless power transfer (Timofeev et al. 2023).

However, TLM has some limitations. For example, the TLM assumes that the electrical properties of the material are constant throughout the device, which may not be the case in real-world devices where the electrical properties may vary with the thickness of the perovskite layer (Yoo et al. 2019). Additionally, the TLM does not account for the effects of interfacial layers, such as the electron transport layer (ETL) and the hole transport layer (HTL), which can significantly impact the electrical behavior of the device. (Yoo et al. 2019). Another limitation of the TLM is that it does not account for the effects of recombination, which can significantly impact the efficiency of the device. (Yoo et al. 2019). Recombination is the process by which charge carriers (electrons and holes) combine and lose their energy, results in decrease in the efficiency of the device. Despite these limitations, the TLM is a useful in understanding the electrical behavior of PSCs and for extracting important parameters of the device. However, it is important to keep in mind the assumptions and limitations of the model when interpreting the results. It does not reveal PSCs' time-dependent behavior and transitory processes. The model assumes uniform circuit elements and ignores nonlinear behavior, reducing its accuracy. This model struggles to explain and extract essential phenomena like trap-assisted recombination and ion migration.

Figure 4.4a shows the transmission line model of PSCs, with R_s, R_t, R_{BL} (TiO_2 compact blocking layer resistance), R_r (recombination resistance), and R_{HTL} (hole

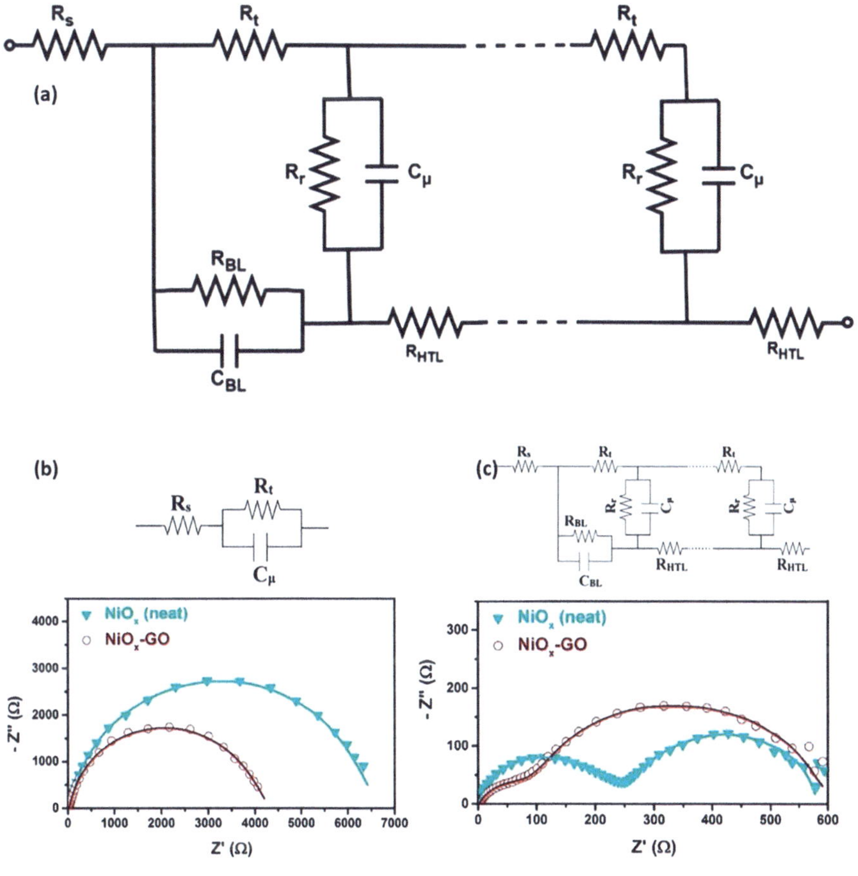

Fig. 4.4 (**a**) Transmission model of PSCs (Nouri et al. 2018). Nyquist plots EEC of the NiO_x- and NiO_x–GO-based PSC devices, these plots and circuits were measured under two different illumination conditions: (**b**) dark and (**c**) AM 1.5G illumination. The bias voltage used was V_{OC}. The fitted curves are represented by solid lines (Nouri et al. 2018)

transport resistance), C_{BL} and $C\mu$ represent blocking layer and chemical capacitance, respectively. Gonzalez-Pedro et al. (2014) conducted a noteworthy study and observed that the transmission line phenomena disappears at greater biases (>0.5 V) in PSCs, and instead, a Gerischer pattern happens. This shift is suggestive of a diminished step of freedom in robust recombination, which ultimately results in a significantly shorter diffusion length (L_d) in comparison to the layer thickness (L) (Gonzalez-Pedro et al. 2014; Bisquert et al. 2014b). Typically, the Gerischer pattern is seen in PSCs when R_{tr} is greater than the R_{rec}. This causes the carrier diffusion length (L_d) to decrease to a point where it is less than the sample size or the layer thickness (L) (Gonzalez-Pedro et al. 2014).

For IS analysis by TLM, we took a study by Nouri et al. (2018) which is represented in Fig. 4.4c. The resistance (R_s) is typically consistent across all devices due to their structural similarities. Yet, a notable imprint was observed in the charge transfer resistance (R_t) values at the interfaces of the electron transport layer (ETL) or hole transport layer (HTL) with the perovskite layer between devices utilizing pristine NiO_x and those engaging enhanced NiO_x–GO layers (specifically NiO_x–3GO). Nyquist plots, depicted in Fig. 4.4, illustrate this difference under dark conditions and at the open-circuit voltage (V_{OC}), revealing internal charge transport and recombination within the PSC devices. The plots display a sole semicircle, signifying the charge transfer resistance (R_t) at the NiO_x or NiO_x–GO and perovskite film interface, with the NiO_x–GO-based devices showing lower R_t, suggesting quicker charge transport at the interface of the HTL and perovskite. This reflection brings into line with deductions drawn from photoluminescence measurements. This observation aligns with conclusions drawn from photoluminescence measurements. Additionally, Nyquist plots under AM 1.5G illumination display high- and low-frequency semicircles, indicative of charge transfer resistance (R_t), series resistance (R_s), dielectric contributions, and recombination resistance (R_r) at the photoactive layer. Fitting the plots with equivalent circuits shows that NiO_x–GO-based devices have smaller R_t and larger R_r, indicating enhanced hole extraction and reduced charge recombination, leading to superior device performance compared to those based solely on pristine NiO_x. This suggests that while the presence of GO does not affect series resistance, it advances the structural body of the HTL, enabling charge transmission and obstructing charge recombination. Additionally, comparing the photoconversion competence over time of PSC devices using up spiro-OMeTAD (as a reference standard HTL) with those engaging NiO_x and NiO_x–3GO, and either Au or carbon back contact electrodes, reveals superior performance of NiO_x–GO-based devices. These comparisons were conducted under controlled humidity and ambient air conditions, highlighting the importance of HTL composition on the photovoltaic performance of PSCs.

In summary, TLM is an EEC used to analyze the electrical behavior of PSCs. The TLM is based on the concept of a series of transmission line segments, each representing a small section of the device. The TLM is used to extract important parameters of the device, such as the R_{TR} for both electrons and holes, and to model the electrical behavior of the device under varying illumination intensity. However, the

TLM has some limitations, such as assuming constant electrical properties throughout the device, not accounting for interfacial layers, and not accounting for the effects of recombination.

Type 4: Diffusion-Recombination Model

Additionally, the Diffusion-Recombination Model was developed with the express purpose of analyzing the behavior of traps in disordered materials. This was accomplished by taking into consideration trapping factors and nonlinear recombination rates. The Transient Line Model was also developed for this purpose. Bisquert et al. have provided the remarkable mathematical expressions for the a new upgraded model with respect to PSCs (Bisquert et al. 2014b). Reader can follow this article about calculation of impedance (Eq. (4.18)). A representation of the kinetics of minority charge carriers is included in this model. This impedance representation can be expressed as follows:

$$z(\omega)=\left(\frac{r_{tr}.r_{rec}}{1+\frac{i\omega}{\omega_{rec}}}\right)^{1/2}\operatorname{cotanh}\left[\left(\frac{r_{tr}}{r_{rec}}\right)^{1/2}\left(1+\frac{i_{\omega}}{\omega_{rec}}\right)^{1/2}\right] \quad (4.18)$$

Within this context, the R_{TR} and the recombination resistance are denoted by the symbols r_{tr} and r_{rec}, respectively. The distinctive angular frequency of recombination is represented by the symbol, ω_{rec}. On the contrary, *cotan*h is a function that characterizes the diffusion length and its effect on the resistance that is given through direct current.

In Fig. 4.5, EEC of transmission line model for diffusion-recombination is represented. When the diffusion resistance is overlooked in a high mobility semiconductor, there is no need to use a transmission line model because the transport features can be misplaced from the commencement. This is how an overall transmission line revealed in Fig. 4.5a can be abridged to the lumped circuit shown in Fig. 4.5c, which is parallel combination of C_d and R_r.

The effect of Warburg impedance is depicted in Fig. 4.3, showcasing the semi-infinite diffusion of chemical species with a single barrier at the electrode/electrolyte interface ($x = 0$), where diffusion extends indefinitely under quiescent conditions as the concentration gradient decreases over time (Fig. 4.5d). This time-dependent behavior is modeled as a transmission line of infinite length with resistors and capacitors representing R_d and C_d (Fig. 4.5e) (Kim and Pyun n.d.). However, in many electrochemical systems, a finite-diffusion model better describes the impedimetric response, where the diffusion length is limited to $x = L$ (Fig. 4.5f). Here, one side of the material is permeable to ions, while the other may be permeable or

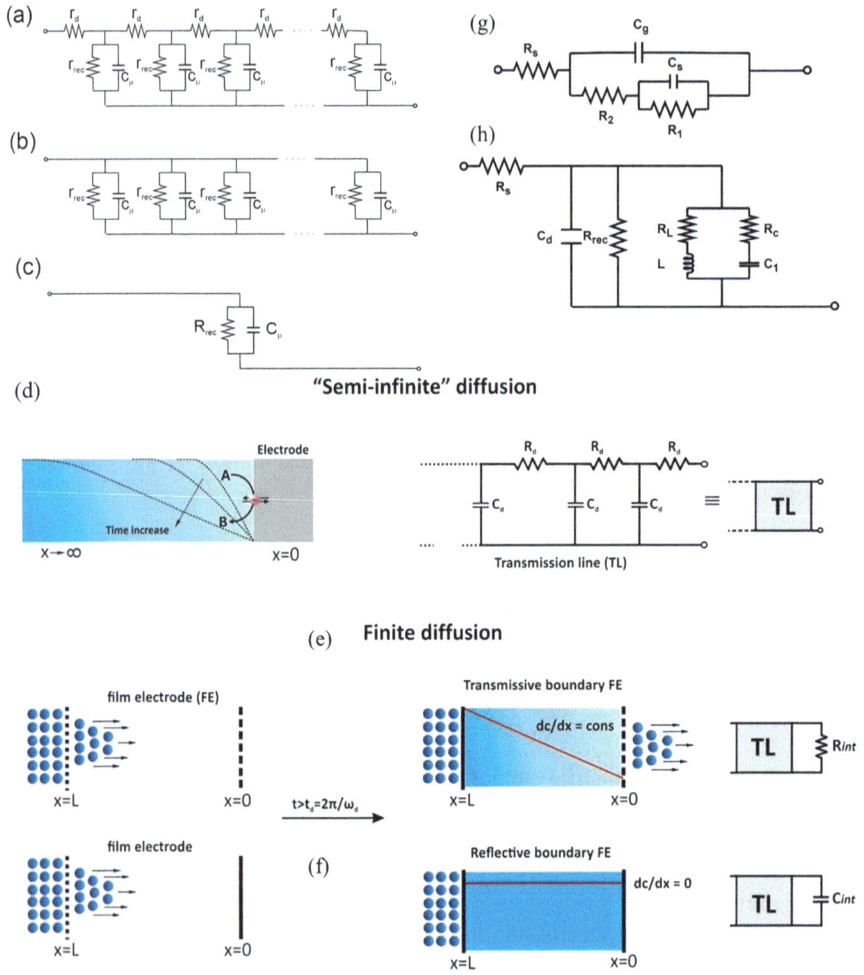

Fig. 4.5 (**a**) EEC of transmission line model for diffusion-recombination (**b**) same when R_{tr} is negligible (**c**) case (**b**) can be written simplified form (Guerrero et al. 2021; Bisquert et al. 2014b). (**d**) The semi-infinite regime of an electrochemical cell with the transmission line (TL) depiction of the semi-infinite regime, finite boundary diffusion at $t < t_d$ spans, (**e**) transmissive, and (**f**) reflective boundary at $t > t_d$ (Lazanas and Prodromidis n.d.)

impermeable. The finite-diffusion region evolves over time until $t > t_d = \frac{2\pi}{\omega_d} = \frac{2\pi}{D/L}$, where D is the diffusion coefficient, beyond which the concentration gradient stabilizes or diminishes. During this period, the system transitions from semi-infinite diffusion to finite-diffusion behavior, governed by the unique geometry of the electrochemical cell.

Essentially, finite diffusion can be divided into two scenarios: (a) When the finite-diffusion zone is permeable to electrochemical cells create a steady-state concentration gradient (dC/dx = constant) as species diffuse, resulting in a current flow. This is a porous or transmissive barrier. (b) Once the finite-diffusion area becomes impermeable to diffusing species, charge transfer is disrupted, and the concentration gradient of the species becomes zero ($dC/dx = 0$). This is a reflecting or impermeable barrier. For gaining much more understanding, reader can refer to reference (Lazanas and Prodromidis n.d.).

Later investigations showed that ionic transport across the interface and within PSCs was important. PSCs were ionic-electronic mixed conductors (Meloni et al. 2016; Eames et al. 2015). Iodine ions made a great effect in the low-frequency region. This reduced low-frequency (lf) dielectric response produces hysteresis induced by current. Therefore, the diffusion-recombination model for PSCs, which involves electrons and holes is inappropriate since it ignored ionic transport and build-up. PSC *lf*-IS spectra show that ETL and HTL selection greatly affects recombination resistance. Photo-generated electron-hole pair recombination diminishes device efficacy. This resistance can be calculated with diffusion length or characteristic times from impedance measurements (Gonzalez-Pedro et al. 2014). Two distinct temporal periods in the lf region suggest various recombination processes (Zhao et al. 2014). Additionally, the recombination method must account for intrinsic ion reorganization in the perovskite layer as well as the lf area. Without HTL material (spiro-OMeTAD), R_{TR} dominates in the high-frequency (hf) region. An interfacial recombination mechanism, not a bulk process, causes this phenomenon, emphasizing charge separation in perovskites at hf (Sanchez et al. 2014; Juarez-Perez et al. 2014b). However, selective interactions must be understood for complete insights.

In summary, both the simplicity of the diffusion-recombination model and the valuable insights it provides into charge carrier movement in semiconductor materials are advantages that the model offers. The fact that it does not consider ionic conductivity and does not consider the intricate recombination routes that are present in particular materials can, however, cause it to deviate from the experimental data. As a consequence of this, the model would not be able to accurately represent the behavior of more complex systems, such as PSC, which feature mixed conductors and various recombination paths. As a result, they are less useful in situations like these. Within the context of addressing these difficulties, the Surface Polarization has emerged as a useful model.

Type 5: Surface Polarization Model

It is a theoretical framework that explains the dynamic hysteresis observed in PSCs. This model suggests that the charge accumulates a large quantity of surface polarization due to the slow transient process associated with the ion migration and charge accumulation at the interfaces (Ravishankar et al. 2017; Ebadi et al. 2019).

The surface polarization model can explain the observed correlation between high apparent capacitances and hysteresis in current-voltage curves, which cannot be used to quantify charge accumulation. The respective resistances, nevertheless, are related to electron and hole injection, recombination, and transport in the solar cell, being primarily determined by the most limiting process. (Ebadi et al. 2019)

The Surface Polarization Model assumes large electric and ionic charge accumulation at the surface and interfaces of the PSCs. This model can be used to understand the origin of apparent light-enhanced and negative capacitance in PSCs, which is due to the surface polarization effect (Ebadi et al. 2019; Ghahremanirad et al. 2017). The polarization field model is another theoretical framework used to describe the behavior of PSCs. This model is built on the Schottky barrier effect and depolarization field-driven photovoltaic effect, which can be used to understand the principles and advances of photoferroelectrics PSCs (Li et al. 2021).

Ravishankar et al. proposed a surface polarization model analysis of dynamic hysteresis with significant transient current to investigate polarization effects (Ghahremanirad et al. 2017). This study showed using fullerene contact instead of TiO_2 as a perovskite layer ETL eliminates PSC hysteresis effects at normal temperature. It was also shown that fullerene-based PSCs lack capacitance, which causes interface polarization, and that the interface charges electronically, creating transient current and hysteresis. These findings match literature (Ravishankar et al. 2017). This study's capacitance-frequency plots show that ionic polarization is low frequency at the perovskite interface while dielectric polarization is high frequency in the bulk.

Another understanding of the surface polarization effect and the behavior of inductive loops in PSCs was achieved by Ghahremanirad and colleagues through the development of a dynamic hysteresis model that included the EIS response (Ebadi et al. 2019). The model takes into account a kinetic relaxation period for ion displacement, which leads to the formation of separate impedance arcs that have their own set of underlying distinctive characteristics. Important kinetic information regarding the interfaces of PSCs, such as delayed surface voltage and its strict dependence on kinetic relaxation time, can be obtained from the inductive loops that have been discovered. These loops indicate positive and negative capacitance.

In order to highlight the dynamic hysteresis behavior of PSCs, Fig. 4.5g, h depicts the EEC of surface polarization model. This model incorporates the electrical elements that are connected with the motions of ions. In this context, the symbol C_d represents the dielectric capacitance that was created in order to specifically account for dielectric bulk/surface phenomena experienced at higher frequencies. In addition, the symbol R_c represents the series resistance in conjunction with the capacitance C_1, and the product of these two variables, denoted as (R_c/C_1), may be used to get the time constant value, which governs the kinetics process within the PSCs. Reader can refer to (Ghahremanirad et al. 2017; Hernández-Balaguera and Bisquert 2024) for calculation of impedance of EEC represented in Fig. 4.5h.

$$Z(\omega) = R_s + \left[j\omega C_g + \frac{1}{R_b} + \frac{1}{R_a + j\omega L_a} + \frac{1}{R_x + \frac{1}{j\omega C_x}} \right] \tag{4.19}$$

The following is possible interpretation of the circuit from a physical standpoint: the parallel combination of R_{rec} and C_d creates a typical circuit for solar cells, which is controlled by recombination. An extra circuit, represented as $R_C - C_1$, is accountable for the charging and discharging of the surface accumulation capacitance. This process is governed by the kinetic relaxation constant, $\tau\text{-}_{kin}$. It is a result of charge build-up and extraction at the surface. The non-standard process $R_L - L$, which is a notable characteristic of the model, is combined in order to account for the temporal delay in surface charging that is produced by the motions of ions. As a result of the presence of an inductive line in the analogous circuit, this delay is controlled by the, V–I. This variable determines the pace at which the interface reaches equilibrium, which in turn leads to the presence of the inductive line.

For better understanding of IS, we took a latest study of 2024 by Balaguera (Hernández-Balaguera and Bisquert 2024) which is denoted in Fig. 4.5. In this, they infer the importance of crucial representation of the frequency-dependent capacitance in the Bode plot illustration. The impedance data discloses a changeover between two capacitive plateaus that is marked by negative values and undershoots in the mid-frequency range and linked to an inductive loop in the impedance. These relaxation processes, tied to underlying physical properties, are demonstrated by the equivalent circuit in Fig. 4.5h, R_s is the parasitic series resistance, R_b and R_a are the two branches of the recombination resistance, and R_x is the resistance for ionic relaxation. The dielectric capacitance is denoted by the symbol C_g, the chemical inductance is denoted by L_a, and the interfacial charging capacitance is the symbol C_x. Since it makes use of the vast array of ambiguous perovskite patterns discovered through IS in a manner that is both universal and physiologically motivated, the EEC depicted in Fig. 4.5h is the most successful equivalent circuit for the impedance response. Impedance is represented by Eq. (4.3). Readers can refer to the reference (Hernández-Balaguera and Bisquert 2024) for much deeper understanding. The surface polarization model was also utilized in research such as (Kim et al. 2013; Ghahremanirad et al. 2017; Fabregat-Santiago et al. 2017) to gain a better understanding of the data that was collected during the creation of positive/negative inductive loops. These investigations shed information on the role that RL plays in determining the cause of the delay in surface charge that is brought about by the passage of ions. The capacitance C_1 is associated to the low-frequency area, while the dielectric capacitance C_d is reflected in the high-frequency portion of the spectrum. To obtain a comprehensive understanding of this phenomena, nevertheless, it is essential to have specific insights into the kinetic process, interfacial chemistry, and the fundamental physical components that impact device performance. This is for the purpose of acquiring a comprehensive understanding of this phenomenon.

In summary, the Surface Polarization Model is a theoretical framework that describes the dynamic hysteresis observed in PSC. This model is built on the idea that the cell accumulates a large amount of surface polarization due to the slow transient course associated with the ion migration and charge accumulation at the interfaces. The polarization field model is one more theoretical framework used to describe the behavior of PSCs, which is grounded on the Schottky barrier effect and depolarization field-driven photovoltaic effect.

Type 6: Neuron Style Model

Gonzales et al. proposed the "Neuron Style Model." It gives a deep insight into the J–V hysteresis behavior and EIS response in Metal Halide PSCs. It also helps us understand negative capacitance (inductive loop) formation at high voltage (Gonzales et al. 2022). This model is essential for better understanding and optimizing the performance of PSCs. They proposed a Neuron Style Model based on differential equations that were valid for any perturbation. According to this model, the total current flowing through combines several components. A portion of the total current (I_{tot}) corresponds to displacement current, which charges Cg. The remaining current (I_v) is then subdivided into three distinct branches: fast current extraction at the contact (I_{cat}), surface capacitance charging (C_1), and slow current extraction through an ion-modulated current (i_d). This systematic division allows for a comprehensive understanding of the current flow dynamics within the device, highlighting the contribution of each component toward its overall operation and performance. These equations define this model:

$$I_{tot} = C_g \frac{\mathrm{d}u}{\mathrm{d}t} + I_v(v) \tag{4.20}$$

$$I_v(v) = \frac{\mathrm{d}Q_s(w)}{dt} + I_c(\omega)^{+i_d} \tag{4.21}$$

$$\tau_d \frac{di_d}{\mathrm{d}t} = I_d(w) - i_d \tag{4.22}$$

In this model, I_{tot} and voltage (u) are exterior variables. The current components $I_v(v)$ and $I_c(\omega)$ are instantaneous roles of the respective voltage, I_d is variable current influenced by ionic effects with a characteristic time τ_d. The surface charge is described by the function $Q_s(w)$ and $I_d(w)$ is the stationary value of the delayed current. Impedance in Laplace form (where $s = i\omega$) of EEC which are represented in Fig. 4.6d can be represented by

$$Z(s) = \left[C_g s + \cfrac{1}{R_3 + \cfrac{1}{C_1 s + \cfrac{1}{R_1} + \cfrac{1}{R_a + L_a s}}} \right] \tag{4.23}$$

A chemical inductor arrangement is represented by R_a and L_a components that describe delayed surface-controlled recombination. The transition from capacitance to inductive behavior is often associated with the plan and configuration of the circuit. The slower ionic movement influences the suggestively faster recombination of electronic current or the transfer of that current toward contacts. This behavior resembles the voltage-gated ion channels in neuron membranes, as described by Foster et al. (1993). Understanding this mechanism offers valuable perceptions into the voltage-gated electronic current observed at the surface of solar cells.

In this, they analyzed the IS into three voltage regions (low, medium, and high). The components of the circuit that are utilized at the various voltage levels (low, transition, and high) that are applied to the MAPBr solar cell are illustrated in Fig. 4.6. The observation of a double RC feature can be seen in Fig. 4.6b, c when

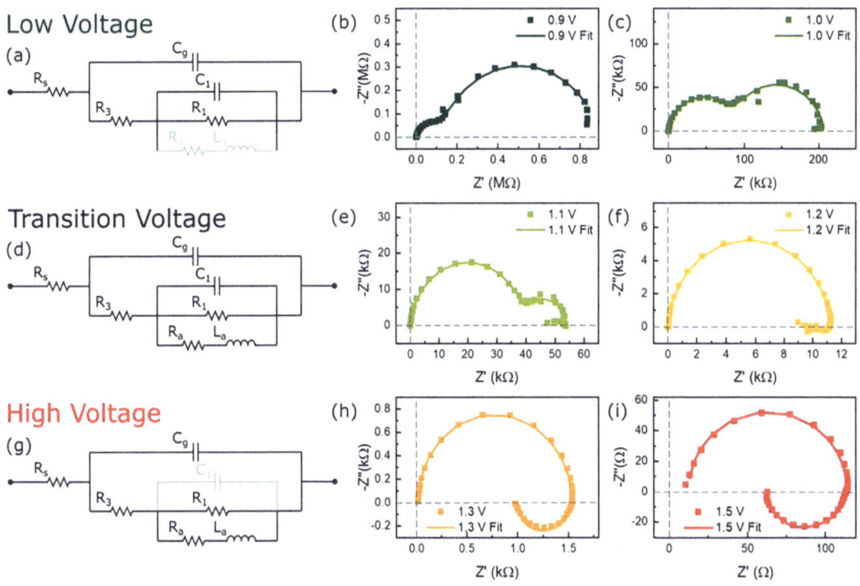

Fig. 4.6 IS examination of MAPBr solar cell under a varying dark condition and applied voltages (V_a). At low V_a (<1.1 V), capacitive behavior is found in circuits (**b**, **c**) that match circuits (**a**). A transition zone may be observed at a voltage range of 1.1 V to 1.2 V (**e**, **f**) associated with circuit (**d**). Circuit (**g**) exhibits inductive behavior when the applied voltage is high (more than 1.2 V) (**h**, **i**) (Gonzales et al. 2022)

the voltages are lower (as shown in Fig. 4.1a, which is caused by the inductor having a high resistance. A low-frequency arc is substituted by an initial inductive feature in the intermediate/transition zone, which is represented by the EEC in Fig. 4.6b. This can be viewed in Fig. 4.6e, f. The capacitive arc that is connected with C_1 is not observed at higher voltages (Fig. 4.6g), and it is replaced by a chemical inductor, which therefore results in a newly observed inductive arc effect (Fig. 4.6h, i).

The advantage of the Neuron Style Model for solar cells, specifically focusing on halide perovskite cells transitioning from capacitive to inductive behavior at varying voltages. Their model accurately replicates experimental features like impedance spectroscopy and hysteresis in current-voltage curves under different sweep rates. They found that capacitive excess current scales with the scan rate, while inductive current inversely scales with it. This model enables precise control over cell kinetics, predicting hysteresis characteristics based on impedance measurements. The model explains the transition of the low-frequency capacitor to a chemical inductor that is regulated by certain time constants and preserves regular hysteresis across the voltage range without negative capacitance. Despite these limitations, the Neuron Style Model is an invaluable technique for investigating and optimizing MHPSCs. (Bisquert 2023; Bou and Bisquert 2021; Hernández-Balaguera et al. 2023; Samardzic et al. 2022)

Suggestions, Conclusion, and Outlook

Typical PSC encompasses complex processes like charge transfer and dissemination of charged species. Additionally, the interaction between different layers and materials adds complexity. So, understanding these processes is essential for analyzing and improving the efficiency of PSCs. Impedance spectra of PSC devices represent a combination of various processes, so interpretation is a big challenge without considering kinetic parameters. Factors like ion migration, electrode porosity, and non-homogeneity must be accounted for in EEC models based on physical origins, enabling precise analysis and insights into PSC behavior.

Physically motivated EECs help uncover the origins of resistances and capacitances in electrochemical systems, eliminating ambiguity through auxiliary techniques or clear argument chains. Before conducting EIS, a theoretical understanding of impedance origins is essential to avoid cognitive bias in interpretation. External parameters like temperature and bias potentials aid in exploring interface impedances, while attention to system non-stationarity prevents impedance response distortions. The multistage intercalation mechanism in PSCs illustrates the superiority of EECs with clear physical meaning, enabling precise interpretation of battery system performance and facilitating the design of new electrode structures.

Research in the emerging field of PSCs is highly interdisciplinary. It draws expertise from diverse domains, e.g., device fabrication, semiconductor physics, chemistry, solid-state physics, material science, and electrical and device engineering. However, beginners often encounter challenges in grasping how IS effectively studies new PSC materials and devices due to its broad scope. In this review, we emphasize the potency of EIS in characterizing interfaces and understanding dynamic processes within PSCs. By analyzing impedance spectra, we can get crucial insights into the various interfaces, such as ITO/FTO contact surfaces, transport layers, and electrode-perovskite interfaces. IS applications have greatly contributed to comprehending charge transport, recombination processes, and interface effects on device performance in PSCs. However, modeling equivalent circuits remains a challenge. This paper presents advancements in equivalent circuit modeling, facilitating the extraction of parameters crucial for understanding electrochemical behavior. This book chapter aims to provide a resource for impedance spectroscopy analysis of PSCs, bridging gaps in the literature by offering insights into equivalent circuit elements not readily available elsewhere.

While significant progress has been made in applying IS to PSCs research, there are still several areas that offer potential for future exploration and advancement. The prospects of IS in the field of PSC are promising and hold great potential. IS can help in improving and characterize interface optimization, involving characterization, defect detection, and stability assessment, remains a focal point for enhancing the effectiveness of solar panels. Advanced circuit models play a pivotal role in accurately analyzing impedance spectra, considering factors like double-layer capacitance and interfacial reactions. Stability and degradation studies utilize IS to monitor long-term performance, identifying degradation mechanisms and enhancing overall stability. Exploration of new materials and architectures through IS provides crucial insights into improving solar panel efficiency. Non-destructive testing and monitoring, coupled with IS, ensures continuous evaluation and reliability of solar systems. The pursuit of low-frequency response analysis and integration of ML aims to overcome challenges and improve predictive capabilities. Additionally, device performance optimization and integration with other techniques underscore the importance of IS in enhancing solar panel efficiency and understanding materials comprehensively. Further exploration and integration of these aspects promise to advance the field of solar energy research toward greater efficiency and stability.

Overall, the future of IS in PSCs research is bright, and research lies in the continued refinement of measurement techniques, the development of advanced circuit models, and the integration of EIS with other characterization methods. It offers opportunities for improved material understanding of device optimization, reliability assessment, and the development of next-generation photovoltaic technologies. By further exploring the electrochemical responses of interfaces and materials, EIS can contribute to developing more efficient, stable, and reliable PSCs. Continued research and technological advancements in EIS will contribute to sustainable solar energy solutions.

References

Akriti, Lin ZY, Park JY, Yang H, Savoie BM, Dou L (2022) Anion diffusion in two-dimensional halide perovskites. APL Mater 10(4). https://doi.org/10.1063/5.0088538

Almora O, Zhao Y, Du X, Heumueller T, Matt GJ, Garcia-Belmonte G, Brabec CJ (2020) Light intensity modulated impedance spectroscopy (LIMIS) in all-solid-state solar cells at open-circuit. Nano Energy 75:104982. https://doi.org/10.1016/J.NANOEN.2020.104982

Alvarez AO, Arcas R, Aranda CA, Bethencourt L, Mas-Marzá E, Saliba M, Fabregat-Santiago F (2020) Negative capacitance and inverted hysteresis: matching features in perovskite solar cells. J Phys Chem Lett 11(19):8417–8423

Barsoukov E, Macdonald JR (2018) Impedance spectroscopy: theory, experiment, and applications, 3rd edn. Wiley, Hoboken

Bisquert J (2023) Electrical charge coupling dominates the hysteresis effect of halide perovskite devices. J Phys Chem Lett 14(4):1014–1021. https://doi.org/10.1021/ACS.JPCLETT.2C03812

Bisquert J, Bertoluzzi L, Mora-Sero I, Garcia-Belmonte G (2014a) Theory of impedance and capacitance spectroscopy of solar cells with dielectric relaxation, drift-diffusion transport, and recombination. J Phys Chem C 118(33):18983–18991. https://doi.org/10.1021/JP5062144

Bisquert J, Mora-Sero I, Fabregat-Santiago F (2014b) Diffusion–recombination impedance model for solar cells with disorder and nonlinear recombination. ChemElectroChem 1(1):289–296. https://doi.org/10.1002/CELC.201300091

Bisquert J, Garcia-Belmonte G, Guerrero A (2016) Impedance characteristics of hybrid organometal halide perovskite solar cells. In: Organic-inorganic halide perovskite photovoltaics: from fundamentals to device architectures. Springer, Cham, pp 163–199. https://doi.org/10.1007/978-3-319-35114-8_7

Bou A, Bisquert J (2021) Impedance spectroscopy dynamics of biological neural elements: from memristors to neurons and synapses. J Phys Chem B 125(35):9934–9949

Bou A, Pockett A, Raptis D, Watson T, Carnie MJ, Bisquert J (2020) Beyond impedance spectroscopy of perovskite solar cells: insights from the spectral correlation of the electrooptical frequency techniques. J Phys Chem Lett 11(20):8654–8659. https://doi.org/10.1021/ACS.JPCLETT.0C02459

Cari C, Khairuddin K, Septiawan TY, Suciatmoko PM, Kurniawan D, Supriyanto A (2018) The preparation of natural dye for dye-sensitized solar cell (DSSC). AIP Conf Proc 2014(1):20106. https://doi.org/10.1063/1.5054510/724338

Castro-Chong A, Riquelme AJ, Aernouts T, Bennett LJ, Richardson G, Oskam G, Anta JA (2021) Illumination intensity dependence of the recombination mechanism in mixed perovskite solar cells. ChemPlusChem 86(9):1347–1356. https://doi.org/10.1002/CPLU.202100233

Chen B, Yang M, Zheng X, Wu C, Li W, Yan Y, Bisquert J, Garcia-Belmonte G, Zhu K, Priya S (2015) Impact of capacitive effect and ion migration on the hysteretic behavior of perovskite solar cells. J Phys Chem Lett 6(23):4693–4700. https://doi.org/10.1021/ACS.JPCLETT.5B02229

Dualeh A, Moehl T, Tétreault N, Teuscher J, Gao P, Nazeeruddin MK, Grätzel M (2014) Impedance spectroscopic analysis of lead iodide perovskite-sensitized solid-state solar cells. ACS Nano 8(1):362–373. https://doi.org/10.1021/NN404323G

Eames C, Frost JM, Barnes PRF, O'Regan BC, Walsh A, Islam MS (2015) Ionic transport in hybrid lead iodide perovskite solar cells. Nat Commun 6(1):1–8. https://doi.org/10.1038/ncomms8497

Ebadi F, Taghavinia N, Mohammadpour R, Hagfeldt A, Tress W (2019) Origin of apparent light-enhanced and negative capacitance in perovskite solar cells. Nat Commun 10(1):1–9. https://doi.org/10.1038/s41467-019-09079-z

Fabregat-Santiago F, García-Cañadas J, Palomares E, Clifford JN, Haque SA, Durrant JR, Garcia-Belmonte G, Bisquert J (2004) The origin of slow electron recombination processes in dye-sensitized solar cells with alumina barrier coatings. J Appl Phys 96(11):6903–6907. https://doi.org/10.1063/1.1812588

Fabregat-Santiago F, Garcia-Belmonte G, Mora-Seró I, Bisquert J (2011) Characterization of nanostructured hybrid and organic solar cells by impedance spectroscopy. Phys Chem Chem Phys 13(20):9083–9118. https://doi.org/10.1039/C0CP02249G

Fabregat-Santiago F, Kulbak M, Zohar A, Vallés-Pelarda M, Hodes G, Cahen D, Mora-Seró I (2017) Deleterious effect of negative capacitance on the performance of halide perovskite solar cells. ACS Energy Lett 2(9):2007–2013. https://doi.org/10.1021/ACSENERGYLETT.7B00542

Fabregat-Santiago F, Barea EM, Giménez S, Bisquert J (2018) Impedance spectroscopy in molecular devices. Springer, Cham, pp 353–384. https://doi.org/10.1007/978-981-10-5924-7_10

Fang Y, Wei H, Dong Q, Huang J (2017) Quantification of re-absorption and re-emission processes to determine photon recycling efficiency in perovskite single crystals. Nat Commun 8(1):1–9. https://doi.org/10.1038/ncomms14417

Foster WR, Ungar LH, Schwaber JS (1993) Significance of conductances in Hodgkin-Huxley models. J Neurophysiol 70(6):2502–2518. https://doi.org/10.1152/JN.1993.70.6.2502

Gateman SM, Gharbi O, Gomes de Melo H, Ngo K, Turmine M, Vivier V (2022) On the use of a constant phase element (CPE) in electrochemistry. Curr Opin Electrochem 36:101133. https://doi.org/10.1016/J.COELEC.2022.101133

Ghahremanirad E, Bou A, Olyaee S, Bisquert J (2017) Inductive loop in the impedance response of perovskite solar cells explained by surface polarization model. J Phys Chem Lett 8(7):1402–1406

Gonzales C, Guerrero A, Bisquert J (2022) Transition from capacitive to inductive hysteresis: a neuron-style model to correlate I-V curves to impedances of metal halide perovskites. J Phys Chem C 126(32):13560–13578

Gonzalez-Pedro V, Juarez-Perez EJ, Arsyad WS, Barea EM, Fabregat-Santiago F, Mora-Sero I, Bisquert J (2014) General working principles of CH3NH3PbX3 perovskite solar cells. Nano Lett 14(2):888–893

Grätzel M (2014) The light and shade of perovskite solar cells. Nat Mater 13(9):838–842. https://doi.org/10.1038/nmat4065

Guerrero A, Garcia-Belmonte GG, Mora-Sero I, Bisquert J, Kang YS, Jacobsson TJ, Correa-Baena J-PP, Hagfeldt A (2016) Properties of contact and bulk impedances in hybrid lead halide perovskite solar cells including inductive loop elements. J Phys Chem C 120(15):8023–8032

Guerrero A, Bisquert J, Garcia-Belmonte G (2021) Impedance Spectroscopy of metal halide perovskite solar cells from the perspective of equivalent circuits. Chem Rev 121:14430

Hernández-Balaguera E, Bisquert J (2024) Time transients with inductive loop traces in metal halide perovskites. Adv Funct Mater 34(6):2308678. https://doi.org/10.1002/ADFM.202308678

Hernández-Balaguera E, Arredondo B, Pereyra C, Lira-Cantú M (2023) Parameterization of the apparent chemical inductance of metal halide perovskite solar cells exhibiting constant-phase-element behavior. J Power Sources 560:232614. https://doi.org/10.1016/J.JPOWSOUR.2022.232614

Herz LM (2017) Charge-carrier mobilities in metal halide perovskites: fundamental mechanisms and limits. ACS Energy Lett 2(7):1539–1548. https://doi.org/10.1021/ACSENERGYLETT.7B00276

Hishikawa Y, Shimura H, Ueda T, Sasaki A, Ishii Y (2016) Precise performance characterization of perovskite solar cells. Curr Appl Phys 16(8):898–904. https://doi.org/10.1016/J.CAP.2016.05.002

Impedance Spectroscopy (2018) Impedance spectroscopy. https://doi.org/10.1002/9781119381860

Jacobs DA, Shen H, Pfeffer F, Peng J, White TP, Beck FJ, Catchpole KR (2018) The two faces of capacitance: new interpretations for electrical impedance measurements of perovskite solar cells and their relation to hysteresis. J Appl Phys 124(22). https://doi.org/10.1063/1.5063259/155643

Johnston MB, Herz LM (2016) Hybrid perovskites for photovoltaics: charge-carrier recombination, diffusion, and radiative efficiencies. Acc Chem Res 49(1):146–154. https://doi.org/10.1021/ACS.ACCOUNTS.5B00411

Juarez-Perez EJ, Sanchez RS, Badia L, Garcia-Belmonte G, Kang YS, Mora-Sero I, Bisquert J (2014a) Photoinduced giant dielectric constant in lead halide perovskite solar cells. J Phys Chem Lett 5(13):2390–2394. https://doi.org/10.1021/JZ5011169

Juarez-Perez EJ, Wußler M, Fabregat-Santiago F, Lakus-Wollny K, Mankel E, Mayer T, Jaegermann W, Mora-Sero I (2014b) Role of the selective contacts in the performance of

lead halide perovskite solar cells. J Phys Chem Lett 5(4):680–685. https://doi.org/10.1021/JZ500059V
Khan MT, Huang P, Almohammedi A, Kazim S, Ahmad S (2021) Mechanistic origin and unlocking of negative capacitance in perovskites solar cells. IScience 24(2):102024
Kim J-S, Pyun S-I (n.d.) Comparison of transmissive permeable and reflective impermeable interfaces between electrode and electrolyte. J Solid State Electrochem. https://doi.org/10.1007/s10008-011-1393-9
Kim HS, Mora-Sero I, Gonzalez-Pedro V, Fabregat-Santiago F, Juarez-Perez EJ, Park NG, Bisquert J (2013) Mechanism of carrier accumulation in perovskite thin-absorber solar cells. Nat Commun 4(1):1–7
Kim B, Kim M, Kim H, Jeong S, Yang J, Jeong MS (2022) Improved stability of MAPbI3Perovskite solar cells using two-dimensional transition-metal dichalcogenide interlayers. ACS Appl Mater Interfaces 14(31):35726–35733. https://doi.org/10.1021/ACSAMI.2C08680
Kojima A, Teshima K, Shirai Y, Miyasaka T (2009) Organometal halide perovskites as visible-light sensitizers for photovoltaic cells. J Am Chem Soc 131(17):6050–6051. https://doi.org/10.1021/ja809598r
Lazanas AC, Prodromidis MI (n.d.) Electrochemical impedance spectroscopy–a tutorial. ACS Measure Sci 3(3):162–193
Li H, Li F, Shen Z, Han ST, Chen J, Dong C, Chen C, Zhou Y, Wang M (2021) Photoferroelectric perovskite solar cells: principles, advances and insights. Nano Today 37:101062. https://doi.org/10.1016/J.NANTOD.2020.101062
Liu Y, Renna LA, Thompson HB, Page ZA, Emrick T, Barnes MD, Bag M, Venkataraman D, Russell TP, Liu Y, Page ZA, Emrick T, Russell TP, Renna LA, Thompson HB, Barnes MD, Bag M (2017) Role of ionic functional groups on ion transport at perovskite interfaces. Adv Energy Mater 7(21):1701235. https://doi.org/10.1002/AENM.201701235
MacDonald DD (2006) Reflections on the history of electrochemical impedance spectroscopy. Electrochim Acta 51(8-9):1376–1388. https://doi.org/10.1016/J.ELECTACTA.2005.02.107
Mahapatra A, Mandal M, Mahapatra AD, Anilkumar V, Nawrocki J, Chavan RD, Yadav P, Prochowicz D (2024) Mechanochemically-assisted synthesis of 3D, 2D and quasi 2D lead halide perovskites for supercapacitor applications. Mater Adv. https://doi.org/10.1039/D3MA01144E
Mahjabin S, Haque MM, Sobayel K, Jamal MS, Islam MA, Selvanathan V, Assaifan AK, Alharbi HF, Sopian K, Amin N, Akhtaruzzaman M (2020) Perceiving of defect tolerance in perovskite absorber layer for efficient perovskite solar cell. IEEE Access 8:106346–106353. https://doi.org/10.1109/ACCESS.2020.3000217
Meloni S, Moehl T, Tress W, Franckeviius M, Saliba M, Lee YH, Gao P, Nazeeruddin MK, Zakeeruddin SM, Rothlisberger U, Graetzel M (2016) Ionic polarization-induced current–voltage hysteresis in CH3NH3PbX3 perovskite solar cells. Nat Commun 7(1):1–9. https://doi.org/10.1038/ncomms10334
Nouri E, Mohammadi MR, Lianos P (2018) Construction of perovskite solar cells using inorganic hole-extracting components. ACS Omega 3(1):46–54. https://doi.org/10.1021/ACSOMEGA.7B01775
Obrero-Perez JM, Contreras-Bernal L, Nuñez-Galvez F, Castillo-Seoane J, Valadez-Villalobos K, Aparicio FJ, Anta JA, Borras A, Sanchez-Valencia JR, Barranco A (2022) Ultrathin plasma polymer passivation of perovskite solar cells for improved stability and reproducibility. Adv Energy Mater 12(32):2200812. https://doi.org/10.1002/AENM.202200812
Parida B, Singh A, Oh M, Jeon M, Kang JW, Kim H (2019) Effect of compact TiO2 layer on structural, optical, and performance characteristics of mesoporous perovskite solar cells. Mater Today Commun 18:176–183. https://doi.org/10.1016/J.MTCOMM.2018.12.007
Pascoe AR, Duffy NW, Scully AD, Huang F, Cheng YB (2015) Insights into planar CH3NH3PbI3 perovskite solar cells using impedance Spectroscopy. J Phys Chem C 119(9):4444–4453. https://doi.org/10.1021/JP509896U
Prochowicz D, Tavakoli MM, Alanazi AQ, Trivedi S, Tavakoli Dastjerdi H, Zakeeruddin SM, Grätzel M, Yadav P (2019) Charge accumulation, recombination, and their associated time scale in

efficient (GUA) x(MA)1-xPbI3-based perovskite solar cells. ACS Omega 4(16):16840–16846. https://doi.org/10.1021/ACSOMEGA.9B01701

Ravishankar S, Almora O, Echeverría-Arrondo C, Ghahremanirad E, Aranda C, Guerrero A, Fabregat-Santiago F, Zaban A, Garcia-Belmonte G, Bisquert J (2017) Surface polarization model for the dynamic hysteresis of perovskite solar cells. J Phys Chem Lett 8(5):915–921

Samardzic NM, Bajic JS, Sekulic DL, Dautovic S (2022) Volatile Memristor in leaky integrate-and-fire neurons: circuit simulation and experimental study. Electronics 11(6):894. https://doi.org/10.3390/ELECTRONICS11060894

Sanchez RS, Gonzalez-Pedro V, Lee JW, Park NG, Kang YS, Mora-Sero I, Bisquert J (2014) Slow dynamic processes in lead halide perovskite solar cells. Characteristic times and hysteresis. J Phys Chem Lett 5(13):2357–2363. https://doi.org/10.1021/JZ5011187

Shao Y, Xiao Z, Bi C, Yuan Y, Huang J (2014) Origin and elimination of photocurrent hysteresis by fullerene passivation in CH3NH3PbI3 planar heterojunction solar cells. Nat Commun 5(1):1–7. https://doi.org/10.1038/ncomms6784

Suresh MS (1996) Measurement of solar cell parameters using impedance spectroscopy. Sol Energy Mater Sol Cells 43(1):21–28. https://doi.org/10.1016/0927-0248(95)00153-0

Tavakoli MM, Tsui KH, Zhang Q, He J, Yao Y, Li D, Fan Z (2015) Highly efficient flexible perovskite solar cells with antireflection and self-cleaning nanostructures. ACS Nano 9(10):10287–10295. https://doi.org/10.1021/ACSNANO.5B04284

Timofeev MI, Guarnieri FV, Huddy JE, Scheideler WJ (2023) Engineering perovskite solar cells for efficient wireless power transfer. APL Energy 1(3):36104. https://doi.org/10.1063/5.0169827

Touré S (2022) Effect of the dynamic resistance on the maximum output power in dynamic modelling of photovoltaic solar cells. Open J Model Simul 10(1):48–57. https://doi.org/10.4236/ojmsi.2022.101003

Von Hauff E, Klotz D (2022) Impedance spectroscopy for perovskite solar cells: characterisation, analysis, and diagnosis. J Mater Chem C 10(2):742–761. https://doi.org/10.1039/d1tc04727b

Wei N, Chen Y, Wang X, Miao Y, Qin Z, Liu X, Wei H, Zhao Y, Wei N, Wei H, Chen Y, Wang X, Miao Y, Qin Z, Liu X, Zhao Y (2022) Multi-level passivation of MAPbI3 perovskite for efficient and stable photovoltaics. Adv Funct Mater 32(16):2108944. https://doi.org/10.1002/ADFM.202108944

Wu Y, Shen H, Walter D, Jacobs D, Duong T, Peng J, Jiang L, Cheng YB, Weber K (2016) On the origin of hysteresis in perovskite solar cells. Adv Funct Mater 26(37):6807–6813. https://doi.org/10.1002/ADFM.201602231

Xu C, Alvar MS, Wetzelaer GJAH, Blom PWM (2023) Analysis of the ionic and dielectric properties of perovskites by impedance spectroscopy. J Appl Phys 133(4):45501. https://doi.org/10.1063/5.0123547/2866736

Yadav P, Alotaibi MH, Arora N, Dar MI, Zakeeruddin SM, Grätzel M (2018) Influence of the nature of a cation on dynamics of charge transfer processes in perovskite solar cells. Adv Funct Mater 28(8). https://doi.org/10.1002/ADFM.201706073

Yoo SM, Yoon SJ, Anta JA, Lee HJ, Boix PP, Mora-Seró I (2019) An equivalent circuit for perovskite solar cell bridging sensitized to thin film architectures. Joule 3(10):2535–2549. https://doi.org/10.1016/J.JOULE.2019.07.014

Zhao Y, Nardes AM, Zhu K (2014) Solid-state mesostructured perovskite CH3NH3PbI3 solar cells: charge transport, recombination, and diffusion length. J Phys Chem Lett 5(3):490–494. https://doi.org/10.1021/JZ500003V

Zimmermann E, Wong KK, Müller M, Hu H, Ehrenreich P, Kohlstädt M, Würfel U, Mastroianni S, Mathiazhagan G, Hinsch A, Gujar TP, Thelakkat M, Pfadler T, Schmidt-Mende L (2016) Characterization of perovskite solar cells: towards a reliable measurement protocol. APL Mater 4(9). https://doi.org/10.1063/1.4960759/121499

Chapter 5
Multi-Physics Studies of Perovskite Solar Cells

Yogesh Sharma, Mirza Tanweer Ahmad Beig, Rajiv Vatsa, Akanksha Namdeo, Shikha Shukla, A. N. Upadhyay, and Sanjeev K. Srivastava

Abstract Perovskite solar cells (PSCs) have emerged as a prominent photovoltaic technology due to it's high-power conversion efficiency, flexible material properties, and inexpensive production costs. However, maximising their efficiency and ensuring long-term stability require a thorough understanding of the interconnected physical processes at work. This chapter examines multi-physics phenomena in PSCs, including electrical charge transport, light absorption and optical behaviour, thermal management, and mechanical stability. The topic incorporates important modelling methodologies, including drift-diffusion models for charge dynamics, optical simulations for light control, and electro-thermal-mechanical coupling for device stability under real-world operating conditions. By merging these physical characteristics into unified multi-physics models, we may improve device performance, optimise material and device designs, and solve efficiency, degradation, and scalability issues. The chapter finishes with a discussion of developing trends in multi-physics simulations and their implications for the future development of PSC technologies.

Y. Sharma (✉) · A. Namdeo
Department of Physics & Environmental Sciences, Sharda School of Engineering & Science (SSES), Sharda University, Greater Noida, Uttar Pradesh, India

M. T. A. Beig
Department of Physics, School of Basic and Applied Sciences, SGT University, Gurugram, Haryana, India

R. Vatsa
Department of Physics, Hindu College, Sonipat, Haryana, India

S. Shukla
Department of Physics, AKS University, Satna, Madhya Pradesh, India

A. N. Upadhyay
Department of Physics, Lady Irwin College, University of Delhi, New Delhi, India

S. K. Srivastava
Department of Physics, Amity Institute of Applied Sciences, Amity University Uttar Pradesh, Noida, India

M. Kumar (ed.), *Computational Analysis of Perovskite Solar Cells*, Springer Tracts in Electrical and Electronics Engineering,
https://doi.org/10.1007/978-981-95-6521-4_5

Keywords Perovskite solar cells · Finite element method · Band gap · Multi-physics · Photovoltaic technology

Introduction

Perovskite solar cells (PSCs) have emerged as highly efficient next-generation photovoltaics due to their excellent light absorption, tunable bandgap, and low-cost fabrication. Their efficiency has rapidly increased, surpassing 25%, making them competitive with silicon-based solar cells. The remarkable power conversion efficiencies (PCEs), low manufacturing costs, and adaptable material features of PSCs have made them a game-changer in the photovoltaics industry. PSCs have developed rapidly since their inception in 2009, when used methylammonium lead halide ($CH_3NH_3PbI_3$) in a liquid-electrolyte system to yield a low efficiency of 3.8% (Kojima et al. 2009). Laboratory-scale devices have attained 25.8% efficiency in just over 10 years, comparable to conventional silicon-based solar cells. Figure 5.1 shows a chart of the highest confirmed conversion efficiencies for research cells for a range of photovoltaic technologies, plotted from 1976 to the present maintained by Nrel (2023).

The distinctive optoelectronic characteristics of perovskite materials, such as their long charge carrier diffusion lengths, low exciton binding energies, tunable bandgaps, and high absorption coefficients, are responsible for this outstanding advancement (Ramirez et al. 2018). An ideal platform for effective light harvesting and charge transport is provided by the perovskite structure, which is commonly represented by the ABX_3 formula, where A is a monovalent cation (such as

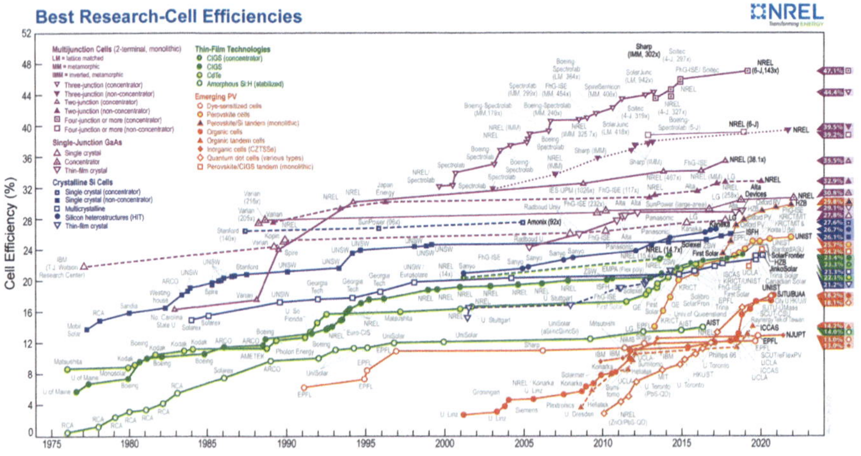

Fig. 5.1 Best research-cell efficiency chart (Nrel 2023)

methylammonium, formamidinium, or cesium), B is a divalent metal cation (such as lead or tin), and X is a halide anion (such as iodide, bromide, or chloride). However, sophisticated modelling techniques are required to comprehend and maximise PSC performance and stability entirely because of the simultaneous interaction of several physical processes, including light absorption, charge carrier dynamics, ion migration, and thermal and mechanical effects.

A vital tool for deciphering the intricate dynamics regulating PSCs, multi-physics modelling offers a comprehensive framework for examining how they behave in actual operational environments. Charge transport and recombination mechanisms have been clarified by electrical modelling, especially the negative consequences of defects that serve as non-radiative recombination centres, such as vacancies, interstitials, and grain boundaries (Yang et al. 2015). By reducing reflection losses and increasing photon absorption in the multilayer architecture, optical modelling has proven crucial for improving light control in PSCs (Zhou et al. 2015). Since excessive heat accelerates degradation processes, including ion migration and phase segregation, thermal modelling simultaneously addresses essential challenges related to heat generation and dissipation (Berhe et al. 2016). Mechanical modelling supports these efforts by clarifying the consequences of mechanical stresses and thermal expansion mismatches, which can result in cracking, delamination, and, ultimately, device failure (Tress et al. 2018). By combining these disparate fields, multi-physics studies offer a thorough understanding of the interdependencies among electrical, optical, thermal, and mechanical processes, allowing researchers to pinpoint degradation pathways and performance bottlenecks.

Beyond basic understanding, multi-physics research has influenced the development of sophisticated PSCs and the development of creative methods to improve their stability and performance. For example, defect passivation approaches have been developed to reduce non-radiative recombination losses and increase device lifetimes, including the use of additives and surface treatments (Noel et al. 2014). Developing tandem solar cells, which combine perovskite layers with silicon or other materials to overcome the Shockley-Queisser efficiency limit, has also significantly benefited from multi-physics modelling (Rolston et al. 2018). Since high-efficiency PSCs are often made using methods like spin-coating, which are challenging to scale for industrial manufacturing, scalability is still a significant issue. By simulating layer uniformity, thickness control, and interfacial characteristics, multi-physics models have proven crucial in enhancing scalable fabrication techniques, including roll-to-roll printing and slot-die coating (Hou et al. 2020). There are intriguing prospects to speed up PSC research by combining multi-physics modelling with cutting-edge technologies such as artificial intelligence (AI) and machine learning (ML). New perovskite compositions and device combinations can be identified more easily, thanks to AI-driven methods that evaluate large datasets to forecast material characteristics and performance trends (Ren et al. 2020; AlHajri et al. 2012; Ren et al. 2017). Insights for creating more reliable and effective solar solutions can be gained by simulating the long-term performance and degradation of PSCs under various climatic circumstances using coupled multi-physics and AI frameworks. The need for dependable, environmentally friendly, and highly efficient solar energy solutions is growing as the world's energy needs continue to rise (Freire et al. 2023). The PSCs research community is in a strong position to tackle current issues and realise the full potential of this game-changing technology by utilising sophisticated multi-physics modelling and experimental validation, opening the door for its broad use in the renewable energy sector (Boyd et al. 2019).

Perovskite Material Properties

Crystal Structure and Composition

Perovskite materials are named after the mineral calcium titanate ($CaTiO_3$), which has a distinct crystal structure as shown in Fig. 5.2. The general formula for perovskites is ABX_3, where A larger cation, typically an alkali, alkaline earth, or rare earth metal (e.g. Ca^{2+}, Sr^{2+}, Ba^{2+}, or Cs^{+}), B: A smaller cation, typically a transition metal or a metal from group IV (e.g. Ti^{4+}, Sn^{4+}, or Pb^{2+}) and X: An anion, usually oxygen (O^{2-}) in oxides or halides (Cl^{-}, Br^{-}, I^{-}) in halide perovskites. If we consider the ideal perovskite structure with cubic symmetry ($Pm\bar{3}m$), the A cation occupies the corners of the cube, the B cation resides at the body centre, and the X anions are positioned at the face centres, and the X anions at the face centres but Octahedral Coordination: The B-site cation is surrounded by an octahedron of X anions, forming BX_6 units (Ramirez et al. 2018; Rolston et al. 2018).

Perovskite materials can be categorised into various types based on their composition and applications. Oxide perovskites (ABO_3) are commonly used in functional ceramics like $BaTiO_3$ and $SrTiO_3$, with applications in dielectrics, ferroelectrics, superconductors, and catalysts. Halide perovskites (ABX_3) include organic-inorganic hybrids where A is an organic cation (e.g., $CH_3NH_3^+$ or Cs^+), B is Pb^{2+} or Sn^{2+}, and X is I^-, Br^-, or Cl^-, as well as all-inorganic variants with similar compositions. These materials are extensively used in solar cells, light-emitting diodes (LEDs), and photodetectors. Double perovskites ($A_2BB'X_6$) contain two different

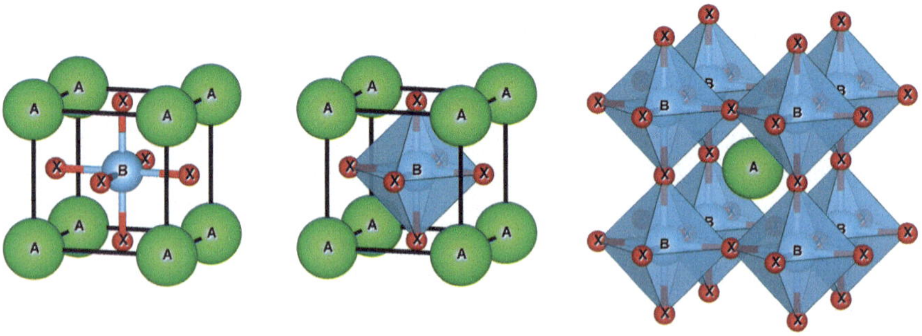

Fig. 5.2 Perovskite ABX_3 lattice (a) large A and small B cations with X anions

B-site cations (B and B′), offering enhanced tunability for electronic and optical properties. Finally, layered perovskites, such as Ruddlesden-Popper phases, exhibit reduced dimensionality and find applications in optoelectronics and quantum materials (Nkele et al. 2020; Banoo 2016).

Electronic Properties

Perovskite materials, with the general chemical formula ABX_3, are well-known for their wide range of tunable and diverse electronic properties. This makes them exceptionally versatile and desirable for a wide range of applications, particularly electronics, optoelectronics, and energy conversion devices. These materials possess a unique crystal structure, which is crucial in determining their electronic characteristics (Habte et al. 2015). The arrangement and interactions between the different ions in the perovskite lattice, which are the A-site cation, B-site cation, and X-site anion, are key factors that influence their electronic behaviour. The flexibility of perovskites lies in their ability to modify their electronic properties by altering the composition of the ions in the structure, allowing for precise tuning of their conductivity, bandgap, and optical properties. The A-site cation, typically a larger ion such as an alkali or alkaline earth metal, can be substituted with various elements, enabling control over the overall charge distribution and lattice dynamics. Meanwhile, the B-site cation, typically a transition metal, directly influences the electronic band structure and the material's electronic conductivity. Moreover, perovskite materials exhibit remarkable flexibility in their electronic properties, such as their high charge carrier mobility, making them suitable for devices such as solar cells, light-emitting diodes (LEDs), and photodetectors. In addition, their ability to support both p-type and n-type doping makes them adaptable to various device architectures, contributing further to their widespread use in next-generation electronic and optoelectronic devices (Gummel 1964; Scharfetter and Gummel 1969). The interactions between ions in the crystal lattice, including the roles of defects, strain, and lattice distortions, also contribute to the unique electronic properties of perovskites. These properties, along with their ease of fabrication and low cost, continue to drive research into new and innovative applications in areas such as renewable energy, optoelectronics, and advanced electronics (Ma et al. 2016; Park 2015).

Bandgap Tunability, Electronic States, and Hybridisation

The electronic bandgap of perovskites can range from insulating to semiconducting or even metallic, depending on the composition and structure.

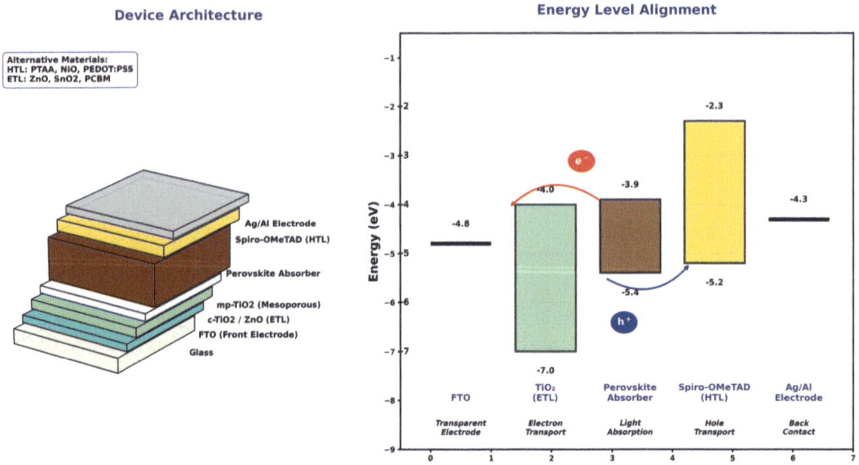

Fig. 5.3 Structure and energy band diagram of a perovskite solar cell

The left side of Fig. 5.3 shows the layered structure of a perovskite solar cell, including glass, electrodes, transport layers, and the light-absorbing perovskite layer. The right side illustrates the energy band alignment, highlighting charge generation and transport. Sunlight excites electrons in the perovskite, which are collected at the electrodes to generate electricity. The conduction band primarily comprises B-site cation orbitals (e.g. Pb 6p or Ti 3d). The valence band is typically dominated by X-anion orbitals (e.g. I 5p or O 2p). Strong hybridisation among A-site, B-site, and X-site ions influences band alignment and carrier dynamics. Halide perovskites, such as $CH_3NH_3PbI_3$, typically possess direct bandgaps ranging from approximately 1.2 eV to 2.3 eV, making them ideal for applications in solar cells and LEDs. Their bandgap can be tuned through compositional adjustments, such as substituting halides like I^-, Br^-, or Cl^-. In contrast, oxide perovskites, such as $SrTiO_3$, often exhibit larger bandgaps exceeding 3 eV, which makes them more suitable for dielectric and photocatalytic applications.

Charge Carrier Properties

Halide perovskites are primarily known for their high charge carrier mobility, a significant factor in their excellent performance in optoelectronic applications such as solar cells, LEDs, and photodetectors. This exceptional mobility is mainly due to their low effective masses of charge carriers, coupled with long carrier lifetimes. Low effective masses allow the carriers to move more freely through the material, while long lifetimes ensure they remain active long enough to contribute to device performance. These characteristics make halide perovskites highly efficient in converting light to electrical energy, which is essential for their role in energy harvesting and light-emitting technologies. However, the performance of perovskite materials is not solely dependent on their inherent properties. Defects and grain boundaries are significant contributors to structural imperfections, and these imperfections can be substantial in limiting charge carrier mobility. Grain boundaries act as traps in polycrystalline films, commonly used in many applications, to hinder the motion of charge carriers and thus contribute to lower device efficiency. These defects can also lead to non-radiative recombination, where carriers recombine without emitting light, reducing the material's effectiveness in optoelectronic devices. Therefore, the quality of the perovskite material becomes critical for optimising charge carrier mobility. High-quality single crystals, for example, often exhibit much higher mobilities than polycrystalline films due to the absence of grain boundaries and defects. In addition, the deposition methods used to form the perovskite films can significantly impact their structural quality. Processing strategies aimed at reducing defect density, promoting crystallisation, and suppressing grain boundary formation are essential for enhancing the performance of halide perovskites in practical applications (Li et al. 2018).

Unique Properties of Perovskites

Perovskites exhibit various unique properties that make them highly versatile for multiple applications. Their high absorption coefficient makes them particularly effective for photovoltaics, while their tunable bandgap allows for optimisation by altering composition and structure.

Figure 5.4 shows diversified applications of the perovskite materials such as solar cells, LEDs, photodetectors, neuromorphic devices, lasers, pressure-induced

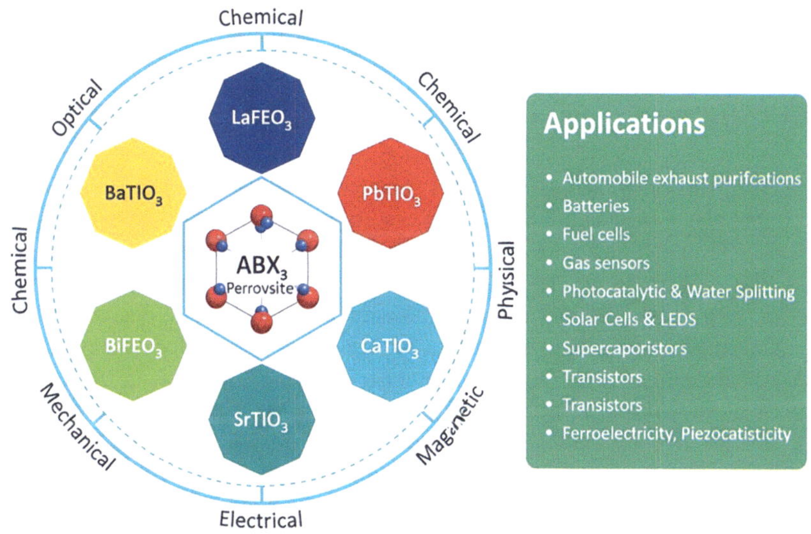

Fig. 5.4 Diverse applications of perovskite materials in modern technologies

emission, and many others. One distinctive optoelectronic property perovskite offers is high optical absorption and well-tunable bandgap, enabling efficient solar light collection, emission, and advanced detection. Additionally, perovskites' ferroelectric and piezoelectric characteristics are valuable for sensors and actuators. Certain types, such as cuprate and doped perovskites, even exhibit superconductivity, further expanding their potential applications.

Charge Transport Dynamics

Charge Generation and Recombination

The charge generation and recombination processes occurs in several distinct steps. Upon excitation, photons are absorbed, causing electrons to transition from the valence band to the conduction band and creating electron-hole pairs, known as excitons. During charge separation, excitons in certain perovskites dissociate into free charge carriers due to the material's low exciton binding energy, a process facilitated by intrinsic electric fields or heterojunctions within the device. These free carriers, comprising electrons and holes, are transported through the perovskite layer. The transport of carriers is influenced by factors such as carrier mobility, grain boundaries, defects, and the dielectric properties of the material. Following this, recombination occurs, which can be either radiative, where electrons recombine with holes and emit photons, as in LEDs and lasers or non-radiative, where energy is lost as heat, often at defect sites or grain boundaries. Finally, the charge carriers are collected during extraction at the interfaces with transport layers, such as electron or hole transport layers. Efficient extraction reduces recombination losses and enhances the overall performance of the device.

Drift-Diffusion Models

The drift-diffusion model is a helpful tool for understanding carrier dynamics, which includes transport, recombination, and collection of charge carriers, as well as simulating the efficiency of solar cells (SCs). It accounts for recombination losses in multilayered structures. This model includes the fundamental physics of charge transport and recombination, revealing how material parameters, device structure, and operating circumstances influence the efficiency and performance of PSCs. Detailed balancing models cannot optimise performance or investigate device physics in carrier dynamics. The drift-diffusion model optimises PSCs' performance by considering their practical device structures. Drift-diffusion models may accurately capture carrier dynamics, including defects, mobility, and bulk/surface recombination. The Drift-diffusion Model quantifies efficiency degradation due to several loss processes, including SRH recombination, surface recombination, and injection/extraction barriers (Ren et al. 2017).

Impact of Defect

Defects in PSCs affect their performance, stability, and degradation mechanisms. The performance of PSCs is highly affected by defect density.

Figure 5.5 describes the basic structure of a perovskite solar cell, which contains glass, an ITO layer, an electron extraction layer, a perovskite absorber, a hole extraction layer, and a metal electrode. Incident photons are absorbed by the perovskite, generating electron-hole pairs (Scharfetter and Gummel 1969; Usmani 1994). The energy diagram explains how electrons and holes move through the material, indicating possible charge traps. Efficient charge extraction at respective layers enhances the overall performance of the solar cell. High defect densities can result in non-radiative recombination, which reduces the charge extraction efficiency and overall performance of the device. An appropriate dopant concentration can mitigate these defects, which enhances charge transport and improves efficiency. These defects, which can occur at the atomic, grain, and interface levels, serve as charge trapping

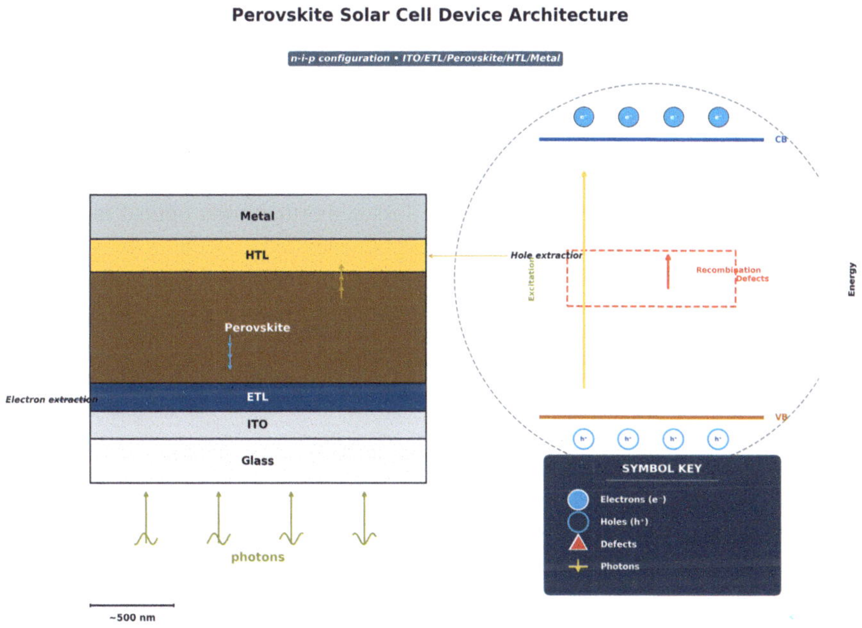

Fig. 5.5 Structure and energy diagram of a PSC showcasing photon absorption and charge extraction in the presence of a defect

and non-radiative recombination centres, lowering power conversion efficiency (PCE) (Pimcharoen 2023; Stein et al. 2019).

Intrinsic defects, such as vacancies, interstitials, and antisite defects, generate localised energy states within the bandgap, allowing carriers to recombine while limiting charge extraction. Grain boundaries, ubiquitous in solution-processed perovskite films, worsen these effects by acting as ion migration channels, leading to hysteresis and long-term instability. Furthermore, interface imperfections at the perovskite-transport layer and perovskite-electrode junctions impair charge transport and encourage undesired chemical processes, especially when exposed to high temperatures, light, and humidity for extended periods. Multi-physics investigations that include electrical, thermal, mechanical, and chemical models provide valuable insights into the intricate interaction between defect dynamics and PSC performance. Such studies enable the design of targeted strategies, including defect passivation, grain boundary engineering, and interface optimisation, to mitigate defect-induced losses and enhance device longevity.

Optical Modelling

Optical modelling is an integral part of comprehending and improving the workings of PSCs since it gives information about light absorption, reflection, scattering, and overall photonic activity within the device. Perovskite materials have unique optoelectronic features, such as a variable bandgap, a high absorption coefficient, and long carrier diffusion lengths, which make them highly efficient at capturing solar energy. Multi-physics optical models replicate light propagation and interaction throughout PSCs' layered architecture, accounting for interference, parasitic absorption, and photon recycling. Such models frequently use FDTD simulations, TMM, and ray-tracing techniques to optimise each layer's thickness and refractive index, improving light capture and reduce losses. Furthermore, optical modelling aids in the design of surfaces of textured or nanostructured materials, anti-reflective protective coatings, and tandem arrangements to enhance light management. Researchers may gain a comprehensive understanding of how light absorption influences charge carrier dynamics and thermal stability by combining optical models with electrical and thermal simulations, paving the way for more efficient and robust PSC designs.

Light Absorption and Management

Given PSCs, light absorption and management play an essential role in understanding the behaviour of light passing through the different layers of the device and, therefore, help calculate its efficiency and find possible causes of energy losses. Besides, the optical behaviour of each layer—dealer, perovskite, and contacts—indeed contributes objectively to the device's performance. Some concepts can be found in PSC optical modelling. For light absorption in the Perovskite Layer, the actual absorption efficiency of light matters for the initiation of charge carriers. Effective anti-reflection and light-trapping strategies are paramount to the device's high performance. Therefore, they work by minimising loss by reflection and keeping light in the active layers for relatively long periods, so that the chances for absorption may increase. The design of the layers thickness of the perovskite it's transport layers, and configuration of optical interfaces enters much into the light behaviour within the device. There are two aspects of light entering the perovskite layer: how much is being reflected from every evident interface and that primarily depends on the refractive indices of the various layers and the thickness of those layers. The refractive index determines how much the interaction between light and the different materials bends an incoming ray. At the same time, the layer thickness measures how much light travels in that layer before either getting absorbed or reflected.

This cell design optimisation minimises reflection and maximises transmission to increase the usable light for conversion into electrical energy. Optical modelling can optimise the interfaces by fine-tuning the refractive indices and thicknesses of each layer to reduce reflections and enhance light transmission, for improved PSC efficiency.

Optoelectronic Simulation

Optoelectronic simulations are great tools to improve the efficiency of PSCs and gain deeper insights into their device physics. Transfer Matrix Method (TMM), one of the most common methods, will help calculate light absorption in thin layers. This numerical method solves Maxwell's equations for complex nanoscale designs, allowing for reasonable accuracy in calculating how light interacts with the layers of the solar cell. TMM is well suited for issues related to multilayers, which usually characterise PSCs since it can consider phenomena like interference, refraction, and absorption on various interfaces within the materials.

Ray-tracing models, unlike other models, simulate the true path of individual photons through devices. This entails tracking the movement of photons through the various layers of the device, including their interaction with surfaces by means of reflection, refraction, and scattering (Iqbal et al. 2023). Considering these processes at each optical interface within the ray-tracing model gives a detailed picture of light behaviour within the device, especially for geometries that are more complicated, such as textured or patterned layers. Such detailed simulations are useful in optimising light management schemes, thus improving the performance of the solar cell by maximising light absorption and minimising reflection losses (Jung et al. 2023). In this case, the study option is favourable for optical phenomena occurring at a larger scale, wherein light-scattering, reflection, and diffraction come into play. Ray-tracing tracks photons through the layers of the solar cell, depending on whether they are scattered or reflected by different surface features. This gives insight into how the geometry of the cell impacts light processing and, ultimately, solar energy conversion efficiency (Kim et al. 2020).

In conjunction, these simulation techniques and ray-tracing optimise the optical designs of PSCs, maximising light absorption and conversion to electrical energy. Each method has unique capabilities: TMM excels with thin films, while ray-tracing provides a comprehensive picture of photon dynamics through complex, textured structures (Xu et al. 2022).

Thermal Effects and Stability

Thermal Degradation Mechanisms

At elevated temperatures, organic-inorganic perovskites undergo intrinsic instabilities, causing thermal degradation of the perovskite materials. This constraint on thermal degradation impedes long-term operating stability in PSCs (Raju et al. 2025). The perovskite materials undergo many detrimental processes under thermal stress, including ion migration, loss of volatile matter, and other degradation mechanisms that culminate in the unstable power output (SPO) of PSCs in real-world conditions, very much aggravated by temperature fluctuations, where the inherent sensitivity to heat causes accelerated degradation (Yang et al. 2024).

Thermal degradation is a primary cause of the deterioration of perovskite materials through several mechanisms. One mechanism is the decomposition of organic acts within the perovskite structure (Zhou et al. 2023). For instance, thermal stress causes methylammonium (MA) ions to decompose into smaller molecules, like methylamine and hydrogen iodide. This decomposition results in phase instability and may subsequently lead to the formation of non-perovskite phases that further degrade the solar cell efficiency (Senocrate et al. 2019). Another type of significant degradation involves the loss of halide ions like iodine, which are essential for the material's stability. As the halide ions migrate through the material, they create vacancies and defects that disrupt the crystallinity. These defects not only reduce the stability of the material but also increase carrier recombination losses, further reducing the efficiency of the solar cell. Improvement in the thermal stability of perovskite materials is thus vital for the long-term performance and reliability of PSCs, considering that in real-world situations studied, temperature fluctuations are unavoidable (Elangovan et al. 2024).

Thermal Management Strategies

For several reasons, it is critical to understand PSCs' thermal stability and their temperature variation. First, many perovskite films and contacts require an annealing stage for proper formation. Second, module encapsulation procedures often

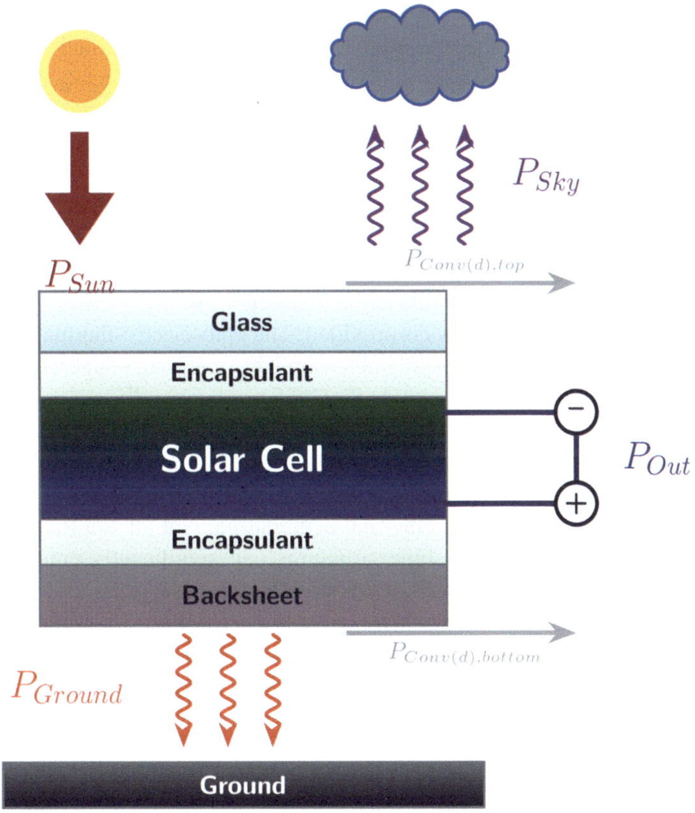

Fig. 5.6 Diagram of a terrestrial photovoltaic module, showing the energy fluxes coming in and flowing out

involve short-term exposure to temperatures exceeding 140 °C. Third, solar cells routinely reach 65 °C during operation in regions with high ambient temperatures. Finally, high-temperature testing accelerates chemical reactions and degradation processes that may take years to manifest under field conditions. Figure 5.6 shows diagram of a terrestrial photovoltaic module. The solar spectrum and the PV's absorptivity determine the absorbed solar irradiance, or P_{Sun}. The sky cooling, or P_{Sky}, is achieved by radiative energy exchange with the atmosphere from the side facing the sky, as well as by energy transfer to the ground, or P_{Ground} from the back-side, convective cooling by air at the top and bottom surfaces, conductive heat transfer through the aluminium frames, or (*d*), *top*/*bottom*. Most importantly, the output power delivered by PV modules to the external load, or P_{Out}.

Energy-balanced equations: To achieve equilibrium, the incoming and outgoing energy fluxes for a thermodynamic system in the steady state must balance out; specifically,

$$P_{Sun} = O_{Sky} + P_{Ground} + P_{Conv(d),\,top} + P_{Conv(d),\,bottom} + P_{Out} \tag{5.1}$$

For solar modules on land, keep in mind that each energy flux in (5.1) depends on the external environment, the PV modules' thermal condition, and their optical characteristics. Therefore, in order to determine the steady-state temperature of PV modules, (5.1) must be solved opto-electro-thermally and continuously. For example, optically, we compute P_{Sun} by combining the sun spectrum with the observed absorptivity. The temperature of PV modules T_{PV}, the ambient temperature, T_A, the emissivity of PV modules, and the atmospheric transmittance in the infrared (IR) region all affect the thermal P_{Sky}. Electrically, the output power, P_{Out}, varies depending on the PV technology and is temperature dependant. Lastly, energy flows must be provided by the computed equilibrium temperature. An overview of the formulas used to determine the energy flows is provided. The opto-electro-thermal simulation framework in this study can physically determine the operating temperature of modules with different solar absorbers (e.g., Si, CIGS) and under diverse condition, circumstances without the need for fitting parameters, in contrast to empirical methodologies (Sah et al. 1957; Faiman 2008).

Techniques such as compositional engineering, interface engineering, and encapsulation are employed to mitigate thermal degradation. Not all additional layers in PSCs impair thermal stability; some structures can significantly enhance it (Ke 2021). For example, thermally stable oxide transport layers like TiO_2 or NiOx, combined with stable contact or barrier layers atop the perovskite, help prevent decomposition and reactions with metal electrodes. Protecting the edges and top surface of the perovskite layer is crucial to prevent the escape of organic and halide species, which are prone to thermal breakdown (Khatoon et al. 2023). This is particularly vital in modules, as scribe lines allow moisture and oxygen ingress while facilitating the escape of volatile decomposition products (Pei et al. 2021).

One practical approach involves introducing silicon dioxide particles at the perovskite/HTL interface. These particles act as heat dissipation materials due to their high thermal conductivity and as surface passivators by coordinating with under-coordinated lead centres in the perovskite. This strategy has yielded devices with significantly improved long-term stability and a power conversion efficiency (PCE) of 22.29% for thermally stable compositions (Njema et al. 2024). Additionally, mixed-cation and mixed-anion perovskites offer a balance between structural and decomposition stability while also enhancing optoelectronic properties. Passivation plays a vital role by preventing moisture and oxygen ingress, metal-induced degradation, and surface-mediated breakdown through precise control of interfaces and contact layers. These measures ensure the long-term stability of PSCs, even under high-temperature conditions (Boyd et al. 2019).

Mechanical and Structural Stability

Mechanical Stresses

It describes the devices' ability to retain structural strength and functional performance under mechanical, environmental, and operational stresses. This notion is crucial to assuring PSCs' long-term stability and endurance, particularly in real-world applications (Tian et al. 2024). The residual stresses in perovskite films are tensile and can surpass 50 MPa, enough to deform copper. These stresses are a significant driving cause for fracture. When films are subjected to heat or humidity, they become more unstable under tensile stress and more stable under compressive stress. Because of perovskites' high thermal expansion coefficient, increasing the temperature of the formation of perovskite films results in higher residual stresses (Rolston et al. 2018). Specifically, this tensile stress arises after cooling to ambient temperature, as the substrate restricts the perovskite from shrinking (Rolston et al. 2018).

Common causes of mechanical stress in PSCs arise from various intrinsic, thermal, environmental, and external factors. Intrinsic stresses include lattice mismatch, which occurs due to differences in the thermal expansion coefficients of the perovskite film and adjacent layers, such as the substrate, electron transport layer (ETL), and hole transport layer (HTL), leading to lattice strain. Additionally, crystallisation stress can develop from volumetric fluctuations and strain during the formation of perovskite crystals. Thermal stresses arise from temperature changes, causing thermal expansion and contraction in the layers, which can lead to delamination or cracking over time. Environmental exposure, such as exposure to moisture, alter the stress distribution within the perovskite layer and compromise its stability. Furthermore, external pressures from wind loads, hail, encapsulation processes, or device handling can induce stress, negatively impacting its performance (Dailey et al. 2021).

Perovskite absorbers, with a linear thermal expansion coefficient (TEC) of $45 \times 10^{-6}{}^{\circ}C^{-1}$, are expected to contract more than glass substrates when cooled to room temperature. However, because they are attached to the substrate, they cannot shrink as much once formed, making them susceptible to breaking, delamination, and deterioration from moisture, heat, or light due to high tensile stress (Castriotta et al. 2025). To mitigate these issues, researchers have developed strategies such as producing perovskites at lower temperatures using a bath conversion process or employing plastic substrates with thermal expansion coefficients similar to the perovskite layer. Studies have shown that perovskite films exhibit better heat resistance when deposited on polymeric substrates compared to glass substrates, as tensile stress is significantly reduced in the former (Kosasih et al. 2022; Liu et al. 2024).

Finite Element Modelling

Finite element modelling (FEM) for multi-physics studies of PSCs is a powerful and specialised technique that enables the simultaneous simulation of multiple physical phenomena, including electrical, optical, thermal, and mechanical behaviours. FEM plays a crucial role in providing these insights, enabling detailed analysis of how these interactions affect the efficiency and durability of PSCs (Zandi and Razaghi 2019). By discretising the geometry of the PSCs into smaller, finite elements, FEM enables a detailed analysis of the electrical, thermal, mechanical, and optical processes occurring within the device. This method makes it possible to model the multilayered structure of the solar cell, which typically includes layers such as the perovskite light absorber, electron transport layers, and hole transport layers (Zhou et al. 2018). Each layer exhibits different material properties and behaves differently under various operational conditions. FEM allows for the detailed study of charge carrier dynamics, including transport, recombination, and the influence of defects and impurities. Understanding these dynamics is essential for improving the efficiency of PSC's and addressing performance-limiting factors such as charge carrier losses due to recombination or trapping (Li et al. 2022). Thermal management is another critical aspect of PSCs performance, and FEM can simulate how heat is generated and dissipated within the solar cell under real-world operational conditions. Excessive heat can lead to material degradation, impacting efficiency and long-term stability. FEM also helps analyse mechanical behaviour by simulating the mechanical loads that arise due to environmental factors, such as temperature fluctuations or mismatches in the thermal expansion coefficients of different layers (Tang et al. 2024). These mechanical stresses can lead to problems like delamination, cracking, and other structural failures, which can significantly reduce the reliability and lifespan of the device. FEM also facilitates the optimisation of device architecture by modelling the interactions among material properties, layer thicknesses, and environmental stimuli such as light and temperature. By adjusting these parameters, researchers can identify the most effective configurations for improving performance and durability. For example, by simulating the effect of varying the thickness of the perovskite layer or modifying the composition of transport layers, it is possible to predict how changes to the design can impact efficiency and stability. When used in combination with experimental verification, FEM becomes an indispensable tool for advancing the development of next-generation perovskite solar cells. It provides valuable insights into potential performance limitations, guides the improvement to material and device designs, and helps optimise manufacturing processes. FEM also helps in overcoming the challenges faced by PSCs, contributing to the development of more reliable, efficient, and cost-effective solar technologies for the future (Alghamdi et al. 2024).

Key Benefits of FEM in PSC Studies

FEM offers numerous advantages in the study of PSCs. It facilitates optimisation by refining design parameters such as layer thickness and material selection to enhance efficiency and stability. FEM is instrumental in predicting device behaviour under varying lighting conditions, temperatures, and mechanical stresses, providing valuable insights into performance. It also supports degradation analysis, enabling the modelling of potential failure mechanisms, such as perovskite layer degradation due to moisture, thermal stress, or mechanical strain. Additionally, FEM aids in design improvements by elucidating trade-offs among different layers and materials, guiding material selection and fabrication processes. Furthermore, FEM enhances cost-efficiency by enabling simulations before physical fabrication, minimising the need for resource-intensive trial-and-error testing.

The multi-physics modelling process in FEM involves several key steps:

1. Material properties: Each material in the solar cell, such as the perovskite, electrodes, and transport layers, must be characterised for properties like permittivity, conductivity, optical absorption coefficient, thermal conductivity, and Young's modulus.
2. Governing equations: Each physical phenomenon has corresponding equations, including Poisson's equation, drift-diffusion equations, and current continuity equations for electrical behaviour; Maxwell's equations for optical interactions; the heat conduction equation for thermal processes; and elasticity equations for mechanical stress and strain.
3. Meshing: The solar cell structure is divided into small finite elements, with refined meshing near interfaces—such as between the perovskite and transport layers—to capture sharp gradients in material properties.
4. Boundary and initial conditions: Appropriate boundary conditions are applied for each physics domain. For instance, electrical boundaries include electrode voltage and current density; optical boundaries account for incident light spectrum and reflection; thermal boundaries define temperature and heat flux; and mechanical boundaries incorporate constraints due to packaging or external loads.
5. Coupling multi-physics: FEM handles coupled systems for multi-physics problems. For example, in electrical-optical coupling, light absorption generates electron-hole pairs tracked in the electrical model. Electrical-thermal coupling addresses heat generation from current flow and its effects on material stability. In contrast, thermal-mechanical coupling evaluates thermal stresses leading to deformation or failure (Morrone et al. 2018).
6. Solver and post-processing: Appropriate solvers compute the system of coupled equations, while post-processing tools visualise critical parameters such as temperature profiles, electric potential distribution, stress/strain maps, and charge transport phenomena. FEM's comprehensive approach enables an in-depth understanding of the complex interactions in PSCs, aiding in the development of efficient and stable solar cell designs.

Coupled Multi-Physics Modelling

Coupled multi-physics modelling is crucial for comprehending and improving PSCs' lifetime, stability, and performance by simulating the interactions between several physical processes. Because of the strong interdependence of electrical, thermal, optical, and mechanical processes in a PSC, it is challenging to derive insights from single-physics models alone (Nasu et al. 2022). This cutting-edge method combines mechanical, chemical, thermal, and electrical processes to fully capture the intricate relationships affecting PSC stability, performance, and degradation mechanisms. For instance, the heat produced by resistive losses speeds up ion migration and chemical deterioration, whereas defect-induced charge recombination affects electrical performance. Device dependability is eventually decreased by mechanical loads brought on by external factors or thermal expansion mismatches, which also aid in creating cracks and delamination. Multi-physics models recreate real-world operating situations by linking these physical domains, uncovering failure reasons and performance bottlenecks that are otherwise concealed. Device architecture, material properties, and fabrication methods can all be optimised with the help of techniques like multi-domain simulation frameworks and FEM. In addition to helping to comprehend degradation pathways, this comprehensive modelling technique facilitates the construction of strong and effective PSCs for long-term operational stability.

Key Physical Phenomena in Perovskite Solar Cells

PSCs exhibit several key physical phenomena that influence their performance and stability. Electrical behaviour is dominated by the generation, transport, and collection of charge carriers (electrons and holes) across various layers, including the perovskite layer, the electron transport layer (ETL), and the hole transport layer (HTL). Factors like material defects, interface properties, and electric fields impact voltage distribution, current density, and charge carrier recombination. Optical behaviour involves efficient light absorption of perovskite materials across a wide wavelength range, influenced by the thickness of the perovskite layer, light incidence angle, and the optical properties of surrounding layers. This absorption generates electron-hole pairs, with exciton dissociation contributing to charge generation. Thermal behaviour is characterised by heat generation from electrical current and light absorption, with temperature distribution affecting electrical conductivity and material stability. Elevated temperatures can accelerate degradation processes, including photodegradation and thermal and moisture-induced degradation. Mechanical behaviour is critical, as stresses from thermal expansion, material mismatches, or delamination can compromise stability and performance. Mechanical issues like crack formation, layer delamination, and strain-induced damage may lead to device failure over time. Understanding these phenomena is essential for optimising PSC design and ensuring long-term durability.

Coupled Multi-Physics Phenomena in PSCs

Coupled multi-physics modelling for PSCs involves analysing interactions between electrical, optical, thermal, and mechanical effects simultaneously. These highly complex interactions are often non-linear, posing significant challenges for accurate simulations. Electro-optical coupling occurs as optical absorption in the perovskite layer generates electron-hole pairs, separated by the solar cell's electric field. The electric field distribution, influenced by optical absorption and incident light intensity, affects charge transport properties such as mobility and recombination rates. Additionally, reflection and scattering at interfaces, like those between the perovskite layer and the electrode, shape the light absorption profile. In thermo-electrical coupling, electric currents through the perovskite material generate heat via Joule heating, altering the cell's internal temperature. Elevated temperatures impact electrical conductivity, reducing charge mobility and affecting material stability and lifetime. The resulting thermal gradients further influence charge transport, as electron and hole mobilities are temperature-sensitive. Electro-mechanical coupling arises from thermal expansion differences between layers, such as the perovskite, transport layers, and electrodes, causing mechanical stresses. These stresses may lead to deformation, delamination, or cracks. Such mechanical deformations can disrupt interfacial properties, reducing charge extraction efficiency and overall performance of the solar cell. Opto-thermal coupling involves the conversion of absorbed photons into electrons and heat within the perovskite layer. Heat can alter optical properties, such as refractive index and absorption coefficients, potentially reducing optical absorption efficiency or exciton dissociation efficiency.

Finally, long-term degradation integrates thermal cycling, mechanical stresses, and moisture absorption effects, leading to perovskite layer degradation. Such degradation impacts electrical conductivity and optical absorption. Multi-physics models incorporating these degradation processes simulate the device's long-term stability under various environmental conditions, such as humidity, temperature variations, and UV exposure. Understanding these coupled phenomena is crucial for optimising PSC design, improving efficiency, and enhancing durability.

Steps for Implementing Coupled Multi-Physics Modelling for PSCs

Modelling PSCs in conjunction with multiple nonlinearities permits more than a comprehensive approach; it implies the interaction of the domains above in an interdisciplinary way, modelling the complexity therein. The first step is to map out the fundamental physics and coupling mechanisms between them. Next, the governing equations of each domain will be set out. For example, electrostatics (charge distribution) within the solar cell are modelled with Poisson's equation (Rismanchi et al. 2023). In contrast, the conduction of heat is explained by some formulation of the heat conduction equation within the materials. Also, the behaviour of light, how light propagates, reflects, and is absorbed in the cell, is modelled by Maxwell's equations. In contrast, Navier-Cauchy's equations are fundamental in understanding the mechanical strains and stresses developed within the device's working. Material characterisation comes next after defining the physics. Here, data about the properties of each layer in the solar cell, namely the perovskite absorber, transport layers, electrodes, and encapsulation materials, are gathered. Relevant properties include electrical conductivity, absorption coefficients, thermal conductivity, and mechanical properties such as Young's modulus. These values are crucial for predicting the materials' responses to light, heat, and mechanical stress. The next step is mesh generation. This is the stage where the physical domain is subdivided into small finite elements to simplify the solution. Care has to be taken in the mesh refinement at places where the physical properties change rapidly, like material interfaces with high charge density or high-temperature gradients. This is crucial since the model can accurately capture the sharp changes in phenomena such as electrical charge and temperature in those critical points with a finer numerical mesh (AlHajri et al. 2012).

Once the mesh has been created, the equations of all the domains-electrical, thermal, optical, and mechanical—are coupled and solved simultaneously. This is where the real magic happens, for the phenomena in these domains are all integrated (Aalloul et al. 2024). For example, light absorption produces charge carriers; however, the conductivity of materials varies with temperature due to Joule heating. Conductivity, in turn, impacts the movement of charges within the cell. Finally, mechanical deformations from thermal expansion or external forces would, in principle, change the geometries of the layers, which may affect charge collection efficiency and, in the long term, lead to the deterioration or extinction of the device (Giebink et al. 2010). The other fascinating outcome from this coupling is that one can obtain a much more realistic picture of how Joule heating raises the temperature and that this heating then affects the electrical behaviour of the materials (Nguyen 2015). Thus, simulating thermal gradients versus the electrical currents will allow us to present a more realistic picture of how the PSC will behave under different environmental conditions. Another critical step is defining the boundary conditions. These would encompass the parameters, providing limits to the model to represent real-world conditions. Electrical boundary conditions may define either voltages or current densities applied across specific layers, and thermal boundary conditions would specify heat fluxes or ambient temperatures around the cell. Lastly, optical boundary conditions describe incoming light, wherein intensity, wavelength, and angle of incidence are defined (Kim et al. 2015). In contrast, mechanical boundary conditions would represent external forces or stresses, for instance, the forces acting on the device or the stresses induced by thermal mismatches between different layers. So, boundary conditions are essential to keep the model realistic, representing varying light intensities, temperatures, and mechanical stresses during operation (Singh et al. 2021; Sagar et al. 2019).

Finally, once the equations are solved, the post-processing tools come in. Post-processing tools are used for analysis and extracting significant performance metrics such as power conversion efficiencies, current-voltage characteristics, temperature distributions, and mechanical stresses on the device (AlHajri et al. 2012; El-Naggar et al. 2012). These results would provide insights into the behaviour of PSCs under different conditions helping the researchers optimise design choices and material selections to improve the performance and durability of the device (Green et al. 2014).

Software for Multi-Physics Modelling of Perovskite Solar Cells

Several software tools are available for conducting multi-physics simulations of perovskite solar cells, each offering unique capabilities tailored to specific modelling aspects (Kowsar et al. 2025). COMSOL Multiphysics is a leading platform that supports simulations for electrical, thermal, and mechanical analyses. Its ability to seamlessly couple these physical models makes it particularly effective for studying perovskite solar cells.

Table 5.1 summarises various software tools used for photonic and photovoltaic simulations, highlighting their full forms, company names, and whether they are open-source, free, or commercial (Fan et al. 2021; Wang and Zhang 2021; Pan et al. 2021). This classification helps researchers select suitable tools based on their accessibility and application requirements. Using these simulation tools, most advanced multi-physics platforms nowadays, like ANSYS, COMSOL Multiphysics, Abaqus, or Open FOAM, iteratively solve the coupled systems of equations for improving accuracy using feedback between different domains (Sah et al. 1957; Alanazi et al. 2017). ANSYS is renowned for its expertise in structural and thermal

Table 5.1 Summary of software for photonic and photovoltaic simulations (Kowsar et al. 2025; Fan et al. 2021; Wang and Zhang 2021; Pan et al. 2021)

Software name	Full form	Company/ developer	Licence type
TranSIESTA	Transport-Spanish initiative for electronic simulations with thousands of atoms	SIESTA consortium	Open source
GPVDM	General purpose photovoltaic device model	University of Sheffield	Open source
OpenSolar	Open solar design and simulation software	OpenSolar	Free licence
COMSOL Multiphysics®	N/A (proprietary software)	COMSOL multiphysics	Commercial
MEEP	MIT electromagnetic equation propagation	MIT	Open source
PVcase	Photovoltaic case simulation software	PVcase	Open source
SCAPS	Solar cell capacitance simulator	University of Ghent	Restricted academic licence

simulations and also provides tools like ANSYS Maxwell or ANSYS Electronics Desktop to model electrical properties. In terms of licensing for tools, SCAPS is free for academic use but is not open-source because the source code is not available for modification or redistribution (Aalloul et al. 2024). FDTD tools such as MEEP and MPB are fully open-source under the GNU General Public Licence (GPL), allowing unrestricted modification and distribution (Nkele et al. 2020). TranSIESTA and GPVDM are also open-source, providing flexibility for academic and research purposes (Liang 2014; Rajasekar and Neeraja 2013). OpenSolar is under a free licence and offers similar freedoms for use, modification, and redistribution. However, COMSOL Multiphysics is commercial software requiring a paid licence, although community-driven open-source extensions are available for specific applications. In a nutshell, the above tools—FDTD (MEEP/PVcase), TranSIESTA, GPVDM, and OpenSolar are entirely open-source and can provide immense flexibility for the researchers. SCAPS is available for academic purposes with restrictions. Finally, post-processing and analysis are conducted to examine how interactions between physical phenomena affect PSC performance and stability (El-Naggar et al. 2012). Key metrics include power conversion efficiency, current density, voltage profiles, temperature distribution, and mechanical stresses.

Additionally, simulations under varying operating conditions (e.g. temperature, light intensity, or mechanical forces) provide insights for optimising material lifetime and overall device performance. For advanced material modelling, ABAQUS excels in simulating mechanical and electrical interactions within the solar cell stack, particularly for complex geometries and stress analyses. OpenFOAM, an open-source option, is primarily designed for fluid dynamics but can be adapted for thermal and electromagnetic modelling in coupled multi-physics systems (Sah et al. 1957; Alanazi et al. 2017). These tools collectively enable a comprehensive analysis of perovskite solar cells, addressing various physical interactions that influence their performance and durability. In conclusion, coupled multiphysics modelling is indispensable for understanding and optimising PSC performance. By simulating the interactions between electrical, optical, thermal, and mechanical phenomena, researchers can gain critical insights into efficiency, stability, and degradation mechanisms, paving the way for more efficient and durable devices (Slotboom 1973).

Example Studies Done by COMSOL

1. Numerical modelling and simulation of a 1-D Silicon solar cell (Nayak et al. 2021)

Key findings

(a) Efficiency Optimisation with Doping Concentration: The study demonstrates that the efficiency of a Silicon 1D solar cell is significantly influenced by the doping concentration. A donor concentration level yields an efficiency of 16.97%, while the highest efficiency of 18.78% and a fill factor of 92% are achieved at specific acceptor and donor concentration levels.

(b) Impact of Resistance on Electrical Parameters: The variation in doping concentration leads to changes in resistance, which directly affects key electrical parameters such as open circuit voltage and fill factor. These changes play a crucial role in determining the overall performance of the solar cell.

2. COMSOL simulation of heat distribution in perovskite solar cells: coupled optical–electrical–thermal 3-D analysis (Saxena and Gorji 2019).

Key Findings

(a) For the first time, a three-dimensional (3D) simulation of heat distribution in perovskite solar cells has been conducted using COMSOL Multiphysics. The study couples optical, electrical, and thermal modules to analyse temperature variations and heat dissipation across the cell structure.

(b) The results reveal that heat dissipation is most significant at the metallic contact, particularly where PbI_2 defects form due to moisture-induced oxidation or decomposition. Heat accumulation occurs at junctions and electrode sides, affecting device performance. The study suggests using highly thermally conductive metallic contacts to improve heat dissipation and enhance cell reliability under normal operating conditions.

Outlook and Future Directions

The development of coupled multi-physics modelling for PSCs presents several challenges and opportunities. Accurate material characterisation is critical, as precise input data for properties like charge mobility and bandgaps are essential but often difficult to obtain. The complexity of multi-physics models poses another challenge, as these simulations can become computationally expensive, demanding high computational power and sophisticated meshing strategies. Validation of FEM predictions against experimental results is crucial, particularly when exploring new materials or advanced device configurations. Additionally, integrating long-term stability models remains an ongoing challenge, as perovskites are prone to degradation over time due to environmental factors. Addressing these issues will enhance simulations' reliability and predictive capabilities, driving PSC design and performance advancements. To overcome the major obstacles of performance, stability, scalability, and sustainability challenges, multi-physics modelling will be essentialfor developing PSCs. Even though PSCs have produced impressive PCEs of over 25%, more advancements in this field necessitate a better comprehension of the complex interactions among the device's electrical, thermal, mechanical, and chemical processes. By facilitating high-resolution simulations of actual operating conditions, multi-physics techniques can direct efforts to improve defect passivation, optimise charge carrier dynamics, and reduce non-radiative recombination losses. Future research should improve these models to incorporate more realistic depictions of dynamic processes essential to long-term stability, such as moisture ingress, phase segregation, and ion movement. Another area where multi-physics modelling can yield insightful information is scalability, especially when spotting possible stress areas and performance losses while producing large-area modules.

Predictive simulations could help improve layer uniformity, thickness, and interface quality in emerging fabrication processes, including roll-to-roll printing and blade coating. Tandem devices, which combine perovskite layers with silicon or other materials to obtain PCEs above the Shockley-Queisser limit, are also gaining popularity. In this case, multi-physics modelling can assist in optimising the thermal and optical coupling between subcells, guaranteeing optimal stability and performance in various environmental conditions. To develop devices that minimise lead toxicity, eliminate energy-intensive fabrication stages, and promote recycling, it is also necessary to consider sustainability by integrating environmental and life-cycle assessments into modelling frameworks.

One interesting approach to speeding up PSC research is the combination of multi-physics modelling with AI and ML. AI-driven models significantly reduce experimental trial and error by analysing big datasets and forecasting trends in material and device performance. The next generation of high-efficiency, stable, and sustainable solar technologies will be made possible by multi-physics modelling, cutting-edge experimental methods, and AI. In conclusion, even though PSCs have made remarkable strides, their widespread adoption depends on resolving critical issues.

Acknowledgments Yogesh Sharma gratefully acknowledges the DST Science and Engineering Research Board (SERB) for financial support under the Core Research Grant (CRG) through Research Grant No. CRG/2022/006820.

References

Aalloul R, Adhiri R, Benlattar M, Elaissaoui A (2024) Coupled electrical–thermal modelling of photovoltaic modules: an overview. Thermal Sci Eng Progr 55:102962

Alanazi HO, Abdullah AH, Qureshi KN (2017) A critical review for developing accurate and dynamic predictive models using machine learning methods in medicine and health care. J Med Syst 41(4):69. https://doi.org/10.1007/s10916-017-0715-6

Alghamdi, H., Maduabuchi, C., Okoli, K., Alobaid, M., Alghassab, M., Alsafran, A. S., Makki, E., & Alkhedher, M. (2024). Latest Advancements in Solar Photovoltaic-Thermoelectric Conversion Technologies: Thermal Energy Storage Using Phase Change Materials, Machine Learning, and 4E Analyses. International Journal of Energy Research, 2024(1). https://doi.org/10.1155/2024/1050785

AlHajri MF, El-Naggar KM, Al Rashidi MR, Al-Othman AK (2012) Optimal extraction of solar cell parameters using pattern search. J Renew Energy 44:245

Williams, C. M., & Cristian, J. B. (2016). Hydrodynamic modeling of hot-carrier effects in a PN junction solar cell. https://repositorio.uchile.cl/handle/2250/146249

Berhe TA et al (2016) Organometal halide perovskite solar cells: degradation and stability. Energy Environ Sci 9(2):323–356

Boyd CC, RongrongCheacharoen TL, McGehee MD (2019) Understanding degradation mechanisms and improving stability of perovskite photovoltaics. Chem Rev 119(5):3418–3451

Castriotta LA, Uddin MA, Jiao H, Huang J (2025) Transition of perovskite solar technologies to being flexible. Adv Mater. https://doi.org/10.1002/adma.202408036

Dailey, M., Li, Y., & Printz, A. D. (2021). Residual Film Stresses in Perovskite Solar Cells: Origins, Effects, and Mitigation Strategies. ACS Omega, 6(45), 30214–30223. https://doi.org/10.1021/acsomega.1c04814

Elangovan NK, Kannadasan R, Beenarani BB, Alsharif MH, Kim M-K, Hasan Inamul Z (2024) Recent developments in perovskite materials, fabrication techniques, band gap engineering, and the stability of perovskite solar cells. Energy Rep 11:1171–1190. https://doi.org/10.1016/j.egyr.2023.12.068

El-Naggar KM, Al Rashidi MR, AlHajri MF, Al-Othman AK (2012) Simulated annealing algorithm for photovoltaic parameters identification. Sol Energy 86(1):266–274

Faiman D (2008) Assessing the outdoor operating temperature of photovoltaic modules. Prog Photovolt Res Appl 16(4):307–315. https://doi.org/10.1002/pip.813

Fan C, Chen M, Wang X, Wang J, Huang B (2021) A review on data preprocessing techniques toward efficient and reliable knowledge discovery from building operational data. Front Energy Res 9. https://doi.org/10.3389/fenrg.2021.652801

Freire P, Manuylovich E, Prilepsky JE, Turitsyn SK (2023) Artificial neural networks for photonic applications—from algorithms to implementation: tutorial. Adv Opt Photon 15(3):739. https://doi.org/10.1364/AOP.484119

Giebink, N. C., Lassiter, B. E., Wiederrecht, G. P., Wasielewski, M. R., & Forrest, S. R. (2010). Ideal diode equation for organic heterojunctions. II. The role of polaron pair recombination. Physical Review B, 82(15). https://doi.org/10.1103/physrevb.82.155306

Green MA, Ho-Baillie A, Snaith HJ (2014) The emergence of perovskite solar cells. Nat Photonics 8(7):506–514

Gummel HK (1964) A self-consistent iterative scheme for one-dimensional steady state transistor alculations. IEEE Trans Electron Devices 11(10):455–465

Habte A, Sengupta M, Lopez A (2015) Developing the next generation of gridded tmys

Hou Y et al (2020) Efficient tandem solar cells with solution-processed perovskite on textured crystalline silicon. Science 367(6482):1135–1140

Iqbal, M. A., Malik, M., Anwar, N., Bakhsh, S., Javeed, S., Sarah Maidin, S., Morsy, K., Capangpangan, R. Y., Alguno, A. C., & Choi, J. R. (2023). Basic concepts, advances and emerging applications of nanophotonics. Arabian Journal of Chemistry, 16(9), 105040. https://doi.org/10.1016/j.arabjc.2023.105040

Jung S-K, Park N-G, Lee J-W (2023) Light management in perovskite solar cells. Mater Today Energy 37:101401. https://doi.org/10.1016/j.mtener.2023.101401

Ke H (2021) Compositional engineering of perovskite materials. In: Hybrid perovskite composite materials. Elsevier, Amsterdam, pp 203–224. https://doi.org/10.1016/B978-0-12-819977-0.00009-3

Khatoon S et al (2023) Perovskite solar cell's efficiency, stability and scalability: a review. Mater Sci Energy Technol 6:437–459. https://doi.org/10.1016/j.mset.2023.04.007

Kim M, Choi W, Choi Y, Yoon C, Choi W (2015) Transmission matrix of a scattering medium and its applications in biophotonics. Opt Express 23(10):12648. https://doi.org/10.1364/OE.23.012648

Kim MS, Lee JH, Kwak MK (2020) Review: surface texturing methods for solar cell efficiency enhancement. Int J Precis Eng Manuf 21(7):1389–1398. https://doi.org/10.1007/s12541-020-00337-5

Kojima A, Teshima K, Shirai Y, Miyasaka T (2009) Organometal halide perovskites as visible-light sensitizers for photovoltaic cells. J Am Chem Soc 131(17):6050–6051

Kosasih FU, Erdenebileg E, Mathews N, Mhaisalkar SG, Bruno A (2022) Thermal evaporation and hybrid deposition of perovskite solar cells and mini-modules. Joule 6(12):2692–2734. https://doi.org/10.1016/j.joule.2022.11.004

Kowsar A et al (2025) An overview of solar cell simulation tools. Solar Energy Adv 5:100077. https://doi.org/10.1016/j.seja.2024.100077

Li Z et al (2018) Scalable fabrication of perovskite solar cells. Nat Rev Mater 3(4):18017

Li D et al (2022) Metal halide perovskite/electrode contacts in charge-transporting-layer-free devices. Adv Sci 9(36). https://doi.org/10.1002/advs.202203683

Eftaiha, A. F., Sun, J.-P., Hill, I. G., & Welch, G. C. (2014). Recent advances of non-fullerene, small molecular acceptors for solution processed bulk heterojunction solar cells. J. Mater. Chem. A, 2(5), 1201–1213. https://doi.org/10.1039/c3ta14236a

Liu J et al (2024) Evolutionary manufacturing approaches for advancing flexible perovskite solar cells. Joule 8(4):944–969. https://doi.org/10.1016/j.joule.2024.02.025

Ma J, Man KL, Guan SU, Ting TO, Wong PW (2016) Parameter estimation of photovoltaic model via parallel particle swarm optimization algorithm. Int J Energy Res 40(3):343–352

Morrone M, Garion C, Aurisicchio M, Chiggiato P (2018) A coupled multiphysics FEM model to investigate electromagnetic, thermal and mechanical effects in complex assemblies: the design of the high-luminosity large hadron collider beam screen. Appl Math Model 57:280–301. https://doi.org/10.1016/j.apm.2017.12.031

Nasu NJ, Al Mahbub MA, Zheng H (2022) A new coupled multiphysics model and partitioned time-stepping method for the triple-porosity-stokes fluid flow model. J Comput Phys 466:111397. https://doi.org/10.1016/j.jcp.2022.111397

Nayak J, Mishra DK, Pattanaik P (2021) Numerical modelling and simulation of a 1-D silicon solar cell. Mater Today Proc 41:451–454. https://doi.org/10.1016/j.matpr.2020.10.233

Nguyen XH (2015) Matlab/Simulink based modeling to study effect of partial shadow on solar photovoltaic array. J Environ Syst Res 4(1):20

Njema GG, Kibet JK, Ngari SM (2024) A review of interface engineering characteristics for high performance perovskite solar cells. Measure Energy 2:100005. https://doi.org/10.1016/j.meaene.2024.100005

Nkele AC, Ike IS, Ezugwu S, Maaza M, Ezema FI (2020) An overview of the mathematical modelling of perovskite solar cells towards achieving highly efficient perovskite devices. Int J Energy Res 45(2):1496–1516

Noel NK et al (2014) Lead-free organic-inorganic tin halide perovskites for photovoltaic applications. Energy Environ Sci 7(9):3061–3068

Nrel (2023) Best research-cell efficiency chart. National Renewable Energy Laboratory, Washington

Pan X, Wang X, Tian B, Wang C, Zhang H, Guizani M (2021) Machine-learning-aided optical fiber communication system. IEEE Netw 35(4):136–142. https://doi.org/10.1109/MNET.011.2000676

Park N-G (2015) Perovskite solar cells: an emerging photovoltaic technology. Mater Today 18(2):65–72

Pei, F., Li, N., Chen, Y., Niu, X., Zhang, Y., Guo, Z., Huang, Z., Zai, H., Liu, G., Zhang, Y., Bai, Y., Zhang, X., Zhu, C., Chen, Q., Li, Y., & Zhou, H. (2021). Thermal management enables more efficient and stable perovskite solar cells. ACS Energy Letters, 6(9), 3029–3036. https://doi.org/10.1021/acsenergylett.1c00999

Szulc, K., Pydzińska-Białek, K., & Ziółek, M. (2023). Modeling of charge injection, recombination, and diffusion in complete perovskite solar cells on short time scales. Materials, 16(22), 7110. https://doi.org/10.3390/ma16227110

Krishnakumar, N., Venugopalan, R., & Rajasekar, N. (2013). Bacterial foraging algorithm based parameter estimation of solar PV model. In 2013 Annual International Conference on Emerging Research Areas and 2013 International Conference on Microelectronics, Communications and Renewable Energy (pp. 1–6). IEEE. 2013 Annual International Conference on Emerging Research Areas (AICERA) - 2013 International Conference on Microelectronics, Communications and Renewable Energy (ICMiCR). https://doi.org/10.1109/aicera-icmicr.2013.6575948

Raju TD, Murugadoss V, Nirmal KA, Dongale TD, Kesavan AV, Kim TG (2025) Advancements in perovskites for solar cell commercialization: a review. Adv Powder Mater 4:100275. https://doi.org/10.1016/j.apmate.2025.100275
Ramirez C, Yadavalli SK, Garces HF, Zhou Y, Padture N (2018) Thermo-mechanical behavior of organic-inorganic halide perovskites for solar cells. Sci Mater 150:36–41
Ren X, Wang Z, Sha WE, Choy WC (2017) Exploring the way to approach the efficiency limit of perovskite solar cells by drift-diffusion model. ACS Photon 4:934–942
Ren Y, Yang Z, Yu Z (2020) Machine learning-enabled discovery and design of inorganic solar cell materials. Adv Mater 32(44):2002146
Rismanchi A, Mottaghitalab V, Rahimpour Soleimani H (2023) The modelling and simulation of perovskite solar cell consisting textile-based electrodes. Results Surf Interfaces 10:100095. https://doi.org/10.1016/j.rsurfi.2022.100095
Rolston, N., Bush, K. A., Printz, A. D., Gold-Parker, A., Ding, Y., Toney, M. F., McGehee, M. D., & Dauskardt, R. H. (2018). Engineering Stress in Perovskite Solar Cells to Improve Stability. Advanced Energy Materials, 8(29). https://doi.org/10.1002/aenm.201802139
Sagar MAK et al (2019) Optical fiber-based dispersion for spectral discrimination in fluorescence lifetime imaging systems. J Biomed Opt 25(1):1. https://doi.org/10.1117/1.JBO.25.1.014506
Sah C-T, Noyce R, Shockley W (1957) Carrier generation and recombination in P-N junctions and P-N junction characteristics. IEEE Proc IRE 45(9):1228–1243
Saxena P, Gorji NE (2019) COMSOL simulation of heat distribution in perovskite solar cells: coupled optical–electrical–thermal 3-D analysis. IEEE J Photovolt 9(6):1693–1698. https://doi.org/10.1109/JPHOTOV.2019.2940886
Scharfetter DL, Gummel HK (1969) Large-signal analysis of a silicon read diode oscillator. IEEE Trans Electron Devices 16(1):64–77
Senocrate A, Kim GY, Grätzel M, Maier J (2019) Thermochemical stability of hybrid halide perovskites. ACS Energy Lett 4(12):2859–2870. https://doi.org/10.1021/acsenergylett.9b01605
Singh Y, Paswan MK, Raghuwanshi SK (2021) Sensitivity enhancement of SPR sensor with the black phosphorus and graphene with bi-layer of gold for chemical sensing. Plasmonics 16(5):1781–1790. https://doi.org/10.1007/s11468-020-01315-3
Slotboom JW (1973) Computer-aided two-dimensional analysis of bipolar transistors. IEEE Trans Electron Devices 20(8):669–679
Stein HS, Guevarra D, Newhouse PF, Soedarmadji E, Gregoire JM (2019) Machine learning of optical properties of materials—predicting spectra from images and images from spectra. Chem Sci 10(1):47–55. https://doi.org/10.1039/C8SC03077D
Tang L, Luo Y, Yin L, Li J, Cao X (2024) Thermal-management performance of phase-change material on PV modules in different climate zones. Energies 17(23):6200. https://doi.org/10.3390/en17236200
Tian R, Zhou S, Meng Y, Liu C, Ge Z (2024) Material and device design of flexible perovskite solar cells for next-generation power supplies. Adv Mater 36(37). https://doi.org/10.1002/adma.202311473
Tress W et al (2018) Interpretation and evolution of open-circuit voltage, recombination, and sub-gap states during reversible light-soaking and irreversible degradation of perovskite solar cells. Energy Environ Sci 11(1):151–165
C. S. Ponseca Jr. et al., "Organometal Halide Perovskite Solar Cell Materials Rationalized: Ultrafast Charge Generation, High and Microsecond-Long Balanced Mobilities, and Slow Recombination," J. Am. Chem. Soc., vol. 136, no. 14, pp. 5189–5192, 2014.
Wang D, Zhang M (2021) Artificial intelligence in optical communications: from machine learning to deep learning. Front Commun Netw 2. https://doi.org/10.3389/frcmn.2021.656786
Xu L et al (2022) Accurate optical modeling of monolithic perovskite/silicon tandem solar cells and modules on textured silicon substrates. PRX Energy 1(2):023005. https://doi.org/10.1103/PRXEnergy.1.023005

Yang WS et al (2015) High-performance photovoltaic perovskite layers fabricated through intramolecular exchange. Science 348(6240):1234–1237

Yang C et al (2024) Achievements, challenges, and future prospects for industrialization of perovskite solar cells. Light Sci Appl 13(1):227. https://doi.org/10.1038/s41377-024-01461-x

Zandi S, Razaghi M (2019) Finite element simulation of perovskite solar cell: a study on efficiency improvement based on structural and material modification. Sol Energy 179:298–306. https://doi.org/10.1016/j.solener.2018.12.032

Zhou H et al (2015) Interface engineering of highly efficient perovskite solar cells. Science 345(6196):542–546

Zhou D, Zhou T, Tian Y, Zhu X, Tu Y (2018) Perovskite-based solar cells: materials, methods, and future perspectives. J Nanomater 2018:1–15. https://doi.org/10.1155/2018/8148072

Zhou J et al (2023) Modulation of perovskite degradation with multiple-barrier for light-heat stable perovskite solar cells. Nat Commun 14(1):6120. https://doi.org/10.1038/s41467-023-41856-9

Chapter 6
Theoretical Study of All-Inorganic and Organic-Inorganic Mixed Halide Perovskite Solar Cells

Chandra Shekhar Verma, Vinay K. Verma, Rajaneesh K. Verma, Manish Kumar, and Ram C. Singh

Abstract Perovskite solar cell (PSC) is a transformative step in advancement of photovoltaic technology as it presents the high power conversion efficiencies, cost-effectiveness, and customizable material properties. The theoretical investigations into mixed halide-based PSCs through computational perspectives are focusing on structural, electronic, and optical characteristics. Advanced simulation techniques, DFT (Density Functional Theory), SCAPS-1D, etc., have provided critical insights into the behavior of both all-inorganic and organic-inorganic hybrid perovskites under various conditions. For all-inorganic systems, simulations demonstrate exceptional thermal stability, tunable band gaps, and efficient charge carrier mobility, which are essential for developing durable and efficient solar cells. In organic-inorganic hybrids like $MAPbI_3$, theoretical studies elucidate the stabilizing role of organic molecules and the impact of compositional engineering at the A-site and X-site on performance and stability. Key areas explored include defect formation, excitons dynamics, and degradation mechanisms that influence long-term device reliability. By integrating theoretical predictions with experimental data, computational studies play a prime role in material screening and performance optimization, paving the route for coming generation PSCs with enhanced efficiency, durability,

---

C. S. Verma · V. K. Verma
Department of Chemistry and Biochemistry, School of Basic Sciences & Research, Sharda University, Greater Noida, India

R. K. Verma
Department of Chemistry, Indian Institute of Technology, Delhi, New Delhi, India

M. Kumar
Experimental Research Laboratory, Department of Physics, ARSD College, University of Delhi, New Delhi, India

R. C. Singh (✉)
Department of Physics, School of Basic Sciences & Research, Sharda University, Greater Noida, India
e-mail: rcsingh@sharda.ac.in

M. Kumar (ed.), *Computational Analysis of Perovskite Solar Cells*, Springer Tracts in Electrical and Electronics Engineering,
https://doi.org/10.1007/978-981-95-6521-4_6

and scalability. This review underscores the importance of theoretical research in advancing mixed halide PSC technologies for sustainable energy solutions.

Keywords Mixed halide · Organic-inorganic · All-inorganic · SCAPS-1D · Perovskite solar cell · Stability

Introduction

The regular use of energy, owing to gradually expanding population and modernization of society, has been gradually increasing all over the world. Due to limited sources of fossil fuels, there is huge demand for research towards the alternative energy resources, i.e., renewable and sustainable. The characteristics such as clean, safe, and widely availability have make solar energy one of the best alternative to fulfill the future requirements of energy in coming years without creating bad effects on global climate (Osman et al. 2021; Wang et al. 2019; Liang et al. 2021). The solar cell basically follows the principle of Photovoltaic Effect wherein solar cell generates DC voltage when it comes in contact with sunrays. The efficacy of a solar cell is measured by its basic characteristics such as Short Circuit Current, Open Circuit Voltage, Fill-Factor, and Power Conversion Efficiency (Denoted by I_{sc}, V_{oc}, FF, and η, respectively) (Dambhare et al. 2021). The technology of photovoltaic can directly generate electricity by utilizing the eco-friendly renewable energy resource (Wang et al. 2019). Although many technologies such as solar heating, photovoltaics, photo-catalysis, and solar architecture have been developed for extracting the energy, the unique strength of photovoltaic technology, i.e., to convert sunlight into electricity directly, has attracted attentions of research community worldwide which ultimately increased the research interests.

Generations in Solar Technology

The generation of the solar technology is basically relying on the kind of materials used in developing of solar devices. In market, single crystalline silicon-based solar cells have occupied maximum space. The silicon is also used in the amorphous as well as polycrystalline solar cells. The materials, used as a fine layer in solar devices, are generally amorphous silicon (a-Si), CdTe (Cadmium Telluride), Cadmium Sulpfide (CdS), and CIGS (Copper Indium Gallium Di-Selenide). In the solar cell technology of third generation, advanced technology is used in solar cells to utilize the maximum solar energy so that optimum efficiency could be obtained. The most popular types of solar cells in this category are Dye-Sensitized, Quantum Dot, Tandem Cells, etc. (Dambhare et al. 2021). Solar cell technology can be grouped into the following three generations based on the materials and process involved for solar cell devices:

First-Generation of Solar Cell

It is one of the most-established and successful technologies for photovoltaic applications due to its good stability and efficiency offered by the materials. In first generation of photovoltaic technology, crystalline silicon wafers are used so it is also called as wafer-based solar cells. Based on the type of silicon materials (single or poly crystalline) used in solar cells, silicon wafer-based solar cells could be categorized into sub-categories, namely:

- Single (Mono) crystalline silicon-based solar cell.
- Poly (Multi) crystalline silicon-based solar cell.

About 80% market of solar cells are covered by the silicon-based solar cell. The monocrystalline-based solar cells are costly but more efficient (Bertolli 2008) while the polycrystalline silicon-based solar cells are cheaper but low efficient (Jayakumar 2009) as compared to monocrystalline silicon (Choubey et al. 2012; McEvoy et al. 2012; Bagher et al. 2015; Srinivas et al. 2015).

Second Generation of Solar Cell

The silicon wafer-based solar cells are costly as it uses the pure crystals of silicon which synthesis process is very complex and takes high heat energy. To reduce the solar cells' expenses, thin film layered solar cells have been adopted by the researchers. In this new technology, thin film (1 μm thick) of silicon is used in solar cell's devices. Thin film-based solar cells are mainly including a very thin film of materials such as amorphous Cadmium Telluride (CdTe), amorphous silicon, and Copper Indium Gallium di-Selenide. In this way, the cost and efficiency are comparatively less to the previous wafer-based technology (Chopra et al. 2004).

Third Generation of Solar Cell

The solar devices of first and second generation have some limitations to achieve optimum efficiency. The third-generation solar cells are mainly based on the CZTS, Dye-Sensitized (DS or DSSC), Quantum Dot (QD or QDSC), Organic, Perovskite materials. Theoretically, the efficiency of the first generation's solar cells which mainly included monocrystalline and polycrystalline silicon solar cells has shown efficiency nearly to 29% and 27%, respectively. Whereas the second generation's solar cells which are mainly based on the thin film of CdTe, CIGS and amorphous silicon have shown efficiency nearly 32%, 33%, and 15–20%, respectively. In addition, more advanced technology of solar cells, i.e., third-generation solar cell the includes QDSC, DSSC, Organic, Perovskite, and Tandem, have shown better efficiencies to near about 44%, 31%, 22–25%, 30–35%, and 45%, respectively. The efficiencies details of the various solar cells, according to generation, have been summarized in Table 6.1.

Table 6.1 Various types of solar cells and their efficiencies

Generation of solar cell's technology	Solar cells	Theoretical efficiency	References
First generation (wafer-based solar cell)	Monocrystalline-Si	~29%	Shockley and Queisser (1961)
	Polycrystalline-Si	~27%	Richter et al. (2013)
Second generation (thin film-based solar cell)	CdTe	~32%	Sites and Pan (2007)
	CIGS	~33%	Gloeckler and Sites (2005)
	a-Silicon	~15% to 20%	Carlson and Wronski (1976)
Third generation of solar cell	QDSC	~44%	Nozik (2002)
	DSSC	~31%	O'Regan and Grätzel (1991)
	Organic	~22% to 25%	Vandewal et al. (2014)
	Tandem	>45%	Geisz et al. (2020)
	Perovskite	~30% to 35%	Zhao and Zhu (2016)

Perovskite Solar Cell

Perovskite-based solar cells have been extensively popularized and emerged out as a better eco-friendly alternative, particularly for their efficiency output, in comparison to well-established silicon containing solar cells and solar devices made of materials such as GaAs, CdTe, dye-sensitized, and solution-processable solar cells (organic solar cells) and recent popular perovskite materials (Liang et al. 2021). The economically and environmentally feasible nature has promoted the PSCs globally to minimize the challenges of energy production and environmental issues (Wang et al. 2019). Recent years have seen intense research in the perovskite solar cells for their high efficiency and low cost fabrication, making them most valuable candidate for large scale commercialization (Liang et al. 2021). First time in 2009, the Scientist, Miyasaka, and coworkers had used perovskite materials in solar cell and had exposed its applications for the development of solar cells, reporting the PCE of 3.81% which now have crossed the 25.7%, near about the same to the PCE of silicon solar cells (Liang et al. 2021; Dambhare et al. 2021).

Perovskite Materials

In perovskite solar cell, the perovskite material acts as the active layer and plays the crucial role in extraction of sunlight and generation of electricity (Zhou et al. 2018). The properties of perovskite materials such thermal, optical, and electronic have attracted great attention at global level. The perovskite is coming under the category of minerals and was discovered by a mineralogist L.A. Perovski and thereafter this kind of mineral is named as "Perovskite" by Gus-tav Rose in 1839 with structure of

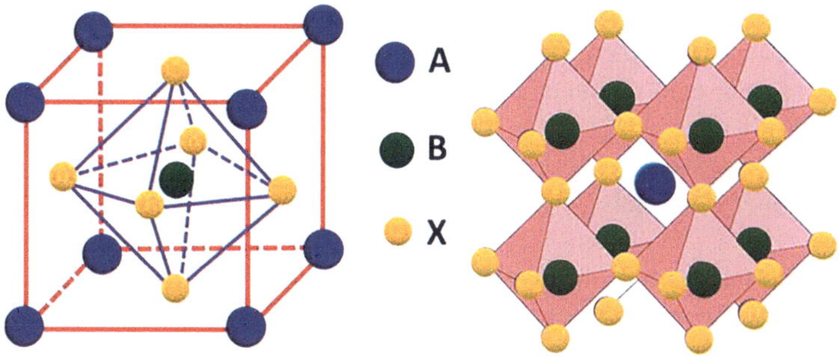

Fig. 6.1 Perovskite crystal structure (Aslam et al. 2021). (Copyright 2020. Reproduced with permission from Book's Corresponding Author)

ABX_3 (Kheralla and Chetty 2021). In ABX_3 formula, "A" denotes a large cation, particularly organic or inorganic such as formamidinium (FA^+, $CH(NH_2)_2^+$) or methylammonium (MA^+, $CH_3NH_3^+$) (Liang et al. 2021; Geisz et al. 2020). As far as Group "B" it is denoting to the inorganic cation (Cu_2^+, Sn_2^+, Pb_2^+) which is lower in size while "X_3" is denoting to the halide ion such as Cl^-, Br^-, and I^- that binds the A and B (as shown in Fig. 6.1). For the solar devices, the most commonly used perovskite materials are generally $CH_3NH_3PbI_3$, $CH_3NH_3PbBr_3$, and the mixed halide structure, $CH_3NH_3PbI_{3-x}Cl_x$ or $CH_3NH_3Pb_xBr_x$ (Osman et al. 2021; Liang et al. 2021; Hamed and Mola 2020; Wu et al. 2024; Jannat et al. 2021).

Electron Transport Materials

In Perovskite-based solar cells, the second crucial layer, i.e., Electron Transporting Layer (ETL), is mainly playing role in extraction and transportation of electron, generated by absorption of light by Perovskite layers, to the respective electrode. So a suitable ETL improves the extraction efficiency of electrons and reduces the migration of holes towards counter electrode, separates effectively electrons and holes, reduces recombination rate, and ultimately resulting in improvement of conversion efficiency. TiO_2 is the electron transporting material, with wide band gap, has been used in most of the electron transport layer in solar devices based on perovskite absorbing materials. Despite the TiO_2, some other n-type semiconductor materials such as ZnO are also useful for the electron transport layers. In electron transporting materials, mainly impacts of the materials and their structures on the efficiency results have been focused. Electron transport semiconducting materials (n-type) should have high carrier mobility with broad band gap and transparent to the visible light. The materials should have broad band gap as well as transparent to visible light. The procedure of preparation should be simple and materials should be easily available, even on low temperatures. The band structure of the materials should be compatible to the band gap of perovskite materials. With the aforesaid characteristics, the electron transport materials lead to the high conversion efficiency. It means that selection of materials is not responsible for low conversion efficiency of solar cells (Zhou et al. 2018).

Hole Transport Materials

Another important layer of the perovskite solar cells is Hole Transport Layer (HTL) in which p-type of materials, known as hole transport material, are used. As the electron transporting materials, the basic function of hole transporting materials are to facilitate in transportation of holes, generated in perovskite layer due to transportation of electrons. At the nearly same energy level of valence band of perovskite materials and highest occupied molecular orbit of hole transport material, the device shows optimum results. The composition of Hole Transport Materials (HTMs) may be either organic or inorganic. In most of the solar devices, Spiro-OMeTAD is widely used as hole material. Spiro-OMeTAD is an organic material and its energy band level satisfying very well with the valence level of perovskite materials, while due to its slow hole mobility, results are decreased to some level. Another commonly using hole transport material in photovoltaic applications is PTAA (poly[bis(4-phenyl) (2,4,6-trimethylphenyl) amine]). The hole mobility of PTAA is good as compared other hole transport materials. An $MAPbI_3$-based perovskite solar cell, developed by Jeon et al., has shown efficiency to 16.2% and 21.3 mA cm^{-2} was the short circuit current in that cell. Other important materials such as P3HT, PEDOT:PSS [Poly(3,4-ethylenedioxythiophene):poly(styrenesulfonate)], PCDTBT [Poly[N-9′-heptadecanyl-2,7-carbazole-alt-5,5-(4′,7′-di-2-thienyl-2′,1′,3′-benzothiadiazole)]] and PCPDTBT [Poly[N-9′-heptadecanyl-2,7-carbazole-alt-5,5-(4′,7′-di-2-thienyl-2′,1′,3′-benzothiadiazole)]] are being used extensively in hole transport layer of the PSCs. Another perovskite solar cell study, by using P_3HT as hole transport material, was carried out by the Zhang et al. in which it has been observed that P_3HT is responsible for reduction in resistance from the hole transfer impedance that ultimately results in better device performance and stability. Another commonly studied hole transport materials is PEDOT:PSS and generally has been used with flexible substrate in solar cells (Zhou et al. 2018).

Whether among the various Hole Transport Materials, the Spiro-OMeTAD is the most considered material for the PSC, but it is too costly. The HTMs has been categorized into either organic HTMs like Spiro-OMeTAD, P_3HT, PEDOT:PSS, DPBTTT-14 (Poly[2,5-bis(3-tetradecylthiophen-2-yl)thieno [3,2-b]thiophene)], etc. or inorganic HTMs such as CuSCN, CuI, NiO, and Cu_2O. The back contact, along with ETL, HTL, and absorber layer, has role in SC performances. The most considered back contact for solar cells is Gold (Au), but it is also very expensive materials and that is the reason, there are various efforts have been made to replace Au by low costly back contacts such as Nickel, Palladium, Platinum, Copper, and Silver, mostly in HTM-free solar cells (Osman et al. 2021).

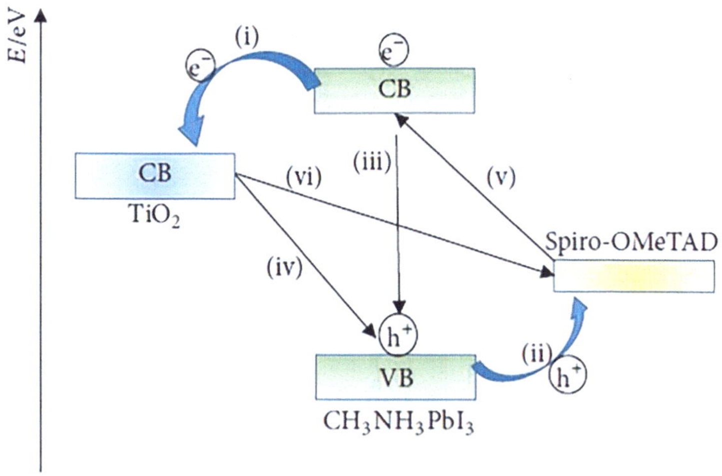

Fig. 6.2 Working mechanism of PSCs (Zhou et al. 2018). (Copyright 2018. Reproduced with permission from Journal of Nanomaterials)

Working Mechanism of Perovskite Solar Cell

The Perovskite Solar Cells function on the principle of photovoltaic effect. When the solar cells come in contact with light of sun, by absorbing light photon the perovskite layer get excited and the formation of excitons and thereafter the differences in binding energy promotes formation of free electron and free hole carriers and movement towards counter electrode resulting current/electricity (Kheralla and Chetty 2021; Zhou et al. 2018). The perovskite layer, when exposed in sunlight, absorb the light photons resulting the electron and hole pairs (excitons) which later separated as free electrons and holes because of low/variation of binding energy of excitons. These free electrons and holes are attracted towards ETM and HTM layers, respectively, and finally with the support of metal electrodes current is obtained through external circuit. The working process of perovskite materials, HTM and ETM of the PSC are represented in the Figs. 6.2 and 6.3 (Zhou et al. 2018).

Architecture of Perovskite Solar Cell

The architecture of PSCs contains an intrinsic semiconductor perovskite material-based absorbing layer positioned between the Electron-Transport-Layer for separation of electrons and Holes Transporting Layer for the separation of holes towards electrodes. The architecture of the PSCs is classified into regular (n-i-p) and inverted

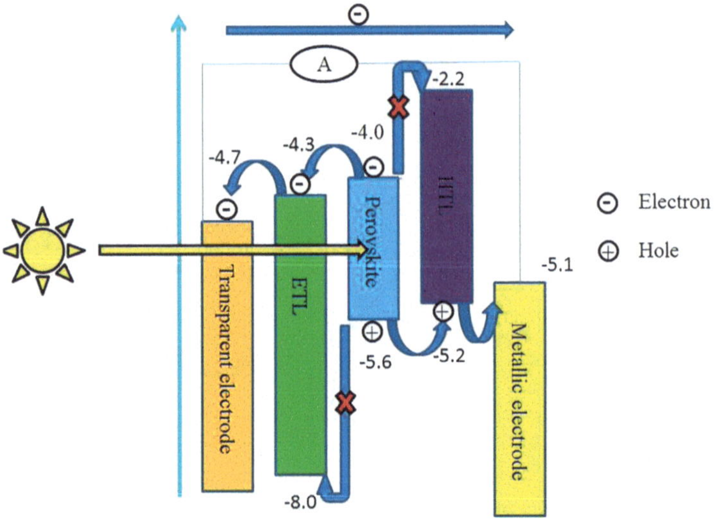

Fig. 6.3 Band structure of PSC (Danladi et al. 2021). (Copyright 2021. Reproduced with permission from East European Journal of Physics)

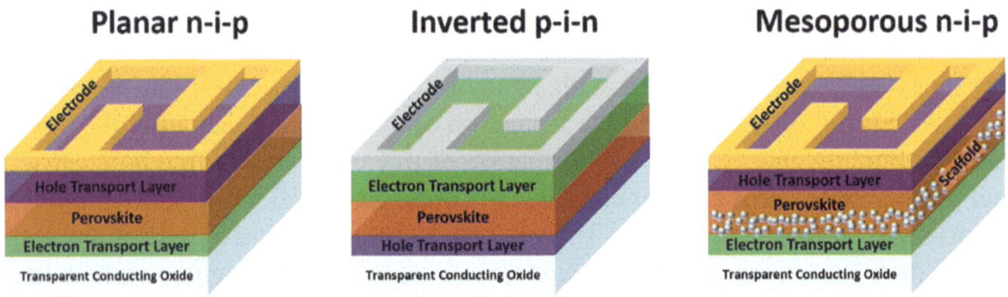

Fig. 6.4 Architecture of perovskite solar cell (Lin et al. 2020). (Copyright 2020. Reproduced with permission from Journal of Advanced Functional Materials)

(p-i-n). The PSCs architecture may be further classified into planar and mesoscopic which is mostly seen in n-i-p type devices (Kheralla and Chetty 2021).

- Regular n-i-p type solar cells
  - Mesoscopic
  - Planar
- Inverted p-i-n type solar cells

Most popular architectures use in the perovskite-based solar devices are mesoscopic and planar. Figure 6.4 is indicating the three type of architecture generally used in solar cells (Osman et al. 2021; Dambhare et al. 2021).

Computational Approach

The computational perspective to photovoltaic technology includes different methods, such as Molecular approaches, Molecular Dynamics, Ab Initio Approaches, Semi-empirical approaches, and Chemical approaches to study about the materials. The aforesaid approaches are helpful to understand the atomic level properties and behavior of the materials. But there are some limitations that these techniques are depending on the external parameters and experimental values. For example, there is need of force field while applying the Molecular Dynamics techniques whereas the force field depends on Ab Initio calculations, a source of actual data atomic and electronic structure. The force field is a set of parameters which define interaction of atoms. A proper force field supports Molecular Dynamic technique in predicting the behavior of molecules structure at different temperature. In recent few years, Molecular Dynamic Technique has been successfully applying during the study of halide perovskites. Recently, Belugas et al. have studied the vibration spectrum behavior of perovskites at different temperature through Molecular Dynamic simulation and obtained accurate reproducible results of vibration spectrum. They have also expressed that by varying the temperature, methyl ammonium lead iodide shows significant structural changes, particularly orthorhombic (at low temperature), tetragonal (at moderate temperature), and at high temperature cubic phase. These structural transformations depend on the movement of molecules within the materials and restriction of their motion. Taufique et al. have also applied Molecular Dynamic simulation tool to predict the accumulation behavior of PCBM, electron transport material used in perovskite solar cells. As far as the Ab Initio Quantum Mechanical Calculations, it is helpful in predicting the energy levels, atomic structure, and optical properties for designing the new materials and saving time, money, etc. during experiment in lab. These calculations are not depending on the external parameters, and their results are very accurate and reliable. By applying these two techniques (Molecular Dynamic and Ab Initio) together, an accurate, reliable, and real insights about structure and electronic nature of a materials can be predicted (Kheralla and Chetty 2021).

Modeling of Perovskite Solar Cells and Their Simulation

With the help of theoretical or numerical study, one can predict the performances of solar cells and by changing the parameters one can optimize parameters to get optimized performance from the solar cells. The theoretical study provides a pre-model of solar cells which are beneficial for understanding the nature and work function of the cells before going for the synthesis. It is very useful to evaluate the physical operation, effects of various changes in the proposed solar cells' performance, and ultimately save the time and money. From time to time, many successful models for simulation have been developed in photovoltaic technology to study the cells. Some of them are: SCAPS (Solar Cell Capacitance Simulator), COMSOL, MULTIPHYSICS, and AMPS (Analysis of Microelectronic and Photonic Structure), TiberCAD, GPVDM. SILVACO, Matlab, and wxAMPS (Osman et al. 2021).

SCAPS-1D Numerical Simulation

The simulation model supports in understanding basic concepts function of the photovoltaic devices and in selecting the major factors that affect the solar cells' performance and their functions (Hossain et al. 2022). Among them SCAPS-1D is one of the simulation tools that are used for simulation of photovoltaic structures with the help of Poisson's equation and continuity equation for the charge carriers. The Poisson's equation to semiconductor has been written as:

$$\nabla_{\psi}^{2} = \frac{q}{\mathrm{e}}\left(n - p + N_{\mathrm{A}} - N_{\mathrm{D}}\right) \tag{6.1}$$

where N_A and N_D are indicating to acceptor concentration and donor concentration, respectively. However, ψ is the symbolized to electrostatic potential.

Whereas the equation of Continuity for the semiconductor can be written as:

$$\nabla^{2}.J_{n} - q\frac{\partial n}{\partial t} = +qR \tag{6.2}$$

$$\nabla^{2}.J_{p} - q\frac{\partial p}{\partial t} = -qR \tag{6.3}$$

In Eqs. (6.2) and (6.3), the current density for electrons and holes are indicated as J_n and J_p, respectively, while R is used for the rate of carrier recombination.

In semiconductor, generally two types of effects occur. One is due to effect of electric field, i.e., flow of carriers and another is due the concentration gradient, i.e., diffusion current. The Drift-Diffusion Current Relations may be represented as in Eqs. (6.4) and (6.5):

$$J_{n} = qn\mu_{n}E + qDn\nabla_{n} \tag{6.4}$$

$$J_{p} = qp\mu_{p}E + qDp\nabla_{p} \tag{6.5}$$

where diffusion coefficient for holes and electrons are represented as D_P and D_n, respectively (Dixit et al. 2019).

Mixed Halide Perovskite

The mixed halide perovskite solar cells include substitution or mixing of either two or more halogen atoms (Cl, Br, I) in proper ratio within a perovskite material. The researchers have indicated that as far as theoretical result by substituting the halogen elements the conversion efficiency of the solar devices have gradually been increased. This is because of the size of halogen atoms, as the size of halogen atoms are increasing, absorption of light along with charge transport characteristics are also improving (Yuan et al. 2024).

The theoretical and simulation study of various parameters such as effects of thickness of layers (Perovskite, hole transport, and electron transport), doping concentration in respect of performance of the solar cells are being done by the researchers efficiently with the help of SCAPS tool. It also helpful in investigating the band offsets such as Valence Band Offset and Conduction Offset, along with the effects of diffusion length and electron affinity of HTMs as well as ETMs on the device performance (Osman et al. 2021).

The mixed halide-based perovskite which can be prepared by combining the halogens elements like $CH_3NH_3PbI_{3-x}Cl_x$ and $CH_3NH_3PbI_{3-x}Br_x$ ($FA_{0.85}MA_{0.15}PbI_{2.55}Br_{0.45}$, $MAPbI_{1-x}Br_x$, $Cs_{0.1}FA_{0.75}MA_{0.15}PbI_{2.49}Br_{0.51}$) (Ono et al. 2017) having a number of good applications. Many studies have indicated that such type of perovskite materials are responsible for reduction in recombination, more smooth movement of carriers which ultimately resulting the higher efficiency and stability of the solar device. The mixed halide plays crucial role in fine tuning or adjusting the band gaps of perovskite materials. $MAPbI_2Cl$ is the first mixed halide-based perovskites which was prepared by reaction of methylammonium halide and lead halide and shown efficiency of 10.9%. Because of the wide band gap, $MAPbBr_{3-x}Cl_x$ and $MAPbI_{3-x}Cl_x$ are most commonly used in photovoltaic technology and have shown high conversion efficiency in solar cells (Hamed and Mola 2020).

The perovskite materials being used in solar applications are basically organic-inorganic mixed halide or all-inorganic halide or mixed halide compound, with perovskite structure ABX_3. So far the perovskite materials having mixed halide in their structure can be categorized into organic-inorganic mixed halide and inorganic mixed halide perovskite (Chopra et al. 2004; Shockley and Queisser 1961).

Experimental Studies

From the reports it has been cleared that the electronic structure of *p* orbit of X and Pb of $MAPbX_3$ have the great impact in photovoltaic performance. The *p* orbit of the halogens, i.e., X plays impactful role in tuning the band gap of $MAPbX_3$ so that it could absorb solar photons in visible range. Noh et al. first time have demonstrated characteristics of $MAPbI_{3-x}Br_x$, mixed halide perovskite, in solar cell and achieved efficiency of the device was about 12.3%. Noh et al. have demonstrated that Br content lower than 10% have shown highest initial device efficiency whereas at more than 20% content of Br, the device shows a good stability (RH 55% for 20 days). Further, it has indicated in the study that the stability of this device is directly related to the phase transformation (tetragonal to pseudo-cubic) of the $MAPbX_3$; the smaller ionic radius Br^- makes the perovskite structure more compact and tightly bounded that increases the tolerance factor and thereby stability of the device. In some extent, the compact structure of perovskite reduces the degradation. In another experimental studies carried out by the Jeon et al. having 18.4% PCE by incorporating $(FAPbI_3)_{1-x}(MAPbBr_3)_x$ ($x = 0$–0.3) as active layer in the perovskite solar cell (Zhou et al. 2018).

Also a study for determination of most stable structure of $MAPbIBr_2$ and $MAPbI_{1/2}Br_{5/2}$ have been carried out by Brivio et al., applying two principles together such as first-principles total-energy calculations and statistical mechanical methods. They have shown that both materials structures are showing natural layered superlattice structures along the [001] direction and revealed that the halogen ion are located at the corner of Pb-X octahedron whereas I ion is positioned at top and bottom site of the octahedron. The reduction of internal strain within the material, due separation of bond lengths of Pb-I and Pb-Br bonds, results in a more stable structure (Zhou et al. 2018).

Stranks et al. in 2013 have studied the mixed perovskite, $MAPbI_{3-x}Cl_x$, in perovskite solar cell. $MAPbI_{3-x}Cl_x$ have been prepared by inserting Cl atoms by substituting I atoms of $MAPbI_3$ in a fixed ratio. In this study, they have affirmed that insertion of Cl atom enhances the stability along with conductivity of the perovskite materials due to increasing of diffusion length of holes and electron in $MAPbI_{3-x}Cl_x$. Whereas in other study of $MAPbI_{3-x}Cl_x$, Docampo et al. have adopted two-step dipping for preparation films of $MAPbI_{3-x}Cl_x$ in the perovskite solar cell and have gained efficiency up to 15.41% along with current density up to 22 mA cm^{-2}. In addition, the electron tunneling effect, developed due to interface connection between the $MAPbI_{3-x}Cl_x$ and TiO_2 layer, also enhances the stability of perovskite materials in air (Zhou et al. 2018).

Similarly, the Cl atom also improved the efficiency of carrier absorptivity that so directly affects the device performance. In the same way, it has been observed that Br atom affects the band gap of MAPb (I, Br, Cl_3) (methyl ammonium lead halide) perovskite materials which ultimately increased the open circuit voltage. Also reported that incorporation of Br enhanced the stability and life span of device. Furthermore, Chiang et al. have used hot solution-based spin coating technique in synthesis of $MAPb(I/Br/Cl)_3$ film. The $MAPbI_{3-x-y}Br_xCl_y$-based perovskite solar cell having PEDOT:PSS and $PC_{61}BM$ as hole transport layer and electron transport layer, respectively, have achieved conversion efficiency up to 16.52% along with J_{sc} up to 19.25 mA/cm^2, V_{oc} up to 1.10 V and FF up to 78% (Zhou et al. 2018).

Theoretical Studies of Mixed Halide Perovskite Solar Cell

In experimental studies, money investment is high, consumes more time in getting results. To address these issues, several endeavors have been carried out through theoretical and simulation method from time to time to predict the performance in a shorter time. Herein, we are trying to provide some important theoretical study by SCAPS-1D simulation techniques about mixed halide-based inorganic and organic-inorganic perovskites solar cells, have been briefly discussed.

Organic-Inorganic Mixed Halide Perovskites

In this section, mainly mixed halide perovskites have been emphasized. The mixed halide perovskites show some good properties applicable for the photovoltaic such as band gap adjustment, light absorption range. These mixed perovskites have greater carrier mobility and diffusion length in comparison to single halide perovskite. Due to wide band gap, these materials absorb light from wide range of spectrum. Generally, Cl and Br are used in mixed halide perovskites. The presence of Cl in mixed perovskite improves the greater quality of film layer while the presence of Br is showing comparatively much lower quality of film layer but better open circuit voltage current (Hamed and Mola 2020).

For tuning the band gap and improving stability, composition of halides (Cl, Br, I) of organic-inorganic perovskite materials have been adjusted. The organic-inorganic mixed halide perovskites (Either A or B is organic or inorganic and X denoting to combination of Cl, Br, I in ABX_3 structure) are the promising materials for the application in LEDs or Light-Emitting Diodes, field-effect transistors, detectors and photovoltaic devices, etc. In the last few years, it has been developed in photovoltaic technology with their Power Conversion Efficiencies (PCEs). It started in 2009 with 3.8% and reached over to an impressive 25.7% certified making them an attractive alternative for new generation technology (Tinedert et al. 2024; Aslam et al. 2021).

Being an impressive efficiency, the organic-inorganic perovskite solar cells are facing various challenges such as thermal instability and environmental degradation. It is highly susceptible to severe decomposition under high temperatures, oxygen, and humidity. The organic–inorganic hybrid perovskite material $MAPbI_3$ readily decomposes into PbI_2 and MAI, with MAI further breaking down into CH_3NH_2 and HI and potentially into CH_3I and NH_3 at high temperatures (Wu et al. 2024). Due to hygroscopic nature of methyl ammonium salt, moisture has also been a major issue for degradation of both organic-inorganic perovskite and organic-inorganic mixed halide perovskites like $MAPbI_3$ and $MAPbI_{3-x}Cl_x$, respectively, along with the toxicity due to the toxic nature of lead (Wang et al. 2019).

There is a continuous progress in the photovoltaic performance of organic and inorganic mixed hybrid perovskites have been researched from more than two decades. Still there is some issues such thermal instability and toxicity due the presence of lead are the main focus of researchers and efforts are being done to short out these issues keeping their main goal on the device conversion efficiency, to be highest. In the studies, it is shown that cesium containing perovskite materials may be a good alternative as it has shown better stability in various environmental situation. Among various studies on cesium-based perovskite materials, a fully inorganic-based mixed halide with low lead content-based solar cell had been designed and studied through SCAPS. Another effort was made to enhance the efficiency and a double layered perovskite-based solar cell was designed for absorption high and low energy photons. In the double-layered perovskite solar cells mostly $CsPb_{0.625}Zn_{0.375}I_2Br$, $CsPb_{0.625}Zn_{0.375}I_2Cl$, and $CsPb_{0.625}Zn_{0.375}IBr_2$ have used as active layer to absorb the low and high energy containing photons. A double-layered architecture such as FTO/TiO_2/$CsPb_{0.625}Zn_{0.375}I_2Cl$/$CsPb_{0.625}Zn_{0.375}IBr_2$/Spiro-MeOTAD/back electrode has shown the highest conversion efficiency up to 30.36% (Rabhi et al. 2024).

To address the lead toxicity in photovoltaic applications, various efforts to investigate the behavior of lead-free perovskite materials have been made in last two decades. In this connection, theoretical investigation of $FASnI_3$-based solar cell by SCAPS tool have been made by Abdelaziz et al. to optimize the performance through varying the electron affinity, band gap, doping content, and thickness of layers of electron transport, hole transport and perovskite and gained efficiency was 14.03% along with 0.92 V open circuit voltage (V_{OC}), 22.65 mA/cm^2 short circuit current density (J_{SC}), and 64.74% fill-factor (FF) (Kumar et al. 2020).

On the same pattern, a theoretical study for perovskites $CH_3NH_3PbI_3$ and $CH_3NH_3SnI_3$, through ATLAS simulation tool, have been carried out by the Hima and his co-researchers to get the optimized device performance. After optimizing the devices, Hima et al. have received improved efficiency (PCE) up to 18.16% and 9.56% for the $CH_3NH_3PbI_3$ and $CH_3NH_3SnI_3$-based solar cell, respectively. Still some gaps are reported that could be fulfilling by proper optimization of parameters for improving the results (Kumar et al. 2020).

Reports have shown that the $MAPbI_{3-x}Cl_x$ films up to 140 °C is thermally stable and are most commonly studied mixed halide perovskite due to the presence of chloride/iodide. The carrier diffusion length of such mixed perovskites is reported better in comparison to normal Cl/I-based perovskite materials such as $MAPbI_3$ and $MAPbCl_3$ (Osman et al. 2021).

Also for enhancing stability and reducing the toxicity due to lead of the $MAPbI_3$ absorber, efforts towards formation of $MA_{1-x}FA_xPb_{1-y}Sr_yI_3$, a hybrid perovskite material by substituting the MA^+ with FA^+ and Pb^{2+} with Sr^{2+} have been successfully processed, proposing Glass/SnO_2:F/TiO_2/$MA_{0.1}FA_{0.9}Pb_{0.85}Sr_{0.15}I_3$/$Cu_2O$/Au architecture. The optimized parameters of the proposed device have resulted optimum efficiency of 29.0% with 24.2 mA/cm^2 short circuit current, 1.37 V open circuit voltage, and 87.49% of fill-factor. Whereas with $MA_{0.5}FA_{0.5}PbI_3$ absorber, the same architecture efficiency was 2.33% lower. The study of operating temperature also suggests more stable to the $MA_{0.1}FA_{0.9}Pb_{0.85}Sr_{0.15}I_3$ absorber than that of $MA_{0.5}FA_{0.5}PbI_3$ absorber (Abena et al. 2022).

The stability (thermal), film-quality, and band gap tunability properties of Cl doped mixed halide, $CH_3NH_3PbI_{3-x}Cl_x$ perovskite have enforced the study of performance of mixed halide perovskite by numerical simulation tool, Solar Cell Capacitance Simulator, following the device architecture: FTO/CdS/ $CH_3NH_3PbI_{3-x}Cl_x$/CuI/Au. In this study, CdS has been used as electron transport material, and CuI has been used as hole transport material. Operating temperature, variation of hole materials and thickness of perovskite layer have been analyzed carefully to optimized the device performance. At optimized parameters of the device, 27.19% of PCE including FF-87.88%, J_{SC}-23.86 mA/cm^2, and V_{OC}-1.29 V was achieved. This results may be used for the experimental purposes to fabricate the film of absorber perovskite, $CH_3NH_3PbI_{3-x}Cl_{x.}$ in solar cell.

In this configuration of solar cell, CdS has been used in the form of ETM in buffer layer, and CuI has been utilized in the form of HTM in hole transport layer. The consequences of operating temperature, absorber layer's thickness, and variation of HTM's in HTL were analyzed on device performance. The optimized parameters of the proposed device have resulted 27.19% PCE along with 87.88% Fill-Factor, 23.86 mA/cm^2 J_{SC}, and V_{OC} of 1.29 V. From the study, it was concluded that the aforesaid results may be useful in fabrication of thin film, particularly in the case of $CH_3NH_3PbI_{3-x}Cl_x$ perovskite-based solar cells. The $CH_3NH_3PbI_{3-x}Cl_x$ perovskite have also shown some other positive characteristics such as low hysteresis due to the presence of chlorine atom, adjustable band gap, carrier movement, good photovoltaic properties, more stability in ambient environment, and can develop into a high quality film (Valeti et al. 2023).

In the year 2016, a study on the electronic properties of mixed perovskites such as $CH_3NH_3PbI_{3-x}Cl_x$ and $CH_3NH_3PbI_{3-x}Br_x$ in tetragonal phase have been analyzed by the Berdiyorov and his coworkers and indicated that the position and value of mixing of halide play a crucial role and affects the electronic characteristics. The doping concentration of chlorine atom makes perovskite more stable crystal that is the reason they have used a mixed halide $CH_3NH_3PbI_{3-x}Cl_x$ in place of $CH_3NH_3PbI_3$ in their proposed solar device to study the performance. A schematic curve for four different perovskites in respect of wavelength versus light absorption efficiency is illustrated in the following Fig. 6.5 (Saffari et al. 2017).

From the figure it is clear that the light absorption efficiency of $CH_3NH_3PbI_3$ is very much lower than the light absorption efficiency of $CH_3NH_3PbCl_3$, whereas the mixed perovskites, $CH_3NH_3PbICl_2$ and $CH_3NH_3PbI_2Cl$, are showing nearly same absorption efficiency as $CH_3NH_3PbI_3$. It is also concluded from the figure that after averaging the light of absorption of each perovskite, it is found that $CH_3NH_3PbCl_3$ is showing lowest absorption while $CH_3NH_3PbICl_2$ has maximum absorption that are 47.64% and 72.21%, respectively (Saffari et al. 2017).

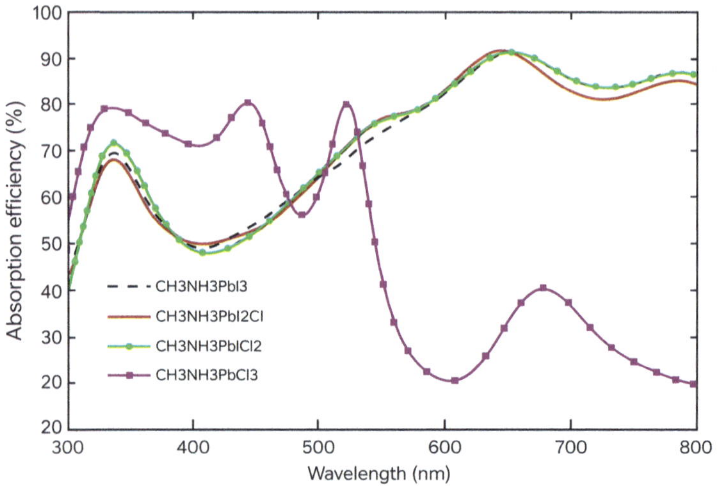

Fig. 6.5 Curve for light absorption efficiency versus wavelength for four perovskites (Saffari et al. 2017). (Copyright 2017. Reproduced with permission from Journal of Physics D: Applied Physics)

Inorganic Mixed Halide Perovskite in Solar Cell

As compared to hybrid perovskite (organic-inorganic materials), completely inorganic-based perovskites have shown the improved thermal, along with improved photo stability in solar cells. Owing to their higher stability, higher conversion efficiency and lower costs, all-inorganic materials-based perovskite solar cells have attracted notices of research community and have become a hot topic of research in the field of photovoltaic applications nowadays (Yuan et al. 2024).

In most of the all-inorganic perovskite materials having general formula ABX_3, cesium is being used at A-site while Sn, Pb, or Ge are mostly being used at B-site and at X-site generally uses either single halogens or mixed halogens during studies. This composition is primarily chosen to maintain the range tolerance factor, i.e., 0.8–1.1, that helps in ensuring the formation of a stable perovskite phase under conventional usage conditions. The tolerance factor (t) can be used for assessing the stability of a particular crystal of perovskite material. Perovskite-type crystals are generally formed when $0.8 < t < 1.0$. If t falls below 0.9, the perovskite crystal structure tends to distort, leading to instability (Wu et al. 2024). Inorganic perovskites generally show different phases of crystal structure depending upon the temperature of environmental, including α-phase (cubic), β-phase (tetragonal), γ-phase (orthorhombic), and δ-phase (orthogonal), with the δ-phase representing a non-perovskite structure (Wu et al. 2024).

By mixing the composition of halide (Cl, Br, I), the properties of perovskites for photovoltaic application can efficiently be regulated. If we consider the case of $CsSnX_3$, when concentration of Cl^- is decreased by I^-, the band gap of the perovskites materials are decreased and absorption also moves towards long wavelength region from ultraviolet/visible region which affects the device performances on various scale (Yuan et al. 2024). Changing the energy band gap of hole and electron transport materials corresponding to the energy band gap of the $CsSnX_3$ absorber also enhances the output of the solar device. Optimization of device parameters, work function, defect density, thickness of active layer, and hole as well as electron transport layer enhance performance of the solar device further. This study also cleared that sequential replacement or insertion of halogens improves the performance of device (Yuan et al. 2024). Although the photovoltaic devices having organic charge transport materials have resulted high PCE, but decomposes easily and have very poor stability. Theoretical studies have concluded that by substituting the halogens in increasing order of their size, the conversion efficiency results of the device also gradually increase because of properties of high light absorption and high charge carrier mobility (Yuan et al. 2024).

In the tandem solar cells as well as in triple junction-based solar cells, mixed halide cesium-based perovskite, $FA_{0.83}Cs_{0.17}$ $PbI_{1.5}Br_{1.5}$, as a top absorber, have attracted attentions of researchers due their suitable photovoltaic properties such as larger diffusion length, high charge carrier mobility, and wide band gap. But generally high valence band offset of hole transport materials limits the performance of $FA_{0.83}Cs_{0.17}$ $PbI_{1.5}Br_{1.5}$-based solar devices. The analysis has confirmed that for $FA_{0.83}Cs_{0.17}$ $PbI_{1.5}Br_{1.5}$ mixed halide perovskite, MoOx as HTM in top cell, is considered to be compatible hole transport materials as it shows low valence band offset with it (Jannat et al. 2021). To increase device stability and their performances in solar cell technology, the mixed perovskite, $FA_{0.85}Cs_{0.15}Pb(I_{0.85}Br_{0.15})_3$, has been studied with various organic and inorganic HTMs, leaving the Spiro-OmeTAD as it causes device instability. Amongst the number of HTMs, Cu_2O is considered most appropriate therefore optimized for $FTO/SnO_2/FA_{0.85}Cs_{0.15}Pb(I_{0.85}Br_{0.15})_3/Cu_2O/Au$ configuration; resulted PCE of 19.38% (Karthick et al. 2022; Bhardwaj et al. 2021).

Kunal Chakraborty and coworkers have proposed the four cell architectures based on the Cesium Titanium perovskite active material ($CuSCN/Cs_2TiX_6/CdS/$Silicon) involving the F, Cl, Br, and I as the halides. All the four cell architectures have been simulated numerically by checking the variation effects on the layer thickness of absorber materials as well as thermal effects of device on the cell made by perovskite materials through simulation tool, SCAPS-1D. It is concluded that optimized thicknesses are 1.5, 1.5, 1.5, and 1.0 (in μm) for the "Cs_2TiI_6," "Cs_2TiF_6," "Cs_2TiCl_6" as well as "Cs_2TiBr_6" active materials, respectively. Also concluded that at the 75, 60, 75, and 80 (in °C), the device has best performance for the Cs_2TiI_6, Cs_2TiCl_6, Cs_2TiF_6, and Cs_2TiBr_6, respectively. The responses towards spectral for these perovskite materials like Cs_2TiBr_6, Cs_2TiI_6, Cs_2TiF_6, and Cs_2TiCl_6 were found at the 750, 650, 625, and 525 (in nm), respectively which indicates that these materials will perform good response in "Visible Spectrum Region" (Chakraborty et al. 2019).

Another effort on mixed halide-based perovskite cell has been made by Karthick et al. (2020), and they had studied the Glass/FTO/SnO_2/FA-Perovskite [$FA_{0.85}Cs_{0.15}$

$Pb(I_{0.85}Br_{0.15})_3]$/Spiro-OMeTAD/Au by SCAPS simulator. In this study, they have proposed three $FAPbI_3$, $FA_{0.85}Cs_{0.15}PbI_3$, and $FA_{0.85}Cs_{0.15}Pb(I_{0.85}Br_{0.15})_3$ perovskite with mixed halides and get efficiency up to 14.7% (Karthick et al. 2020).

Future Perspective

The theoretical and computational studies of perovskite solar cells, comprising of mixed halide-based inorganic and organic-inorganic perovskites materials, may serve as an encouraging step to promote the technology of photovoltaics. In this connection, computational tool, SCAPS-1D, may be a suitable option which is playing an important role in optimizing device architectures and material combinations for predicting the improved efficiency, stability, and scalability. In additional simulation, techniques can explore novel halide compositions and dopants, enabling tailored band gap engineering to enhance light absorption and carrier transport, interface optimization, addressing recombination losses, and enhancing stability in both organic-inorganic and fully inorganic perovskite devices. It may also be helpful for predicting the degradation mechanisms and for the development of materials with reduced defects. Further, simulations can be fruitful techniques for evaluating the performance of environmentally friendly, lead-free perovskites and in optimizing scalable fabrication techniques. Integration with AI techniques will tremendously change the performance efficiency and even future of the theoretical and computational studies.

Conclusion

The theoretical study of all-inorganic and organic-inorganic mixed halide Perovskite Solar Cells are focusing towards the study of effects of mixed halides, organic and inorganic compositions on the performances of perovskite solar cells. It provides insights of the materials properties, working mechanism, challenges regarding stability, strategies for efficiency improvement, electronic and optical properties, etc. For the purposes, numerical simulation tools have contributed a major role, particularly in saving time and money. Simulations techniques help in analyzing the performance of eco-friendly, lead-free perovskites and in optimizing scalable fabrication techniques. With the help of advance simulation tools such as SCAPS-1D, the research community is able to fine tune the perovskite compositions, optimize the parameters and architectures, device stability and performances of the solar cells.

***

References

Abena AMN, Ngoupo AT, Ndjaka JMB (2022) Computational analysis of mixed cation mixed halide-based perovskite solar cell using SCAPS-1D software. Heliyon 8:e11428. https://doi.org/10.1016/j.heliyon.2022.e11428

Aslam M, Mahmood T, Naeem A (2021) Organic inorganic perovskites: a low-cost-efficient photovoltaic material. IntechOpen, London. https://doi.org/10.5772/intechopen.94104

Bagher AM, Vahid MMA, Mohsen M (2015) Types of solar cells and application. Am J Opt Photonics 3:94–113

Bertolli M (2008) Solar cell materials (course: solid state II). Department of Physics, University of Tennessee, Knoxville

Bhardwaj KS, Rai S, Sadanand, Lohia P, Dwivedi DK (2021) Investigating the performance of mixed cation mixed halide-based perovskite solar cells using various hole-transport materials by numerical simulation. Opt Quant Electron 53:602. https://doi.org/10.1007/s11082-021-03262-7

Carlson DE, Wronski CR (1976) Amorphous silicon solar cell. Appl Phys Lett 28(11):671–673. https://doi.org/10.1063/1.88790

Chakraborty K, Choudhury MG, Paul S (2019) Numerical study of Cs_2TiX_6 ($X = Br^-$, I^-, F^-, and Cl^-) based perovskite solar cell using SCAPS-1D device simulation. Sol Energy 194:886–892. https://doi.org/10.1016/j.solener.2019.11.052

Chopra KL, Paulson PD, Dutt V (2004) Thin-film solar cells: an overview. Prog Photovolt 12:69–92

Choubey PC, Oudhia A, Dewangan R (2012) A review: solar cell current scenario and future trends. Rec Res Sci Technol 4:99–101

Dambhare MV, Butey B, Moharil SV (2021) Solar photovoltaic technology: a review of different types of solar cells and its future trends. J Phys Conf Ser 1913:012053. https://doi.org/10.1088/1742-6596/1913/1/012053

Danladi E, Dogo DS, Michael SU, Uloko FO, Salawu AO (2021) Recent advances in modeling of perovskite solar cells using SCAPS-1D: effect of absorber and ETM thickness. East Eur J Phys 2021(4):5–17. https://doi.org/10.26565/2312-4334-2021-4-01

Dixit H, Punetha D, Pandey SK (2019) Improvement in performance of lead-free inverted perovskite solar cell by optimization of solar parameters. Optik 179:969–976. https://doi.org/10.1016/j.ijleo.2018.11.028

Geisz JF, France RM, Schulte KL, Steiner MA, Norman AG, Guthrey HL et al (2020) Six-junction III–V solar cells with 47.1% conversion efficiency under 143 suns concentration. Nat Energy 5(8):326–335. https://doi.org/10.1038/s41560-020-0598-5

Gloeckler M, Sites JR (2005) Efficiency limits for polycrystalline thin-film solar cells. J Appl Phys 98(10):103703. https://doi.org/10.1063/1.2135310

Hamed MSG, Mola GT (2020) Mixed halide perovskite solar cells: progress and challenges. Crit Rev Solid State Mater Sci 45(2):85–112. https://doi.org/10.1080/10408436.2018.1549976

Hossain MK, Rubel MHK, Toki GFI, Alam I, Rahman MF, Bencherif H (2022) Effect of various electron and hole transport layers on the performance of $CsPbI_3$-based perovskite solar cells: a numerical investigation in DFT, SCAPS-1D, and wxAMPS frameworks. ACS Omega 7:43210–43230. https://doi.org/10.1021/acsomega.2c05912

Jannat F, Ahmed S, Alim MA (2021) Performance analysis of cesium formamidinium lead mixed halide based perovskite solar cell with MoOx as hole transport material via SCAPS-1D. Optik 228:166202. https://doi.org/10.1016/j.ijleo.2021.166202

Jayakumar P (2009) Solar energy resource assessment handbook. Renewable Energy Corporation Network for the Asia Pacific

Karthick S, Velumani S, Bouclé J (2020) Experimental and SCAPS simulated formamidinium perovskite solar cells: a comparison of device performance. Sol Energy 205:349–357. https://doi.org/10.1016/j.solener.2020.05.071

Karthick S, Velumani S, Bouclé J (2022) Chalcogenide $BaZrS_3$ perovskite solar cells: a numerical simulation and analysis using SCAPS-1D. Opt Mater 126:112250. https://doi.org/10.1016/j.optmat.2022.112250

Kheralla A, Chetty N (2021) A review of experimental and computational attempts to remedy stability issues of perovskite solar cells. Heliyon 7:e06211. https://doi.org/10.1016/j.heliyon.2021.e06211

Kumar M, Kumar A, Raj A, Sati PC, Sahni M, Anshul A (2020) Organic-inorganic perovskite-based solar cell designs for high conversion efficiency: a comparative study by SCAPS simulation. Mater Today Proc 49. https://doi.org/10.1016/j.matpr.2020.11.035

Liang X, Ge C, Fang Q, Deng W, Dey S, Lin H, Zhang Y, Zhang X, Zhu Q, Hu H (2021) Flexible perovskite solar cells: progress and prospects. Front Mater 8:634353. https://doi.org/10.3389/fmats.2021.634353

Lin L, Jones TW, Yang TC-J, Duffy NW, Li J, Zhao L, Chi B, Wang X, Wilson GJ (2020) Inorganic electron transport materials in perovskite solar cells. Adv Funct Mater 2020:2008300. https://doi.org/10.1002/adfm.202008300

McEvoy A, Castaner L, Markvart T (2012) Solar cells: materials, manufacture and operation, 2nd edn. Elsevier, Amsterdam

Nozik AJ (2002) Quantum dot solar cells. Physica E 14(1–2):115–120. https://doi.org/10.1016/S1386-9477(02)00374-0

O'Regan B, Grätzel M (1991) A low-cost, high-efficiency solar cell based on dye-sensitized colloidal TiO_2 films. Nature 353(6346):737–740. https://doi.org/10.1038/353737a0

Ono LK, Juarez-Perez EJ, Qi Y (2017) Progress on perovskite materials and solar cells with mixed cations and halide anions. ACS Appl Mater Interfaces 9:30197–30246. https://doi.org/10.1021/acsami.7b06001

Osman BA, Abdolkader TM, Ahmed IS (2021) A review of perovskite solar cells. Int J Mater Technol Innov 1(2):48–66. ISSN 2682-4299

Rabhi S, Sekar K, Kalna K, Bouderbala YI, Bouri N, Oueldna N, Belbachir N, Dadda K, Aida MS, Attaf N (2024) Experimental findings and SCAPS-1D simulations for high-efficiency $MAPbI_3$ perovskite solar cells beyond 31%. Opt Quant Electron 56:1372. https://doi.org/10.1007/s11082-024-07282-x

Richter A, Hermle M, Glunz SW (2013) Reassessment of the limiting efficiency for crystalline silicon solar cells. IEEE J Photovolt 3(4):1184–1191. https://doi.org/10.1109/JPHOTOV.2013.2260317

Saffari M et al (2017) DFT analysis and FDTD simulation of CH3NH3PbI3-xClx mixed halide perovskite solar cells: role of halide mixing and light trapping technique. J Phys D Appl Phys 50:415501. https://doi.org/10.1088/1361-6463/aa83c8

Shockley W, Queisser HJ (1961) Detailed balance limit of efficiency of p-n junction solar cells. J Appl Phys 32(3):510–519. https://doi.org/10.1063/1.1736034

Sites JR, Pan J (2007) Strategies to increase CdTe solar-cell voltage. Thin Solid Films 515(15):6099–6102. https://doi.org/10.1016/j.tsf.2006.11.145

Srinivas B, Balaji S, Nagendra Babu M, Reddy YS (2015) Review on present and advance materials for solar cells. Int J Eng Res Online 3:178–182

Tinedert IE, Saadoune A, Hossain MK (2024) A theoretical study of all-inorganic perovskite solar cells: computational modeling of the $CsPbI_3$/$RbGeI_3$ bilayer absorber structure. J Phys Chem Solids 189:111951. https://doi.org/10.1016/j.jpcs.2024.111951

Valeti NJ, Prakash K, Singha MK (2023) Device simulation of $CH_3NH_3PbI_{3-x}Cl_x$ based mixed halide perovskite thin film solar cells. Mater Today Proc. https://doi.org/10.1016/j.matpr.2023.07.057

Vandewal K, Tvingstedt K, Gadisa A, Inganäs O, Manca J (2014) On the origin of open-circuit voltage in polymer-fullerene solar cells. Nat Mater 8(11):904–909. https://doi.org/10.1038/nmat2548

Wang R, Mujahid M, Duan Y, Wang Z-K, Xue J, Yang Y (2019) A review of perovskites solar cell stability. Adv Funct Mater 29:1808843. https://doi.org/10.1002/adfm.201808843

Wu H, Sun Z, Li H, Chen X, Ma W, Li S, Chen Z, Xi F (2024) Review of all-inorganic perovskites and their tandem solar cells with crystalline silicon. Energy Mater Dev 2(3):9370045. https://doi.org/10.26599/EMD.2024.9370045

Yuan S, Li Z, Wang Y, Zhao H (2024) Theoretical study and analysis of $CsSnX_3$ (X = Cl, Br, I) all-inorganic perovskite solar cells with different X-site elements. Molecules 29:2599. https://doi.org/10.3390/molecules29112599

Zhao Y, Zhu K (2016) Organic–inorganic hybrid lead halide perovskites for optoelectronic and electronic applications. Chem Soc Rev 45(11):655–689. https://doi.org/10.1039/C5CS00848D

Zhou D, Zhou T, Tian Y, Zhu X, Tu Y (2018) Perovskite-based solar cells: materials, methods, and future perspectives. J Nanomater 2018:8148072. https://doi.org/10.1155/2018/8148072

Chapter 7
Computational Studies of Tandem Solar Cells

Ipsita Mohanty and Udai P. Singh

Abstract Tandem-based solar cells reflect a significant advancement in photovoltaic (PV) technology, offering the potential for enhanced power conversion efficiency (PCE) compared to the widely used but limited single-junction cells. This chapter provides an overview of computational studies on tandem perovskite solar cells, focusing on their principles, structures, and benefits, and highlighting their potential for high-efficiency solar energy conversion. It explores the working mechanisms and structural advantages of tandem configurations, which improve light absorption and reduce energy losses. The chapter also covers the theoretical foundations of both perovskite-on-perovskite and heterojunction perovskite tandem architectures, emphasizing the critical role of interface engineering in enhancing device stability and reducing degradation. Experimental evidence is presented to support the feasibility of tandem perovskite solar cells, demonstrating recent advancements in performance. Additionally, the limitations of computational models, such as accuracy and resource constraints, are discussed, underscoring the need for a combined approach of computational and experimental research to overcome challenges and drive future advancements in high-efficiency, stable tandem perovskite solar cells.

Keywords Tandem solar cells · Perovskite solar cells · Computational studies · Electron and hole transport layer

---

I. Mohanty
Department of Physics, Dharanidhar University, Keonjhar, Odisha, India

U. P. Singh (✉)
School of Electronics Engineering, Kalinga Institute of Industrial Technology (KIIT), Deemed to be University, Bhubaneswar, Odisha, India
e-mail: singhup@kiit.ac.in

M. Kumar (ed.), *Computational Analysis of Perovskite Solar Cells*, Springer Tracts in Electrical and Electronics Engineering,
https://doi.org/10.1007/978-981-95-6521-4_7

Introduction

The International Energy Outlook Report 2024 marks a historic milestone, projecting global energy investment to exceed USD 3 trillion for the first time, with USD 2 trillion allocated to clean energy technologies and infrastructure (IEA 2024; Solar Power Europe 2024; Kheralla and Chetty 2021). This strong financial commitment reflects the growing global momentum towards sustainable energy solutions. Complementing this trend, the Global Market Outlook for Solar Power 2024 reveals that 2023 witnessed the deployment of 447 GW of new solar PV capacity—a remarkable 87% increase over 2022. This surge represents the fastest growth rate since 2010 when the global solar market was merely 4% of its current scale. Solar cells are at the forefront of this renewable energy transformation, embodying technological innovation with the potential to reduce fossil fuel dependence and mitigate the environmental impacts of traditional energy generation. The successful commercialization of solar cells hinges on achieving a balance of high efficiency, affordability, and durability. These factors are essential for ensuring the widespread adoption of solar technology. Among the various materials available, silicon has emerged as the dominant choice in the modern solar market, thanks to advancements in fabrication processes that have made it abundant and cost-effective. Notably, more than 90% of solar cells today are manufactured using single-crystalline/polycrystalline silicon, a material prized for its exceptional efficiency and long-term reliability (Shrivastav et al. 2023). However, researchers have increasingly focused on perovskite solar cells (PSCs) due to their remarkable efficiency growth, which has surged from 3.9% in 2009 to an impressive 26.7% to date (Mohanty et al. 2024; National Laboratory of the Rockies n.d.). The rapid advancements in PSCs underscore their potential to revolutionize the solar energy industry and accelerate the global transition to renewable energy. Their performance rivals, and in some cases exceeds, that of traditional silicon-based solar cells. Additionally, PSCs can be manufactured using cost-effective, solution-based methods compatible with flexible substrates, opening the door to innovative applications beyond the constraints of rigid silicon cells. These include integrating solar technology into building exteriors, wearable electronic devices, and portable power sources, heralding a new era of solar-powered innovation (Mohanty et al. 2021).

To enhance efficiency and address the stability challenges of PSCs, extensive research has been conducted on various device configurations, material properties, and phase transitions. To further overcome these limitations, tandem solar cell structures have been proposed, achieving notable success in advancing the performance and stability to some extent. According to the National Renewable Energy Laboratory (NREL) reports, perovskite tandem solar cells have achieved an efficiency of 30.1%, as illustrated in Fig. 7.1a. These findings highlight the superior efficiency of tandem structures compared to single-junction solar cells. Furthermore, as depicted in Fig. 7.1b, hybrid tandem solar cells, particularly the perovskite/silicon tandem configuration, have demonstrated an impressive efficiency of 34.6%, underscoring their potential to redefine the benchmarks in solar energy conversion (National Laboratory of the Rockies n.d.).

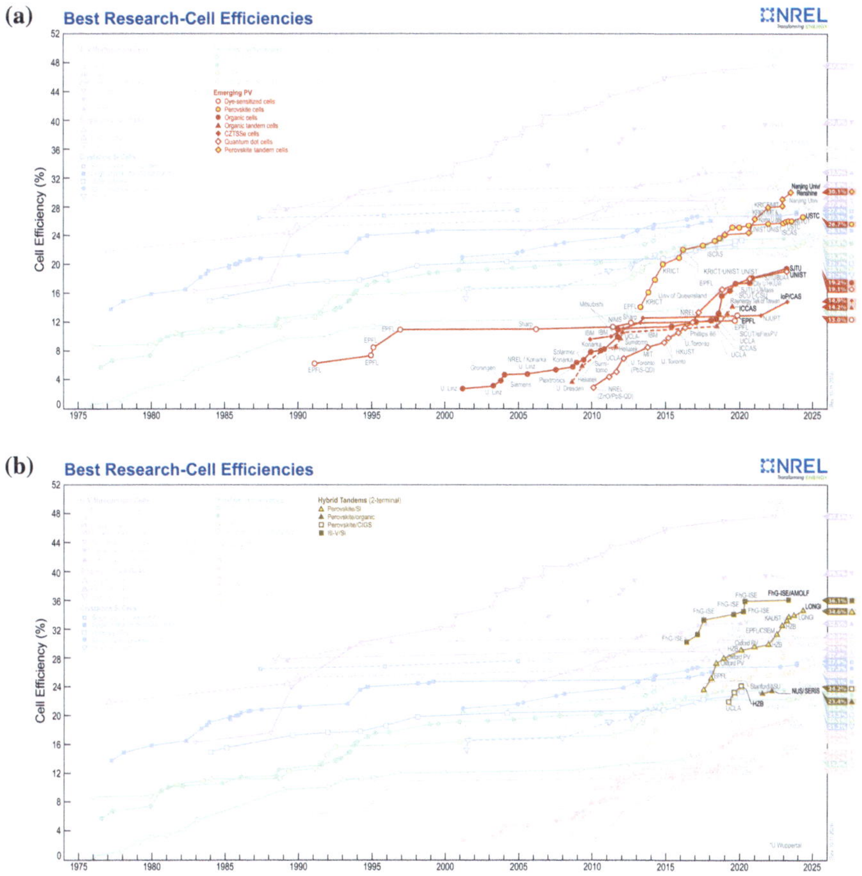

Fig. 7.1 Efficiency chart of (**a**) emerging photovoltaic cells, and (**b**) hybrid tandem cells (National Laboratory of the Rockies n.d.)

Tandem solar cells (TSCs) represent a transformative step in the evolution of photovoltaic technology. By combining multiple layers of materials with distinct bandgap energies, tandem configurations aim to surpass the theoretical efficiency limits of single-junction cells, as described by the Shockley-Queisser (S-Q) limit, i.e. the detailed balance limit that defines the theoretical maximum efficiency of a single-junction solar cell under standard conditions (Li and Zhang 2020). Multijunction TSCs hold significant promise for surpassing the S-Q limit of 33.7%, which represents the theoretical efficiency ceiling for a single-junction solar cell with a bandgap (E_g) of 1.34 eV under 1 sun illumination. By stacking multiple sub-cells with complementary bandgaps, TSCs can harvest a broader spectrum of sunlight, minimizing thermalization and transmission losses that constrain single-junction efficiency. This innovative approach allows TSCs to achieve higher efficiencies, paving the way for transformative advancements in photovoltaic technology (Ho-Baillie et al. 2021).

The earliest tandem solar cells were developed using III–V semiconductors, silicon (Si), and germanium (Ge), fabricated through high-precision, high-vacuum deposition techniques, viz. molecular beam epitaxy (MBE) and metal-organic chemical vapour deposition (MOCVD). While these devices achieved remarkable efficiency and stability, making them ideal for powering spacecraft, their high production costs have limited their adoption in commercial terrestrial applications. This cost barrier has driven research towards alternative materials and fabrication methods to develop more economically viable tandem solar cell technologies (Mazumdar et al. 2023). In recent years, the rapid advancement of PSCs has introduced a cost-effective alternative for tandem solar cells. PSC-based tandems not only offer significant economic advantages due to their low-cost materials and fabrication processes but also show exceptional potential in achieving high efficiencies. These attributes position PSCs as a highly promising technology for next-generation tandem solar cells, combining affordability with performance (Mazumdar et al. 2023).

This chapter offers an overview of the computational study of tandem PSCs, focusing on their principles, structures, and advantages, while highlighting their potential for achieving higher efficiency compared to traditional single-junction cells. It explores the fundamental principles of TSCs, emphasizing their structural design and working mechanisms, as well as the benefits of tandem configurations in enhancing light absorption and minimizing energy losses. The chapter further examines the theoretical foundations of perovskite-on-perovskite and heterojunction perovskite tandem architectures. It also addresses the critical role of interface engineering in improving device stability and preventing degradation, a significant challenge for perovskite-based technologies. Experimental evidence is provided, demonstrating recent advancements and performance gains in tandem perovskite solar cells. Additionally, the chapter discusses the limitations of computational studies, including issues with model accuracy and resource constraints, which can impact predictions and experimental guidance. Finally, it emphasizes the importance of combining computational and experimental approaches to tackle the challenges of developing stable, high-efficiency tandem PSCs, laying the groundwork for future research in this promising field.

Principles of Tandem Solar Cells

Structure of TSCs

TSCs are a cutting-edge advancement in photovoltaic technology, aimed at boosting energy conversion efficiency by stacking various types of solar cells. By incorporating multiple layers, each designed to capture specific portions of the solar spectrum, these systems exceed the efficiency limits of traditional single-junction solar cells (Jošt et al. 2020). These cells are available in two primary configurations: (1)

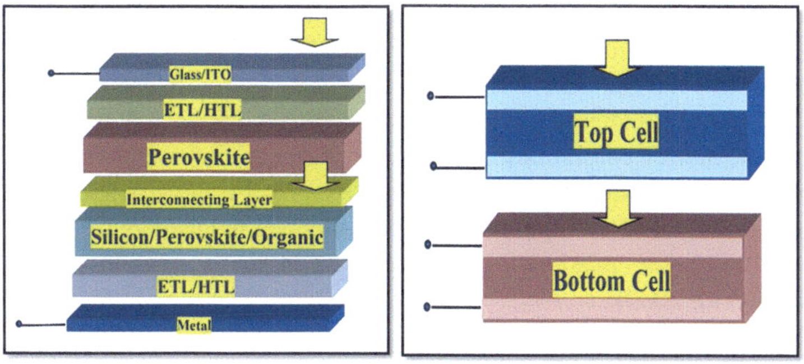

Fig. 7.2 (**a**) Two-terminal (2T) and (**b**) four-terminal (4T) TSCs

two-terminal (2T) and (2) four-terminal (4T). In these cells, the top layer typically features a higher bandgap, effectively capturing high-energy photons, while the bottom layer absorbs the lower energy light that passes through, significantly boosting efficiency compared to single-junction cells. In the 2T design, the cells are connected in series, requiring current matching across the layers, which simplifies the structure but may limit performance due to mismatched current flows. In contrast, the 4T configuration allows each cell to operate independently with separate electrical connections, maximizing power extraction but increasing complexity and component requirements. The selection of these configurations is determined by the specific requirements of the application, including factors such as cost, efficiency, and system integration (Zhou et al. 2024). As shown in Fig. 7.2a, b, each layer in the 2T and 4T TSCs is crucial for optimizing overall performance:

(a) **Transparent Conductive Oxide (TCO)**: As the sunlight needs to reach the perovskite layer, the cathode or anode should be transparent. Hence, TCO like indium-doped tin-oxide (ITO) or fluorine-doped tin-oxide (FTO) is used as-electrode on glass that enables light transmission.
(b) **Hole-Transport Material/Hole Transport Layer (HTM/HTL)**: The basic purpose of the hole-transport layer in PSCs is to behave as p-type semiconductor, increase selectivity and reflectivity of the contact, minimize recombination in an increased open-circuit voltage and higher luminescence efficiency, and accelerate the internal quantum efficiency with reduction of losses of charges.
(c) **Electron Transport Material/Electron Transport Layer (ETM/ETL)**: The electron transport layer acts as an n-type semiconductor, provides a cascading energy level for smooth transportation of electrons from the absorber layer to the cathode, and allows unipolar transference that reduces the charge carrier recombination tendencies.
(d) **Top Layer**: This layer has been designed with a higher bandgap material, it captures high-energy photons (e.g. ultraviolet and visible light) while allowing lower energy photons to pass through. In a 2T configuration, the top layer's current must align with the bottom layer to maintain efficient charge flow since the cells are connected in series. In a 4T configuration, the top layer operates inde-

pendently, optimizing its own power generation without needing to match the bottom layer's performance (Ameri et al. 2009).

(e) **Bottom Layer**: This layer has been composed of a lower bandgap material; it absorbs the lower energy photons (e.g. near-infrared light) that pass through the top layer. In a 2T configuration, the bottom layer complements the top layer by capturing the remaining solar energy, but its output is constrained by the series connection. In a 4T configuration, the bottom layer independently processes the transmitted light, maximizing its contribution to the overall power output (Ameri et al. 2009).

Working Principle of TSCs

In a TSC, single-junction solar cells are layered on top of each other. The top cell, with a wide bandgap, captures high-energy photons, while the bottom cell, with a narrow bandgap, absorbs low-energy photons. This arrangement broadens the range of wavelength absorption, reducing hot carrier thermalization losses and enabling high power conversion efficiency (PCE) (Li and Zhang 2020). In TSCs, the absorption of light is optimized by stacking layers with different bandgap energies. The top layer, with a wider bandgap (~1.7 to 2.0 eV), absorbs high-energy photons from shorter wavelengths, such as blue and ultraviolet light, and converts them into electricity. Lower energy photons, corresponding to longer wavelengths like red and infrared light, pass through the top layer and are absorbed by the bottom layer, which has a narrower bandgap (~1.0 to 1.2 eV). This design minimizes thermalization losses, as each layer captures photons closer to its optimal energy range, thereby reducing heat loss and enhancing overall efficiency. The layers are electrically connected either in series, using tunnel junctions for seamless current flow, or in parallel, combining currents at a shared terminal when voltage levels are compatible. Transparent intermediate layers or materials ensure efficient transmission of unabsorbed light to the lower layers, maximizing the utilization of the solar spectrum (Wali et al. 2018).

The fundamental difference in the working of single-junction and tandem solar cells lies in their approach to photon absorption and energy conversion. In single-junction devices, photons with energy lower than the bandgap cannot be absorbed, while those with energy higher than E_g generate hot carriers. These hot carriers quickly lose excess energy as heat through thermalization due to interactions with phonons, leading to significant energy loss in the form of optical spectral losses. In contrast, tandem solar cells consist of two or more different materials, often from the same family, with varying bandgaps. This configuration enables them to absorb and convert a broader range of photon energies, thereby reducing energy losses and achieving higher efficiency (Wali et al. 2018). The schematic representations of single-junction, tandem solar cells and the response of the top and bottom cells to the spectrum of light are portrayed in Fig. 7.3.

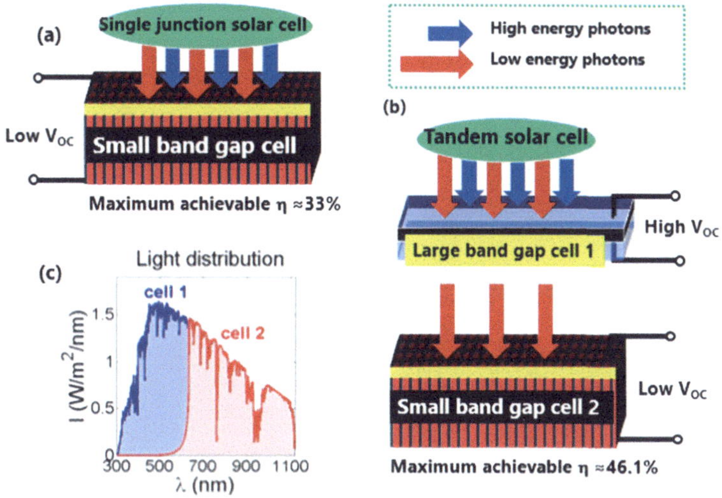

Fig. 7.3 Schematic representation of (**a**) single-junction; (**b**) TSC; and (**c**) response of the top and bottom cells to the spectrum of light (Wali et al. 2018). (Copyright 2018. Reproduced with Permission from Elsevier)

Theoretical Foundations of Tandem Solar Cells

Before designing and simulating a tandem solar cell, it is crucial to identify appropriate materials for the top and bottom cells. Narrow bandgap materials are used in the bottom cell's active layer to capture low-energy photons, while wide bandgap materials are utilized in the top cell's active layer to absorb high-energy photons. Simulation tools are then employed to assess parameters such as defect density and interface properties of the absorber layers in both cells. The standalone performance of each cell is analysed under the AM1.5G spectrum. For tandem configuration, the filtered spectrum from the top cell layers is used to illuminate the bottom cell, ensuring optimal light absorption (Shrivastav et al. 2023). In this context, various tandem structures have been elaborately discussed.

Perovskite-on-Perovskite Tandem Structures

The perovskite-perovskite tandem design provides a viable pathway for developing low-cost, high-efficiency solar cells. Both the top and bottom cells follow similar trends in efficiency improvements, exhibit comparable sensitivities to environmental stressors, and require the same encapsulation and packaging solutions. Additionally, their manufacturing processes can utilize shared infrastructure, reducing capital expenditure (Ho-Baillie et al. 2021). Table 7.1 summarizes different

Table 7.1 Summary of reports related to perovskite-perovskite TSC

Tandem configuration	Top layer	Bottom layer	Simulation software	Efficiency	Year	References
Au/Cu_2O/$FAMASnGeI_3$/ $CH_3NH_3GeI_3$/ZnO/FTO	$FAMASnGeI_3$	$CH_3NH_3GeI_3$	SCAPS-1D	26.72	2020	Singh et al. (2020)
NiO/$Cs_2AgBi_{0.75}Sb_{0.25}Br_6$/ PCBM/$SnO_2$/PEDOT:PSS/ $FACsPb_{0.5}Sn_{0.5}I_3$/C_{60}/BCP	$Cs_2AgBi_{0.75}Sb_{0.25}Br_6$	$FACsPb_{0.5}Sn_{0.5}I_3$	SCAPS-1D	17.3	2020	Madan et al. (2020)
FTO/TiO_2/$MASnIBr_2$/ spiro-OMeTAD/$MASnI_3$/ TiO_2/FTO	$MASnIBr_2$	$MASnI_3$	SCAPS-1D	15.66	2022	Abdelaziz et al. (2022)
FTO/TiO_2/ $Cs_2AgBi_{0.75}Sb_{0.25}Br_6$/ spiro-OMeTAD/ $MAPb_{0.5}Sn_{0.5}I_3$/TiO_2/FTO	$Cs_2AgBi_{0.75}Sb_{0.25}Br_6$	$MAPb_{0.5}Sn_{0.5}I_3$	SCAPS-1D	21.97	2022	Pandey et al. (2021)
Glass/ITO/SnO_2/ $FA_xCs_{1-x}Pb(I_yBr_{1-y})_{3x}$/ PTAA/$SnO_2$/PCBM/ $FA_xMA_{1-x}Pb_ySn_{1-y}I_3$/ PEDOT:PSS/Ag	$FA_xCs_{1-x}Pb(I_yBr_{1-y})_{3x}$	$FA_xMA_{1-x}Pb_ySn_{1-y}I_3$	SCAPS-1D	35	2023	Kumar et al. (2023)
Au/Cu_2O/$FAMASnGeI_3$/ n+/p+/$MAGeI_3$/ZnSe/FTO	$FAMASnGeI_3$	$MAGeI_3$	SCAPS-1D	22	2023	Alsalme and Alsaeedi (2022)
Au/CuO/K_2PtBr_6/IGZO/ ITO/$CuSbS_2$/$MAGeI_3$/ SnO_2/FTO	$MAGeI_3$	K_2PtBr_6	SCAPS-1D	31.68	2023	Jayan (2023)
$Zn_{0.75}Mg_{0.25}O$/$Cs_2AgBiBr_6$/ Cu:$Ni_{1-x}O$/TiO_2/ $Cs_4CuSb_2Cl_{12}$/Cu_2O	$Cs_2AgBiBr_6$	$Cs_4CuSb_2Cl_{12}$	SCAPS-1D	28.08	2024	Kumar and Kumar (2024)

reports on computational study of perovskite-perovskite tandem structures over the years.

In 2020, Singh et al. utilized Solar Cell Capacitance Simulator in One Dimension (SCAPS-1D) software to simulate low bandgap perovskite $FA_{0.75}MA_{0.25}Sn_{0.25}Ge_{0.5}I_3$ ($FAMASnGeI_3$) and wide bandgap perovskite methyl ammonium germanium iodide ($CH_3NH_3GeI_3$) as perovskite-perovskite tandem structure. In this study, five different ETLs were evaluated to determine the optimal ETL for maximizing solar cell efficiency. Additionally, key parameters such as the absorber layer thickness, defect density, and the front electrode's work function were optimized to enhance performance. Following the optimization process, Cu_2O emerged as the best-performing ETL, with an optimized perovskite layer thickness of 200 nm, a defect density of 10^{14} cm^{-3} and a front electrode work function of 4.3 eV. This optimization yielded an output of open-circuit voltage (V_{OC}) = 1.07 V, short-circuit current density (J_{SC}) = 28.36 mA/cm^2, fill factor (FF) = 84.39%, and efficiency (η) = 26.72%. The final simulated *J-V* curve is presented in Fig. 7.4 (Singh et al. 2020). Similarly, in 2023, Kumar et al. demonstrated 2T all perovskite TSC using $FA_xCs_{1-x}Pb(I_yBr_{1-y})_{3x}$ and $FA_xMA_{1-x}Pb_ySn_{1-y}I_3$ and enhanced the PCE by incorporating different ETLs like C_{60}, PCBM, ZnO, WO_3, CdSe, and In_2O_3. After optimization, PCE of over 35% has been obtained by reducing the interface surface defect density and band offsets at the ETL/perovskite interface (Kumar et al. 2023).

However, two primary challenges need to be addressed for successful implementation of 2T perovskite-perovskite tandems: (1) achieving high-efficiency, low-bandgap (typically Sn–Pb-based) perovskite cells, and (2) optimizing the interface between the top and bottom cells. This involves minimizing damage to the bottom cell during the processing of the top cell, reducing interface recombination, and ensuring efficient carrier transport between the sub-cells (Shrivastav et al. 2023).

Heterojunction Perovskite Tandem Structures

Heterojunction perovskite tandem solar cells are a type of tandem solar cell that combines perovskite-based materials with other photovoltaic materials, typically silicon, CIGS, or organic absorber layers. This design exploits the complementary absorption properties of the materials to enhance overall efficiency. Table 7.2 summarizes different reports on computational study of heterojunction perovskite tandem structures over the years. In 2017, Rolland et al. conducted numerical simulations of a hybrid organic–inorganic perovskite-on-silicon tandem solar cell using Silvaco Atlas through drift–diffusion modelling, incorporating trap-assisted recombination effects and a tunnel junction. The tandem cell achieved maximum efficiency of 27%, outperforming the individual efficiencies of silicon (17.3%) and perovskite (17.9%) solar cells (Rolland et al. 2018). Madan et al. in 2021 designed 2T perovskite–lead sulphide colloidal quantum dots (PbS CQD) monolithic TSC using SCAPS-1D software. In this work, before evaluating the performance of the proposed TSC, the performance of the perovskite top cells was optimized under

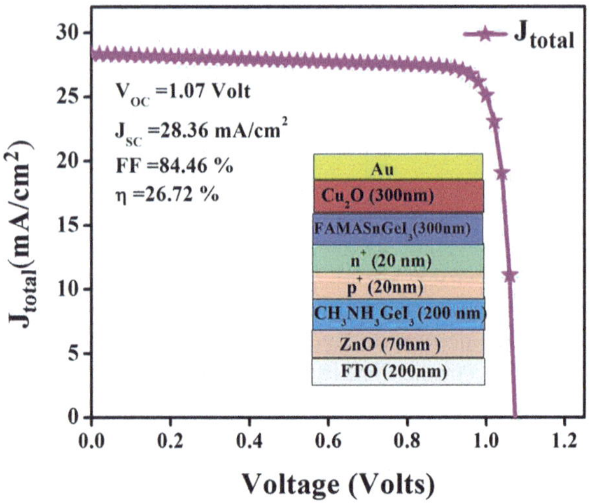

Fig. 7.4 Simulated *J-V* curve for perovskite-perovskite tandem PSC (Singh et al. 2020). (Copyright 2020. Reproduced with Permission from Elsevier)

standalone conditions by varying optical properties, thickness, and interface defect density. After thorough calibration, the TSC yielded an output of 1.79 V (V_{OC}), 16.67 mA cm^{-2} (J_{SC}), 78.3% (FF), and 23.36% (η) (Madan et al. 2021).

In a 2021 simulation study by Shrivastav et al., a perovskite-Si tandem structure achieved an enhanced efficiency of 33%. The research began with the optimization of a perovskite top cell incorporating an Me-4PACz hole transport layer for improved hole extraction, achieving a standalone efficiency of 21.3% and a silicon-based bottom cell with a standalone efficiency of 22.5%. These optimized cells were then integrated to form the PVK–Si TSC. A detailed analysis using the energy band diagram was conducted to investigate carrier dynamics within the designed tandem cell. The PVK–Si tandem solar cell achieved a conversion efficiency of 33%, utilizing absorber layers with thicknesses of 336 nm for the top cell and 150 μm for the bottom cell. Key performance parameters included a J_{SC} of 20.11 mA/cm^2, V_{OC} of 2.02 V, and FF of 81.36%. This comprehensive study highlights a promising pathway for developing high-efficiency tandem solar cells in the future (Shrivastav et al. 2021).

Another promising composition for high-efficiency TSCs with perovskite is Copper Indium Gallium Selenide (CIGS). Among various studies on PVK-CIGS tandem structures, the most recent report by Islam et al. in 2023 demonstrated an impressive efficiency of 38.39%. This work involved designing, simulating, and optimizing an all-inorganic, lead-free tandem PV cell comprising two sub-cells: a $CsSn_{0.5}Ge_{0.5}I_3$ (perovskite) top cell and a CIGS bottom cell. Simulations were performed using the SCAPS-1D simulator, with the optimization focusing on varying the thicknesses of critical layers. Initially, the two sub-cells were simulated independently—the top sub-cell under the standard AM1.5G solar spectrum and the

Table 7.2 Summary of reports related to heterojunction perovskite TSC

Tandem configuration	Top layer	Bottom layer	Simulation software	Efficiency	Year	References
Front contact/HTM/ $MAPb(I_3)_{1-x}(Br_3)_x/TiO_2$/Si n++/Si p++/Si/back contact	Perovskite	Silicon	Silvaco Atlas	27	2017	Rolland et al. (2018)
ITO/perovskite/HIT/ CIGS/PbS/ITO	Perovskite	CIGS	AFORS-HET	20.2	2020	Qasim et al. (2020)
FTO/c-TiO_2/perovskite/ spiro-MeOTAD/MZO/ PbS-CQD/PbS EDT	Perovskite	PbS CQD	SCAPS-1D	23.36	2021	Madan et al. (2021)
C-TiO_2/perovskite/ spiro-MeOTAD/ZnO/ CdS/CIGS	Perovskite	CIGS	SCAPS-1D	23.17	2021	Kumar et al. (2021)
IZO/SnO_2/C_{60}/perovskite/ Me-4PACz/ ITO/n-nc-SiO_x/i-a-Si:H/ n-c-Si/i-a-Si:H/p-a-Si:H	Perovskite	Silicon	SCAPS-1D	33.05	2021	Shrivastav et al. (2021)
FTO/TiO_2/ perovskite/c-Si(n)/c-- Si(p)/c-Si(p+)	Perovskite	Silicon	SCAPS-1D	32.97	2021	Nowsherwan et al. (2021)
FTO/TiO_2/ $MAGeI_3$/c-Si(n)/c-Si(p)/ c-Si	$MAGeI_3$	Silicon	SCAPS-1D	29.85	2021	Sarker et al. (2021)
FTO/TiO_2/ $MAPbI_3$/c-Si(n)/c-Si(p)/ c-Si	$MAPbI_3$	Silicon	SCAPS-1D	33.67	2021	

(continued)

Table 7.2 (continued)

Tandem configuration	Top layer	Bottom layer	Simulation software	Efficiency	Year	References
SnO_2/PCBM/perovskite/NiO/ITO/ZnO/n-SnS/p-SnS	Perovskite	SnS	SCAPS-1D	28.92	2022	Jafarzadeh et al. (2022)
Au/TiO_2/$CH_3NH_3PbI_{3-x}Br_x$/spiro-OMeTAD/a-Si(n)/a-Si(i)/c-Si(n)/a-Si(i)/a--Si(p)/ITO	$CH_3NH_3PbI_{3-x}Br_x$	Silicon	Silvaco ATLAS	31.67	2022	Bacha et al. (2022)
ITO/PEDOT:PSS/perovskite/C_{60}/Ag/ITO/PEDOT:PSS/DPPEZn-TBO:$PC_{61}BM$/PFN/Al	Perovskite	DPPEZn-TBO:$PC_{61}BM$	SETFOS	14	2023	Salem et al. (2023)
ITO/ZnO/$CsSn_{0.5}Ge_{0.5}I_3$/NiO/spiro-OMETAD/P-CIGS/TiO_2/FTO	Perovskite	CIGS	SCAPS-1D	38.39	2023	Islam et al. (2023)

bottom sub-cell using a filtered spectrum. Current matching between the two cells was achieved by fine-tuning the absorber layer thicknesses. The optimized tandem cell structure yielded an outstanding overall PCE of 38.39%, along with FF of 83.4%, V_{OC} of 2.48 V, and J_{SC} of 18.64 mA/cm^2. These results highlight the significant potential of PVK-CIGS tandem structures for next-generation high-efficiency solar technologies (Bacha et al. 2022).

However, heterojunction perovskite TSCs face challenges such as stability issues due to moisture, heat, and UV exposure, which degrade performance over time. Interface engineering is critical to minimize recombination losses and optimize band alignment for efficient charge transfer. Current matching between the top and bottom cells is complex, requiring precise material and thickness control. Additionally, scalability, uniformity in large-area fabrication, and addressing lead toxicity remain significant barriers to commercial viability.

Experimental Evidence of Tandem Perovskite Solar Cells

Experimental research on tandem PSCs has demonstrated substantial progress in enhancing their long-term efficiency and stability over time. These cells typically combine a perovskite cell with a silicon or other low-bandgap material (such as CIGS or perovskite), exploiting the complementary absorption properties of each layer. Table 7.3 summarizes various experimental reports on PVK-PVK, PVK-CIGS and PVK-Si TSCs over the years.

In 2016, Jiang et al. introduced the first bottom-up, solution-processed, two-terminal perovskite/perovskite tandem solar cell by designing a novel charge recombination layer (CRL) consisting of spiro-OMeTAD/PEDOT:PSS/PEI/PCBM:PEI. This CRL effectively collects electrons and holes at its top and bottom interfaces, while also providing sufficient robustness to shield the bottom perovskite layer during the deposition of the top layer. Furthermore, the CRL is fabricated using orthogonal solution-based fabrication under mild thermal conditions, facilitating conformity with the underlying perovskite structure. The PEI/PCBM:PEI component was specifically designed for effective electron collection in both single-junction and tandem PSCs. By optimizing the CRL to connect two $CH_3NH_3PbI_3$ perovskite sub-cells, the TSC achieved an efficiency of 7% (Heo and Im 2015). Another PVK-PVK tandem structure was reported by Yu et.al in 2020 by introducing an interconnection structure of $C_{60}/SnO_{1.76}$. With the device structure of Cu/BCP/C_{60}/$Cs_{0.05}MA_{0.45}FA_{0.5}Pb_{0.5}Sn_{0.5}I_3$/$SnO_{1.76}$/$C_{60}$/$Cs_{0.4}FA_{0.6}PbI_{1.95}Br_{1.05}$/PTAA/ITO/glass and use of inter connection layer, the device efficiency of small-area TSCs (5.9 mm^2) and large-area TSCs (1.15 cm^2) to 24.4% and 22.2%, respectively, while ensuring illumination stability over 1000 h (Yu et al. 2020).

In 2015, Todorov et al. reported PVK-CIGS TSC with each cell specifically tailored for seamless monolithic integration. This study presents a reactor designed for continuous in situ monitoring and precise tuning of the optical properties of the perovskite layer via vapour-phase halide exchange reactions. Utilizing in situ

Table 7.3 Experimental evidences related to TSCs

Tandem configuration	Top layer	Bottom layer	Efficiency	Year	References
TCE/PCBM/perovskite/PEDOT:PSS/ITO/CdS/CIGS/Mo/Si_3N_4/glass	Perovskite	CIGS	15.9	2015	Todorov et al. (2015)
Glass/FTO/bl-TiO_2/perovskite/wet P3HT/PCBM/$MAPbI_3$/PEDOT:PSS/ITO/glass	$CH_3NH_3PbBr_3$	$CH_3NH_3PbI_3$	10.8	2015	Heo and Im (2015)
Glass/FTO/c-TiO_2/m-TiO_2/perovskite/spiro-OMeTAD/PEDOT:PSS/PEI/PCBM:PEI/perovskite/spiro-OMeTAD/Ag	$MAPbI_3$	$MAPbI_3$	7	2016	Jiang et al. (2016)
Glass/ITO/NiO_x/perovskite/C_{60}/Bis-C_{60}/ITO/PEDOT:PSS/perovskite/$IC_{60}BA$/Bis-C_{60}/Ag	$MA_{0.9}Cs_{0.1}Pb(I_{0.6}Br_{0.4})_3$	$MAPb_{0.5}Sn_{0.5}I_3$	18.5	2017	Rajagopal et al. (2017)
MgF_2/ITO/ZnO NPs/perovskite/PTAA/ITO/BZO/iZnO/CdS/CIGS/Mo/glass	Perovskite	CIGS	22.43	2018	Han et al. (2018)
a-Si(p) rear/n-Si/nc-Si:H(nþ)/nc-Si:H(pþ)/Spiro-TTB/$Cs_xFA_{1-x}Pb(I,Br)_3$/LiF/C_{60}/SnO_2/IZO/Ag/MgF	Silicon	Perovskite	25.2	2018	Sahli et al. (2018)

Tandem configuration	Top layer	Bottom layer	Efficiency	Year	References
LiF/IZO/SnO_2/perovskite/PTAA/NiO_x/ZnO/CdS/CIGSe/Mo/glass	Perovskite	CIGSe	21.6	2019	Jošt et al. (2019)
Glass/ITO/PTAA/perovskite/C_{60}/$SnO_{1.76}$/perovskite/C_{60}/BCP/Cu	$Cs_{0.4}FA_{0.6}PbI_{1.95}Br_{1.05}$	$Cs_{0.05}MA_{0.45}FA_{0.5}Pb_{0.5}Sn_{0.5}I_3$	24.6	2020	Yu et al. (2020)
ITO/PTAA/perovskite/PCBM/BCP/Ag/M--PEDOT/PBDB-T:SN6IC4F/Bis-C_{60}/BCP/Ag	$Cs_{0.1}(FA_{0.6}MA_{0.4})_{0.9}Pb(I_{0.6}Br_{0.4})_3$	PBDB-T:SN6IC4F	15.1	2020	Li et al. (2020)
a-Si(pþ) rear/n-Si/a-Si(i)/nc-SiO_x(n)/ITO/PTAA(or SAM)/perovskite/LiF/C_{60}/SnO_2/IZO/LiF	Silicon	$Cs_{0.05}(FA_{0.77}MA_{0.23})_{0.95}Pb(I_{0.77}Br_{0.23})_3$	29.2	2020	Al-Ashouri et al. (2020)
a-Si(pþ)rear/n-Si/ITO/PTAA/$Cs_{0.1}MA_{0.9}Pb(I_{0.9}Br_{0.1})_3$/$C_{60}$/$SnO_2$/ITO/Ag/PDMS	Silicon	$Cs_{0.1}MA_{0.9}Pb(I_{0.9}Br_{0.1})_3$	26.1	2020	Chen et al. (2020)
a-Si(pþ) rear/n-Si(CZ)/a-Si(i)/nc-SiO_x(n)/ITO/2PACz/perovskite/LiF/$C6_0$/SnO_2/IZO/LiF	Silicon	$Cs_{0.05}(FA_{0.77}MA_{0.23})_{0.95}Pb(I_{0.77}Br_{0.23})_3$	27.9	2021	Köhnen et al. (2021)
Perovskite/SnO_2/C_{60}/LiF/Me-4PACs/nc-SiO_x:H (n)/TCO	$Cs_{0.05}(FA_{0.79}MA_{0.21})_{0.95}Pb(I_{0.79}Br_{0.21})_3$	Silicon	29.80	2022	Tockhorn et al. (2022)

(continued)

Table 7.3 (continued)

Tandem configuration	Top layer	Bottom layer	Efficiency	Year	References
Ag/MgF_2/IZO/SnO_2/PDAI/LiF/C_{60}/PVK/2PACz/IZO/nc-Si(n)/a-Si(i)/c-Si/a--Si(i)/IZO/MgF_2/Ag	Perovskite	Silicon	32.8	2023	Aydin et al. (2023)
ITO/PEDOT:PSS/Me-4PACs/PTAA/NiO_x/Cz-Si/ITO/glass	Me-4PACs	Czochralski (Cz)-Si	30	2024	Harter et al. (2024)

bandgap engineering, device structure Glass/Si_3N_4/Mo/CIGS/CdS/ITO/PEDOT:PSS/Perovskite/PCBM/TCE was fabricated, and this cell yielded an efficiency of 15.9% (Todorov et al. 2015). Similarly, Han et al. developed a high-performance monolithic perovskite/CIGS tandem solar cell with an efficiency of 22.43%. Their study presents a promising strategy for reducing the environmental impact of cadmium and prolonging the operational lifespan of perovskite/CIGS tandem solar cells. In a similar effort, Han et al. achieved a 22.43% efficiency with a monolithic perovskite/CIGS tandem solar cell, highlighting a strategy that reduces cadmium usage and enhances long-term device stability. Notably, CIGS solar cells can be reused after the degraded perovskite front cell is removed. By employing PTAA as the hole transport layer, the entire perovskite front cell can be dissolved and separated from the CIGS rear cell using a combination of chlorobenzene and *N*,*N*-dimethylformamide (Han et al. 2018).

In 2020, Ashouri et al. reported a monolithic perovskite/silicon TSC achieving a certified PCE of 29.15%. The perovskite absorber, featuring a bandgap of 1.68 eV, exhibited excellent phase stability under illumination, enabled by efficient hole extraction and reduced nonradiative recombination at the hole-selective interface. This performance was achieved through the implementation of a self-assembled, methyl-substituted carbazole monolayer as the hole-selective layer. The improved charge extraction led to a low ideality factor of 1.26 and single-junction FF reaching up to 84%, culminating in a tandem open-circuit voltage as high as 1.92 V. Notably, the tandem device retained 95% of its initial efficiency after 300 h of operation in air without encapsulation (Al-Ashouri et al. 2020). Another remarkable contribution to PVK-Si TSCs was reported by Aydin et al. in 2023. In this study, the perovskite top cell is electrically connected to a silicon heterojunction bottom cell through a self-assembled monolayer (SAM) anchored on a transparent conductive oxide (TCO), facilitating efficient charge transfer between the sub-cells. Through the integration of optical enhancements, including ultrathin IZO rear electrodes and optimized front contact stacks, an independently certified PCE of 32.5% was achieved. This performance stands among the top values reported for perovskite/silicon TSCs (Tockhorn et al. 2022).

Challenges and Limitations of Computational Study

The landscape of computational study for tandem solar cells is fraught with various challenges and limitations that hinder their efficiency and performance optimization. These challenges are crucial for researchers aiming to develop more effective and reliable solar technologies. Accurate modelling of interfaces in tandem solar cells remains a significant challenge due to the complex interactions between different semiconductor materials. These interfaces are critical to device performance, yet traditional models often fail to capture their intricacies, particularly the effects of material property mismatches, such as bandgap and temperature sensitivity. Similarly, simulating electronic properties, including nonradiative recombination and parasitic resistances, is difficult, as current computational methods often rely on idealized conditions that overlook real-world complexities (Harter et al. 2024). The variability of operating conditions, such as changes in temperature and irradiance, further complicates modelling efforts, with many models oversimplifying or ignoring these factors, resulting in unrealistic efficiency predictions. Efficiency optimization is also hindered by trade-offs between physical properties, such as balancing absorption and resistance, which can yield impractical or contradictory outcomes in simulations. Moreover, modelling long-term degradation and stability remains inadequate, as many approaches fail to account for environmental impacts and material fatigue, leading to overly optimistic performance projections and limiting insights into developing durable tandem solar cells (Cheng and Ding 2021).

Recent advancements in TSC concepts, particularly perovskite/silicon tandems, represent a remarkable milestone in the field and highlight the potential for discovering novel tandem configurations. A critical factor in these developments is the compatibility between sub-cells, with the optimal sub-cell interconnection configuration still under debate. Likewise, multiple strategies for tandem module integration are being explored, but no clear frontrunner has emerged at this stage of development. While two-terminal monolithic configurations offer simplicity, their reliance on current matching poses limitations, especially under variable spectral conditions (Martinho 2021).

Conclusion

The growing global investment in clean energy, coupled with the rapid advancements in solar technology, underscores a transformative shift towards a greener and more sustainable energy future. The remarkable progress of perovskite solar cells, particularly when incorporated into tandem configurations, offers promising potential to surpass the efficiency of traditional single-junction cells. As researchers continue to explore innovative strategies to improve the stability and performance of these cells, tandem perovskite solar technologies are essential in decreasing global reliance on fossil fuels, further accelerating the adoption of renewable energy worldwide.

The success of perovskite-based TSCs hinges on overcoming challenges such as device stability and degradation, which require continued collaboration between computational and experimental research. As efforts intensify to refine these technologies, the combination of affordability, high efficiency, and versatility makes them a game-changer for the solar industry. Ultimately advancements in perovskite TSCs lay the foundation for the next generation of solar energy, supporting the global shift towards cleaner and more efficient energy systems.

Acknowledgement One of the authors (UPS) like to thank Department of Science and Technology (DST)–FIST for financial assistance (Grant # SR/FST/ET-1/2021/862).

References

Abdelaziz S, Zekry A, Shaker A, Abouelatta M (2022) Investigation of lead-free $MASnI_3$-$MASnIBr_2$ tandem solar cell: numerical simulation. Opt Mater 123:111893

Al-Ashouri A, Köhnen E, Li B, Magomedov A, Hempel H, Caprioglio P, Albrecht S (2020) Monolithic perovskite/silicon tandem solar cell with >29% efficiency by enhanced hole extraction. Science 370(6522):1300–1309

Alsalme A, Alsaeedi H (2022) Twenty-two percent efficient Pb-free all-perovskite tandem solar cells using SCAPS-1D. Nanomaterials 13(1):96

Ameri T, Dennler G, Lungenschmied C, Brabec CJ (2009) Organic tandem solar cells: a review. Energy Environ Sci 2(4):347–363

Aydin E, Ugur E, Yildirim BK, Allen TG, Dally P, Razzaq A, De Wolf S (2023) Enhanced optoelectronic coupling for perovskite/silicon tandem solar cells. Nature 623(7988):732–738

Bacha M, Saadoune A, Youcef I (2022) Design and numerical investigation of perovskite/silicon tandem solar cell. Opt Mater 131:112671

Chen B, Zhengshan JY, Manzoor S, Wang S, Weigand W, Yu Z et al (2020) Blade-coated perovskites on textured silicon for 26%-efficient monolithic perovskite/silicon tandem solar cells. Joule 4(4):850–864

Cheng Y, Ding L (2021) Perovskite/Si tandem solar cells: fundamentals, advances, challenges, and novel applications. SusMat 1(3):324–344

Han Q, Hsieh YT, Meng L, Wu JL, Sun P, Yao EP et al (2018) High-performance perovskite/$Cu(In,Ga)Se_2$ monolithic tandem solar cells. Science 361(6405):904–908

Harter A, Artuk K, Mathies F, Karalis O, Hempel H, Al-Ashouri A, Wolff CM (2024) Perovskite/silicon tandem solar cells above 30% conversion efficiency on submicron-sized textured Czochralski-silicon bottom cells with improved hole-transport layers. ACS Appl Mater Interfaces 16(45):62817–62826

Heo JH, Im SH (2015) $CH_3NH_3PbBr_3$-$CH_3NH_3PbI_3$ perovskite-perovskite tandem solar cells with exceeding 2.2 V open circuit voltage. Adv Mater 28(25):5121–5125

Ho-Baillie AW, Zheng J, Mahmud MA, Ma FJ, McKenzie DR, Green MA (2021) Recent progress and future prospects of perovskite tandem solar cells. Appl Phys Rev 8(4):041307

IEA (2024) World energy investment 2024. IEA, Paris. https://www.iea.org/reports/world-energy-investment-2024, Licence: CC BY 4.0

Islam MA, Jawad A, Jahan NA, Hossain MM (2023) Outstanding conversion efficiency of 38.39% from a perovskite/CIGS tandem PV cell: a synergic optimization through computational modeling. Heliyon 9(10):e20558

Jafarzadeh F, Aghili H, Nikbakht H, Javadpour S (2022) Design and optimization of highly efficient perovskite/homojunction SnS tandem solar cells using SCAPS-1D. Sol Energy 236:195–205

Jayan KD (2023) Complete modelling and simulation of all perovskite tandem solar cells. Mater Sci Eng B 294:116506

Jiang F, Liu T, Luo B, Tong J, Qin F, Xiong S, Zhou Y (2016) A two-terminal perovskite/perovskite tandem solar cell. J Mater Chem A 4(4):1208–1213

Jošt M, Bertram T, Koushik D, Marquez JA, Verheijen MA, Heinemann MD, Albrecht S (2019) 21.6%-efficient monolithic perovskite/Cu (In, Ga) Se_2 tandem solar cells with thin conformal hole transport layers for integration on rough bottom cell surfaces. ACS Energy Lett 4(2):583–590

Jošt M, Kegelmann L, Korte L, Albrecht S (2020) Monolithic perovskite tandem solar cells: a review of the present status and advanced characterization methods toward 30% efficiency. Adv Energy Mater 10(26):1904102

Kheralla A, Chetty N (2021) A review of experimental and computational attempts to remedy stability issues of perovskite solar cells. Heliyon 7(2):e06211

Köhnen E, Wagner P, Lang F, Cruz A, Li B, Roß M et al (2021) 27.9% efficient monolithic perovskite/silicon tandem solar cells on industry compatible bottom cells. Sol RRL 5(7):2100244

Kumar P, Kumar A (2024) Double perovskite tandem solar cells: design and performance investigation of the use of CABB and CCSC as top and bottom cell absorber materials. J Electron Mater:1–17

Kumar A, Singh S, Mohammed MK, Shalan AE (2021) Computational modelling of two terminal CIGS/perovskite tandem solar cells with power conversion efficiency of 23.1%. Eur J Inorg Chem 2021(47):4959–4969

Kumar D, Bansal NK, Dixit H, Kulkarni A, Singh T (2023) Numerical study on the effect of dual electron transport layer in improving the performance of perovskite–perovskite tandem solar cells. Adv Theory Simul 6(3):2200800

Li H, Zhang W (2020) Perovskite tandem solar cells: from fundamentals to commercial deployment. Chem Rev 120(18):9835–9950

Li Z, Wu S, Zhang J, Lee KC, Lei H, Lin F, Jen AK (2020) Hybrid perovskite-organic flexible tandem solar cell enabling highly efficient electrocatalysis overall water splitting. Adv Energy Mater 10(18):2000361

Madan J, Pandey R, Sharma R (2020) Device simulation of 17.3% efficient lead-free all-perovskite tandem solar cell. Sol Energy 197:212–221

Madan J, Singh K, Pandey R (2021) Comprehensive device simulation of 23.36% efficient two-terminal perovskite-PbS CQD tandem solar cell for low-cost applications. Sci Rep 11(1):19829

Martinho F (2021) Challenges for the future of tandem photovoltaics on the path to terawatt levels: a technology review. Energy Environ Sci 14(7):3840–3871

Mazumdar S, Zhao Y, Zhang X (2023) Comparative architecture in monolithic perovskite/silicon tandem solar cells. Sci China Phys Mech Astron 66(1):217304

Mohanty I, Mangal S, Jana S, Singh UP (2021) Stability factors of perovskite ($CH_3NH_3PbI_3$) thin-films for solar cell applications: a study. Mater Today Proc 39:1829–1832

Mohanty I, Singh UP, Jana S, Mangal S (2024) Influence of annealing on stability and regeneration of degraded $CH_3NH_3PbI_3$ thin films. Appl Phys A 130(2):143

National Laboratory of the Rockies (n.d.) Best research-cell efficiency chart. NREL. https://www.nrel.gov/pv/cell-efficiency.html

Nowsherwan GA, Jahangir K, Usman Y, Saleem W, Khalid M (2021) Numerical modeling and optimization of perovskite silicon tandem solar cell using SCAPS-1D. Sch Bull 7(7):171–184

Pandey R, Sharma S, Madan J, Sharma R (2021) Numerical simulations of 22% efficient all-perovskite tandem solar cell utilizing lead-free and low lead content halide perovskites. J Micromech Microeng 32(1):014004

Qasim UB, Imran H, Kamran M, Faryad M, Butt NZ (2020) Computational study of stack/terminal topologies for perovskite based bifacial tandem solar cells. Sol Energy 203:1–9

Rajagopal A, Yang Z, Jo SB, Braly IL, Liang PW, Hillhouse HW, Jen AKY (2017) Highly efficient perovskite–perovskite tandem solar cells reaching 80% of the theoretical limit in photovoltage. Adv Mater 29(34):1702140

Rolland A, Pedesseau L, Kepenekian M, Katan C, Huang Y, Wang S et al (2018) Computational analysis of hybrid perovskite on silicon 2-T tandem solar cells based on a Si tunnel junction. Opt Quant Electron 50:1–13

Sahli F, Werner J, Kamino BA, Bräuninger M, Monnard R, Paviet-Salomon B, Ballif C (2018) Fully textured monolithic perovskite/silicon tandem solar cells with 25.2% power conversion efficiency. Nat Mater 17(9):820–826

Salem MS, Shaker A, Abouelatta M, Saeed A (2023) Full optoelectronic simulation of lead-free perovskite/organic tandem solar cells. Polymers 15(3):784

Sarker S, Islam MT, Rauf A, Al Jame H, Jani MR, Ahsan S et al (2021) A SCAPS simulation investigation of non-toxic $MAGeI_3$-on-Si tandem solar device utilizing monolithically integrated (2-T) and mechanically stacked (4-T) configurations. Sol Energy 225:471–485

Shrivastav N, Madan J, Pandey R, Shalan AE (2021) Investigations aimed at producing 33% efficient perovskite–silicon tandem solar cells through device simulations. RSC Adv 11(59):37366–37374

Shrivastav N, Madan J, Pandey R (2023) A short study on recently developed tandem solar cells. Mater Today Proc
Singh N, Agarwal A, Agarwal M (2020) Numerical simulation of highly efficient lead-free all-perovskite tandem solar cell. Sol Energy 208:399–410
Solar Power Europe (2024) Global market outlook for solar power 2024–2028
Tockhorn P, Sutter J, Cruz A, Wagner P, Jäger K, Yoo D, Becker C (2022) Nano-optical designs for high-efficiency monolithic perovskite–silicon tandem solar cells. Nat Nanotechnol 17(11):1214–1221
Todorov T, Gershon T, Gunawan O, Lee YS, Sturdevant C, Chang LY, Guha S (2015) Monolithic perovskite-CIGS tandem solar cells via in situ band gap engineering. Adv Energy Mater 5(23):1500799
Wali Q, Elumalai NK, Iqbal Y, Uddin A, Jose R (2018) Tandem perovskite solar cells. Renew Sust Energ Rev 84:89–110
Yu Z, Yang Z, Ni Z, Shao Y, Chen B, Lin Y, Huang J (2020) Simplified interconnection structure based on C_{60}/SnO_{2-x} for all-perovskite tandem solar cells. Nat Energy 5(9):657–665
Zhou Q, Liu X, Liu Z, Zhu Y, Lu J, Chen Z et al (2024) Annual research review of perovskite solar cells in 2023. Mater Futures 3:022102

Chapter 8
Computational Investigations on Stability of Perovskite Solar Cells: A Real Challenge

Utkarsh Tiwari, Abhishek Raj, Sandeep Kumar Pundir, Bhupendra Singh, and Manish Kumar

Abstract The stability of perovskite solar cells remains one of the most significant challenges hindering their commercialization and long-term performance. This chapter investigates the major problem of stability, beginning with a summary of the factors that contribute to the challenges faced by perovskite solar cells. Key causes of instability, including moisture, heat, UV light, oxygen, ion migration, and hysteresis, are discussed in detail, highlighting their impact on the material's degradation and performance. The chapter also explores various strategies to enhance stability, such as compositional engineering, 2D layered materials, nanoengineering, interfacial modification, and encapsulation techniques. Furthermore, it delves into the role of computational studies in understanding and improving the stability of perovskite solar cells, emphasizing the benefits of simulation-based approaches. Key computational methods, such as density functional theory (DFT), molecular dynamics (MD), and ab initio molecular dynamics (AIMD) are discussed along with examples where these methods have been useful.

Keywords Perovskite · Stability · Computation · Photovoltaic · Degradation

---

U. Tiwari · S. K. Pundir (✉)
Department of Physics, University of Lucknow, Lucknow, Uttar Pradesh, India

A. Raj
CSIR-National Environmental Engineering Research Institute (NEERI), Nagpur, India

Academy of Scientific and Innovative Research (AcSIR), Ghaziabad, India

B. Singh · M. Kumar (✉)
Experimental Research Laboratory, Department of Physics, ARSD College, University of Delhi, New Delhi, India
e-mail: mkumar2@arsd.du.ac.in

M. Kumar (ed.), *Computational Analysis of Perovskite Solar Cells*, Springer Tracts in Electrical and Electronics Engineering,
https://doi.org/10.1007/978-981-95-6521-4_8

Introduction

Perovskite solar cells (PSCs) have revolutionized the field of photovoltaics, offering unprecedented potential with their remarkable power conversion efficiencies, tunable bandgaps, and low-cost manufacturing processes (Sandhu and Park 2024). However, as this technology moves closer to large-scale deployment, its stability under real-world operating conditions has emerged as a bottleneck. Achieving long-term stability is essential for PSCs to compete with already existing technologies such as silicon-based solar cells. Solar cells are typically expected to operate reliably for about 25 years (Azzopardi and Mutale 2009), enduring continuous exposure to sunlight, heat, humidity, and mechanical stress. However, the rapid degradation of PSCs under such conditions makes them unsuitable for practical applications unless stability is significantly improved. The highest recorded time duration for a stable PSCs is just 10,000 h or approximately 1 year and as compared to the expected lifetime of a conventional solar cell, the PSCs fall short in this regard (Grancini et al. 2017). Hence, it is quite evident that stability is a major issue for PSCs. Environmental concerns also underscore the importance of stability. Commonly developed PSCs use lead-based perovskite materials, which poses a risk of toxic lead leakage into the environment when the cells degrade (Raj et al. 2022a). Ensuring stability minimizes this risk and makes sure that the PSC technology aligns with the broader goals of clean and sustainable energy.

Unlike traditional photovoltaic materials, perovskites are highly sensitive to environmental factors, including moisture, oxygen, heat, and light, which can trigger a cascade of degradation processes and significantly impair device performance and lifespan (Tiep et al. 2016). Moisture, one of the most severe threats, rapidly degrades perovskites into lead iodide and other by-products, often irreversibly. Similarly, exposure to oxygen, particularly under illumination, leads to photooxidation reactions that compromise the structural and electronic integrity of the active layer. Heat is also such an issue that causes problems. Furthermore, ultraviolet (UV) light absorption by electron transport materials, such as titanium dioxide, can generate reactive oxygen species that attack the perovskite, leading to further degradation (Wang et al. 2016).

These challenges can only be addressed if the search for their solution occurs in different fields of science together. Innovations in material chemistry, such as engineering more stable perovskite compositions and incorporating additives, have shown promise in improving intrinsic stability (Mahapatra et al. 2019). Additionally, improved device architectures, including the use of different charge materials for charge transport layers and interfacial modification, aim to minimize environmental sensitivity of the device (Fatima et al. 2024; Rao et al. 2021; Ali et al. 2020). Encapsulation technologies are also being developed to shield devices from moisture and oxygen, ensuring durability in outdoor conditions (Chu et al. 2023). Therefore, stability is essential for the economic, environmental, and practical success of PSCs, but it remains a significant challenge due to various vulnerabilities they face. Overcoming these challenges requires a multi-faceted approach involving material innovation, device engineering, and protective strategies to ensure that PSCs can deliver reliable and long-lasting performance. Only by addressing these issues can PSCs Cells realize their potential as a transformative renewable energy technology. While the quest for stability remains ongoing, the progress in research and development underscores the determination of the scientific community to bridge this gap. By overcoming these hurdles, PSCs hold the potential to become a transformative technology for sustainable energy generation, paving the way for their integration into the global energy landscape.

This chapter delves into the key stability challenges facing PSCs and explores the innovative strategies being implemented to ensure their long-term performance. The computational analysis focusing on the stability assessments of PSCs is also examined.

Major Causes of Instability of Perovskite Solar Cells

Moisture

Moisture is one of the most significant environmental factors that damage the stability of PSCs. Multiple studies have demonstrated the damaging effects of moisture on PSCs. These cells are made to be used outside in the open environment so that they can easily achieve their intended function. This naturally becomes a problem when the environment outside has considerable amount of moisture due to humidity which makes it a major concern for use of PSCs. $CH_3NH_3PbI_3$ or $MAPbI_3$ is one of most widely used perovskite absorbing material in PSCs due to their superior optical and electrical properties such as high light absorption, ease of fabrication, and impressive photovoltaic efficiency (Wang et al. 2016). However, this perovskite material presents a limitation due to its chemical reaction with moisture. This reaction takes place due to the hydrogen bond formed between the organic and inorganic units. Excess water can help to break this bond (Habisreutinger et al. 2016). This material on reacting with water decomposes into other compounds that limits the stability of the solar cell considerably (Wang et al. 2016).

The reactions that cause this decomposition of the material are given as follows:

$$CH_3NH_3PbI_3 \overset{H_2O}{\Leftrightarrow} CH_3NH_3I(aq) + PbI_2(s) \tag{8.1}$$

$$CH_3NH_3I \Leftrightarrow CH_3NH_2(aq) + HI(aq) \tag{8.2}$$

$$4HI + O_2 \Leftrightarrow 2I_2(s) + 2H_2O \tag{8.3}$$

$$2HI(aq) \overset{h\upsilon}{\Leftrightarrow} H_2 \uparrow + I_2(s) \tag{8.4}$$

After the second equation, there is an equilibrium of CH_3NH_3I and HI. From this step, HI can degrade in two possible ways, one is the third equation, and the other is the fourth equation. The third equation is a redox reaction whereas the fourth reaction is the photochemical reaction under the presence of UV rays. The third reaction shows that the effect of degradation can be increased by the presence of oxygen whereas the fourth reaction shows that the effect of degradation can be increased by the presence of UV rays. It is the third and the fourth equation that are main reasons for the degradation of the PSCs (Niu et al. 2014).

Another way the degradation of the perovskite cell can be represented by the diagram below in the form of a cycle as shown in Fig. 8.1. This cycle shows the formation of PbI_2 as a by-product of the reaction of the PSC with moisture. Hence, it is completely evident that the harmful effect of the moisture on the PSCs does not only limit itself to the degradation of the material but also in the formation of harmful materials such as PbI_2 (Frost et al. 2014). The material PbI_2 strongly impacts the photovoltaic performance of the solar cell. It works as an insulator and causes a hindrance to the efficient working of the solar cell (Wang et al. 2017). Along with the decomposition of the perovskite material and the formation of harmful by-products, there is the case of formation of hydrated compounds decrease the absorption of the films (Christians et al. 2015).

The various impacts of moisture on PSCs work together in a cumulative manner to progressively hinder the functionality of the solar cells. Over time, these effects lead to a steady decline in the stability and performance of the cells, primarily as a result of the degradation of the perovskite material. This degradation compromises the structural and chemical integrity of the solar cells, ultimately rendering them less efficient and reliable for long-term use.

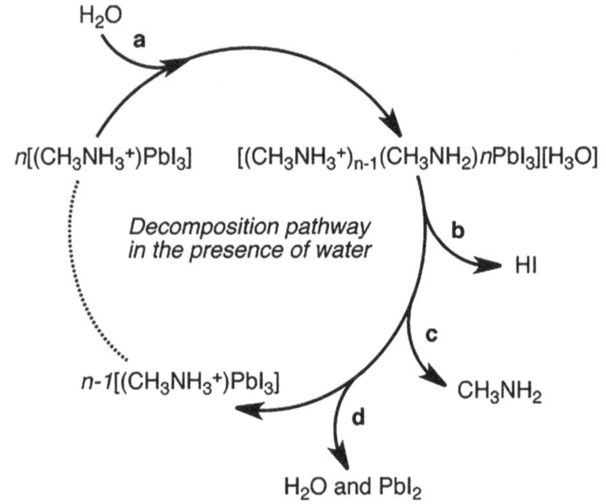

Fig. 8.1 Diagram showing the pathway of decomposition of the perovskite material in the presence of moisture. (Reproduced with permission from Frost et al. 2014 Copyright (2014), American Chemical Society)

Temperature

Annealing is an extremely important step in the fabrication of PSCs and involves heat treatment to modify the properties (mostly physical properties) of a material. In this process a material is heated to a specific temperature above its recrystallization temperature for a specific duration. After the annealing period, the material is then cooled to room temperature. This process leads to various alterations in the properties of a material, depending on the duration for which it is subjected to heat. So, it can be inferred that a particular temperature is associated with the structure and a particular phase of the perovskite material used. At different temperatures the phase of the perovskite material may change (Quarti et al. 2015).

The PSCs are made to be used in environments in which they face the sun constantly for their proper functioning. Naturally it is understandable that the solar cell will be withstanding an environment which involves a lot of heat and high temperatures. Given the importance of the process of annealing and the usage of the PSCs, it is quite clear that the effect of heat on the perovskite material is extremely pertinent.

According to the international standards, it is required that perovskite material must withstand a temperature of 85 °C with considerable stability, this corresponds to 0.093 eV (Kim et al. 2016). So 85° is a good benchmark temperature to measure the stability of the PSCs. Excess temperature is responsible for degrading the perovskite material. As the perovskite material gets heated, it gets decomposed with leaving only PbI_2 behind and the organic part of the material gets evaporated. This happens following the given equation below (Philippe et al. 2015):

$$CH_3NH_3PbI_3 \rightarrow PbI_2 + CH_3NH_2 \uparrow + HI \uparrow \tag{8.5}$$

This equation shows how the perovskite material degrades under the influence of excess heat. As seen above, the international standards give a particular temperature for the proper functioning of photovoltaics, i.e. 85 °C, and experiments have shown that the perovskite material ($CH_3NH_3PbI_3$) degrades when the temperature was increased from room temperature to 100 °C (Philippe et al. 2015). It is seen that the Perovskite material currently used is not very thermally stable, and new methods should be developed to tackle this problem.

UV Light

Just as $CH_3NH_3PbI_3$ is an excellent perovskite material for the application of solar cells, TiO_2 is also an excellent material to be used as an electron transport layer (ETL). The advantages of TiO_2 are many and include good adhesion with the glass substrates, long-term stability of the material, its ability to passivate the surface of perovskite (Fatima et al. 2024). But just as $CH_3NH_3PbI_3$ caused some issues in the presence of moisture, titanium dioxide also proves to be harmful for the perovskite solar cells in the influence of UV light. This can be understood with the help of the photocatalytic effect. The photocatalytic effect of titanium dioxide refers to its ability to act as a catalyst for chemical reactions upon exposure to light. This light can be ultraviolet (UV) light. This is due to the semiconductor property of TiO_2 which allows the compound to absorb photons and generate electron-hole pairs. These generated charge carriers react with the environment (O_2 and H_2O) to form highly reactive radicals and ions which further react with organic materials to decompose them.

The above mechanism is important to understand the effect of UV light on PSCs. As the photocatalytic effect of TiO_2 occurs, the organic compound of the perovskite material decomposes. This decomposition occurs when the photocatalytic effect of TiO_2 extracts electrons from the I^- ions and deconstructs the perovskite material. After this, the $CH_3NH_3^+$ ion maintains an equilibrium reaction between itself and CH_3NH_2 and H^+ ion. Among these two, CH_3NH_2 evaporates and the H^+ ion along with the extracted electrons from the I^- ions reduce the I_2 to form HI which also evaporates. In this way, the decomposition of the perovskite material takes place (Azad et al. 2024; Ito et al. 2014). This can be shown by the following Eqs. (8.6)–(8.8). Figure 8.2 shows the degradation of perovskite material in the presence of light. The diagram shows different products due to various reactions involved in the degradation of the perovskite material. These reactions involve the same Eqs. (8.6)–(8.8) (Ito et al. 2014), Eq. (8.6) occurs at the interface between TiO_2 and $CH_3NH_3PbI_3$.

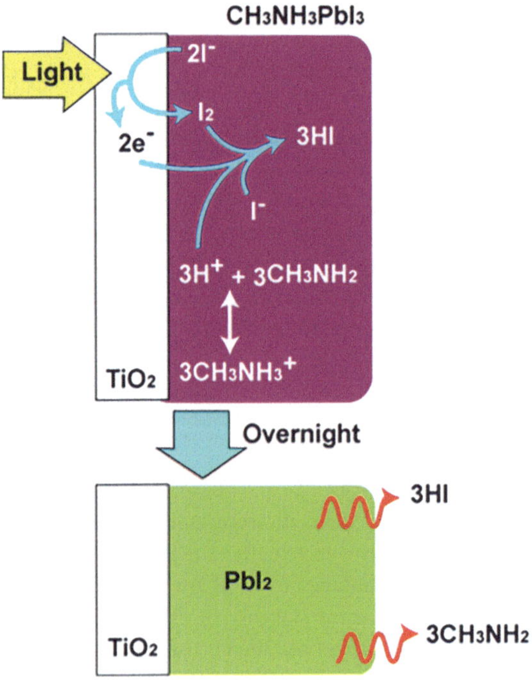

Fig. 8.2 Diagram showing the effect of light on the perovskite material and the degradation following it. (Reproduced with permission from Ito et al. 2014 Copyright (2014), American Chemical Society)

$$2I^- \leftrightarrow I_2 + 2e^- \tag{8.6}$$

$$3CH_3NH_3^+ \leftrightarrow 3CH_3NH_2 \uparrow + 3H^+ \tag{8.7}$$

$$I^- + I_2 + 3H^+ + 2e^- \leftrightarrow 3HI \uparrow \tag{8.8}$$

Oxygen

Oxygen is present in our atmosphere in abundance and therefore, it is reasonable to say that the presence of oxygen and its effect should be a valid concern for the working of the PSCs.

While studying the reasons for degradation and instability of the perovskite material, it become clear that the responsible factors do not work independently. These factors operate simultaneously, influencing one another and impacting each other's responses. One such factor is oxygen. As seen from the effect of moisture on the stability of the PSCs, it is evident that oxygen plays a role in the decomposition of the perovskite material along with moisture. These conditions exist in an ambient environment. The process of degradation of the perovskite material due to moisture, especially Eq. (8.3), shows that in the presence of oxygen the degradation rapidly increases due to the oxidation of HI leading to the formation of PbI_2 (Wang et al. 2016).

Other studies and research have been performed to show that oxygen along with white light can also be a reason for degradation of the perovskite material. One of these ways is the reaction of the superoxide (O_2^-) with the perovskite material. These superoxide ions are generated by the reaction between photoexcited electrons of the perovskite material and oxygen. The reactions of this process are given below (Aristidou et al. 2015),

$$CH_3NH_3PbI_3 \xrightarrow{\text{Light}} CH_3NH_3PbI_3^* \tag{8.9}$$

$$O_2 \xrightarrow{CH_3NH_3PbI_3^*} O_2^- \tag{8.10}$$

$$CH_3NH_3PbI_3 + O_2^- \xrightarrow{\text{Deprotonation}} CH_3NH_2 + PbI_2 + \frac{1}{2}I_2 + H_2O \tag{8.11}$$

Ion Migration

Ion migration is a phenomenon in which ions within a material move under the influence of external force such as electric field, heat, or light. Perovskite materials show ionic properties. As a result, these materials are sensitive to ion migration. In the case of perovskite materials, ion migration occurs under the influence of an electric field and illumination (Zhang et al. 2019). In $CH_3NH_3PbI_3$, Schottky defects are the most common defects (Walsh et al. 2014). These defects play an important role in the ion migration for this perovskite material. For the case of the perovskite material, that is, $CH_3NH_3PbI_3$, it is known that there are three ions in the material; they are I^-, $CH_3NH_3^+$, and Pb^{2+}. It was discovered that out of these three ions, I^- ions are the ions which can most easily migrate in the material. This is due to low activation energy which is around 0.58 eV which implies a much more favourable vacancy assisted ion migration of iodide ions. For the case of $CH_3NH_3^+$ ions, they have slightly higher activation energy and not only that but high level of orientational motion that may limit their transport for long distances. And for the case of Pb^{2+} ions, they have a high activation energy for ion migration that impedes their motion (Eames et al. 2015). As the influence due to the external forces is low, these ions move but they stop at the ETL and the hole transport layer (HTL) which are at relatively low range. But when the intensity of these external factors increases, such as high electric field, high temperature, and illumination, then these ions can travel large range and reach up to the external metal electrodes. These ions may not only interact with the electrodes at the interface and cause reactions that may damage the cell permanently but also these ions may change the electronic nature of the material and may cause hindrance in the intended working of the device (Sakhatskyi et al. 2022).

Ion migration can also cause the stoichiometric ratios of the perovskite material to become unbalanced which may further result in defects such as antisites and metal interstitials. Another effect due to ion migration is the dealloying and the phase segregation, which is the separation of the constituent parts of a mixed quantity. Also, ions from the perovskite layers may migrate to charge transport layers which may negatively affect the transport properties of the solar cell (Bi et al. 2021). Another degrading effect of ion migration is hysteresis which is discussed in the next section.

Hysteresis

Hysteresis is the phenomenon in the PSCs where the forward and reverse current voltage curves do not overlap. This means that the curves change depending on whether it is forward scanning (voltage is increasing) or reverse scanning (voltage is decreasing). Hysteresis also depends on the scan rate interface of material, illumination condition, biasing history, device structure, reversible light and voltage, and bias pre-conditioning not just the scan direction (Khalid and Mallick 2023). A lot of studies have been to discuss the possible reasons due to which this hysteresis occurs. The possible reasons of hysteresis are capacitive effect, trapping of charges, ion migration, and ferroelectric polarization (Chen et al. 2016, 2018).

Another possible reason of hysteresis as discovered by various studies is the grain size of the material. Smaller grain size shows much more pronounced hysteresis (Kim and Park 2014). Hysteresis effect can cause the measurements of different values such as voltage and current to be calculated inaccurately. Not only this, but the calculation of PCE (power conversion efficiency) is also greatly impacted due to this reason. Because of this, it is important to eliminate the effect of hysteresis for the proper working of the PSCs (Tiep et al. 2016).

Techniques to Enhance the Stability of Perovskite Solar Cells

Compositional Engineering of the Perovskite Materials

$CH_3NH_3PbI_3$ as discussed before is a widely used perovskite material in the application of PSCs. But using this material comes with several drawbacks as seen in the previous sections. But studies are being continuously performed to improve the performance and limit the drawbacks of this material. One such method is compositional engineering (Wang et al. 2016).

An example of this type of engineering is using chloride or bromide as an additive. Using Br ions to replace I ions in the perovskite material to form the compound of $CH_3NH_3Pb(I_{1-x}Br_x)_3$ enhances the PCE and stability of the PSCs (Mahapatra et al. 2019). Another perovskite material is $CH(NH_2)_2PbI_3$ or $FAPbI_3$. The bandgap of $FAPbI_3$ is less when compared to $CH_3NH_3PbI_3$ making it easier to absorb wavelengths which are longer (Near IR range) (Koh et al. 2013). Although $FAPbI_3$ shows more stability than $CH_3NH_3PbI_3$, the PCE of $CH_3NH_3PbI_3$ is still more than $FAPbI_3$ (Noman et al. 2023). In $FAPbI_3$, inorganic cation Cs was used in the place of "A" in the perovskite structure (ABX_3) by Li et al. This would result in the structure of $FA_{1-x}Cs_xPbI_3$. It was shown in that study that the stability under moisture increased (Li et al. 2015). Another example was of using Rb cation at the location A of Perovskite structure. This increased the PCE of the solar cell and also made the solar cell more stable against the harmful effects of moisture (Park et al. 2017).

Another compositional engineering approach used to enhance the performance of PSCs involves combining two perovskite materials, namely $FAPbI_3$ and $MAPbBr_3$, to form a new compound $(FAPbI_3)_{1-x}(MAPbBr_3)_x$. This compound was shown to have increased PCE and improved stability especially for $x = 0.15$ (Jeon et al. 2015). The design of this compound was later optimized computationally and with different materials for transport layers and using different parameters to get efficiency as high as 25.86% (Bencherif et al. 2022). Further compositional changes have been made on the compound of $(FAPbI_3)_{1-x}(MAPbBr_3)_x$ by using Cs cations to improve its performance (Saliba et al. 2016).

So, it is seen that compositional engineering is indeed a viable strategy to improve the performance of the PSCs. There have been numerous examples and studies done that prove that just changing the composition of the perovskite material by employing different elements and compounds improves the working of the solar cell.

2D Layered Perovskite Structure

Recently, 2D perovskite materials have emerged as a promising option for use as passivation layers between the 3D perovskite material and the transport layers. It has been found that the 2D perovskite films offer more stability than their 3D counterpart (Cao et al. 2015). Although the efficiency in the case of 2D structures is low as compared to their 3D counterparts (Tsai et al. 2016). This lower efficiency is due to their higher exciton binding energies, higher bandgaps, and poor charge carrier dynamics (Metcalf et al. 2023).

Given their superior stability as compared to 3D perovskite materials and their ability to effectively reduce stability issues at grain boundaries, within the bulk material, and on the surface of 3D perovskites, 2D Perovskite materials may help improve the overall performance and longevity of perovskite-based devices (Mahapatra et al. 2019). Thus, using 2D materials with 3D materials is a viable strategy of improving the overall performance of the solar cell.

An example of this method of the 2D/3D composite material was when the 2D material was the AVAI (Amino valeric acid iodide) or $HOOC(CH_2)_4NH_3I$ and the 3D material was $CH_3NH_3PbI_3$. The interaction between the 2D and 3D perovskite layers led to a widening of the 3D perovskite bandgap in the interface region. This improved the stability of the device (Grancini et al. 2017). New methods for employing this method of using 2D/3D interface of Perovskite materials are being developed. With these developments, new materials that can be used to obtain a 2D layer on the 3D layer's surface are being discovered and tested upon. Also new methods are being experimented which make this process easier and cost-effective. For example, a technique involving dynamic spin coating of PEAI isopropanol (IPA) solution on a Perovskite film of $Cs_{0.1}FA_{0.74}MA_{0.13}PbI_{2.48}Br_{0.39}$ was used instead of the traditional methods which increased stability. Here PEAI is phenylethyl ammonium iodide (Cho et al. 2018). In another study, *n*-Butylamine (BA) was used instead of *n*-Butylammonium iodide (BAI) with $MAPbI_3$ to form 2D layers of $(BA)_2PbI_4$ and improved stability (Lin et al. 2018).

Nanoengineering

Nanostructures are materials or structures that have at least one dimension within the nanoscale range, typically between 1 and 100 nm. These structures exhibit unique physical, chemical, and biological properties that differ significantly from their bulk counterparts due to their small size, increased surface area, and quantum effects (Khan et al. 2017). For nanomaterials, the phenomenon of quantum confinement and carrier multiplication are both observed (Rocks et al. 2018; Nozik 2010). The nanostructured perovskite materials include nanosheets, nanorods, nanoparticles, and quantum dots (McDonald et al. 2019). The quantum confinement effect increases the bandgap of the material (Mello Donegá 2010). Although it might seem

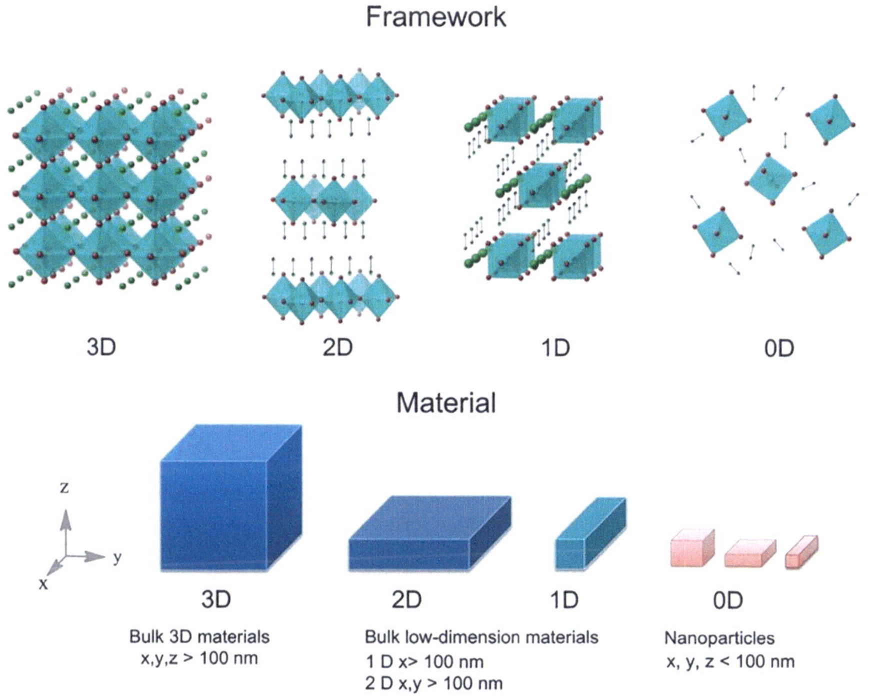

Fig. 8.3 Diagram showing representations of various dimensions of perovskite material (top) and size and scale level for different dimensions (bottom). (Reproduced with permission from González-Carrero et al. 2015 Copyright (2015), Wiley)

as though this is a negative effect for the PSCs, but using the material in a tandem solar cell allows for increased efficiency to balance the effect (Nozik 2010). But comparatively the advantages of the nanostructures far outweigh the disadvantages. In the case of Perovskite material, different dimensionality can change the framework and structure of the material depending upon the arrangement of MX_6 octahedra (González-Carrero et al. 2015). This is shown in Fig. 8.3 which illustrates various dimensional representations of perovskite materials (top) and the corresponding size and scale levels (bottom) (González-Carrero et al. 2015).

Starting with nanosheets or 2D nanostructures, the conjunction of 2D and 3D perovskite materials has been discussed in the previous section. It was noted that the 2D materials showed higher stability as compared to their 3D counterparts whereas the 3D materials showed higher PCE. Their interface provided a good middle ground for good performance and stability working of the solar cell. Other than the 2D/3D layered structure, 2D perovskites themselves alone have been used in the Solar Cells. An example of this is when selenophene-based spacer was developed for the 2D RP Perovskite Solar Cells showing good efficiency and thermal, moisture, and light stability (Fu et al. 2023). Another way 2D structures can be used is through the use of Transition Metal Dichalcogenides (TMDs) which can be used as ETL, HTL, or interfacial layer (Raj et al. 2024).

1D perovskite materials show better performance when compared to 3D Perovskite materials in surface defect passivation, self-healing, and moisture stability (Li et al. 2024a). 1D nanostructures include the structure of nanowires and nanorods. Perovskite nanowires have been developed with good charge collection properties while reducing defects at grain boundary showing a high stability even without encapsulation (Cha et al. 2022). In another case, using the spin coating method, $MAPbI_3$ nanowire PSCs was constructed. This showed improved hole injection from perovskite layer to hole transport layer with an efficiency of 14.71% (Im et al. 2015). Nanorods have been used for the electron transport layer in the form of TiO_2 with impressive performance (Park et al. 2023), and ZnO nanorods were also used as ETL for the $CsPbIBr_2$ PSCs to enhance its performance (Yang et al. 2023). In the case 0D perovskite materials, quantum dots show extreme potential in the field of PSCs. Quantum dots have been used for PSCs by using $MAPbBr_{3-x}I_x$ quantum dots as interface-regulating materials between the perovskite material and HTL to improve the performance (Cha et al. 2016). Perovskite quantum dots (PQDs) were also used on devices based on $FASnI_3$ and $MASnI_3$ showing improved PCE and stability by low hysteresis (Aftab et al. 2023).

As discussed, nanostructures offer significant advancements in the development of high-performance perovskite solar cells. Overall, the integration of nanostructures into perovskite solar cells represents a promising path toward achieving more efficient, stable, and scalable solar technologies.

Interfacial Modification

Grain boundaries and interfaces can be one of the reasons of degradation of the device instead of the absorbing layer (Wang et al. 2018). The characteristics of the interface and its properties can significantly impact the device's performance, as various processes, including recombination and ion migration, occur in these areas, along with the presence of grain defects (Guo et al. 2020). Perovskite films, often made using solution-based methods, are known to have many crystallographic defects on their surfaces. And in recent times with the current advancements in developing high-quality polycrystalline perovskite films, the defects within the material are minimal compared to those on the surface (Lu et al. 2020). In such cases, interfacial modification can prove to be extremely helpful.

The method of interfacial modification deals with improving the contact between the absorber layer and the transport layers. Interfacial modification helps in reducing recombination losses, decreasing defect density, improve the morphology of the layers, suppress ion migration, improve interfacial contact, and improve stability of the device (Ali et al. 2020). Some instances which involved the method of interfacial modification included using Sb_2S_3 between TiO_2 ETL and the perovskite layer showing a reduction in the deterioration rate of the solar cells (Ito et al. 2014).

Another instance was using TPFPB (Tris(pentafluorophenyl)boron) the SnO_2 ETL layer and the perovskite layer, thus improving the efficiency and the stability (Li et al. 2020). Different additives with functional groups were also studied which included amino, carbonyl, Thio, carboxyl, hydroxyl, and cyano groups with each of them providing different advantages and disadvantages (Wang et al. 2020). Also, alkali metal salts (NaBr, KBr, and RbBr) were used between SnO_2 ETL and the Perovskite layer to improve the performance of the device (Zhang et al. 2024). The case of 2D layered perovskite structure that was discussed in a previous section was also a type of interfacial modification. This is because the 2D perovskite layer acts as an interface barrier between the Perovskite layer and the Charge Transport Layer.

In short, the main issues that interfacial modification addresses are poor interfacial contact, defect states, and chemical reactions at these boundaries which can lead to efficiency losses and accelerated degradation under operational conditions.

Encapsulation

It has been already established that the PSCs are highly susceptible to environmental factors such as humidity (moisture and oxygen), heat, and UV light (Sandhu and Park 2024). For these conditions, encapsulation is a crucial technique for the PSCs that involves enclosing the cells with protective materials to shield them from harmful factors. Researchers and industry professionals have introduced a variety of innovative methods and materials aimed at improving encapsulation effectiveness in recent years (Wang et al. 2016).

Encapsulation can be classified into two categories, internal encapsulation (protect the device from processes such as interaction between layers) and external encapsulation (protect it from environmental hazards that exist in the outside word or the environment) (Chu et al. 2023). Internal encapsulation is an excellent strategy for reducing issues such as ion migration, defects, and vacancies that effect the performance of the cell (Xu et al. 2022). In these methods, a layer of additives is added between the charge transport layers and the perovskite layer or the electrodes. An example of internal encapsulation is interfacial modification that is discussed in the previous section. The choice of the additives depends upon the nature of the materials of the device and the situation. Examples of internal encapsulation include using polystyrene-*co*-polyacrylonitrile (SAN) copolymer with the perovskite layer in PSCs (Zhou et al. 2021). And another method is using polytetrafluoroethylene (PTFE) between the perovskite film and the electron transport layer in inverted PSCs (Li et al. 2024b). Coming to the external encapsulation, the protective layer of the encapsulation should provide both electrical insulation and mechanical protection, especially for outdoor applications where conditions can be harsh. This step is not only vital for maintaining the performance and lifespan of the solar cells but also represents a critical requirement for their successful commercialization and widespread use (Ahn and Choi 2023). Examples of external encapsulation to improve the stability of PSCs include using a customized thin-film encapsulation (TFE) method

(Lee et al. 2017), and also using a UV-curable epoxy resin combined with a desiccant sheet and a protective glass (Dong et al. 2016).

These findings highlight the potential of optimized encapsulation to enhance the durability of PSCs, making them more suitable for practical applications despite their sensitivity to environmental factors.

Methods to Investigate Stability of Perovskite Solar Cells

Experimental techniques while conducting studies on PSCs are conventionally used. They may include methods like X-Ray diffraction, scanning electron microscopy, atomic force microscopy, and photoluminescence experiments (Zhu et al. 2019; Valerio et al. 2019; Wu et al. 2022). Testing of the device also provides insights into the performance and stability of the devices in a real environment. Testing also involves exposing the PSCs to different stressors, such as light, heat, moisture, and mechanical stress. The goal is to assess how well the devices maintain their performance over time and identify the mechanisms underlying degradation (Emery et al. 2022; Jošt et al. 2020).

In addition to experimental techniques, computational methods have emerged as powerful tools for understanding and predicting the stability of PSCs. These methods use simulations and modelling to study the intrinsic and extrinsic factors affecting stability. There have been studies done computationally that help researchers discover better solar cells with improved performance (Raj et al. 2022b, 2023;). These methods complement experimental techniques by offering a deeper understanding of degradation mechanisms and guiding the development of more stable PSCs. The main focus of this discussion will be the computational methods.

The computational methods include a number of ways that researchers use to study about the working of the Solar Cells. These methods include DFT, Molecular Dynamics, Ab Initio Molecular Dynamics, etc. (Kheralla and Chetty 2021).

Benefits of Computational Studies

Computational studies offer a number of advantages for understanding and enhancing the stability of PSCs. Initially, computational studies enable the analysis of mechanisms at different length scales from atomic scale all the way up to the scale of the device and different time scales (Zhang et al. 2022). Techniques like density functional theory (DFT) and molecular dynamics (MD) allow researchers to study processes that take place at the atomic level (Ikeuba et al. 2024). These insights are difficult to achieve through experimental methods alone, particularly when dealing with complex, dynamic phenomena. Afterwards, computational studies offer the ability to simulate long-term stability under various environmental conditions, such as heat, light, and humidity, without the need for time-intensive ageing experiments.

Computational tools like Machine Learning can model how different materials and their performance evolve over time, providing a predictive understanding of degradation pathways (Wang et al. 2024). There is also an advantage of the ability to study extreme or inaccessible conditions, such as high temperatures or high pressure, which are challenging to replicate in a laboratory setting. Computational simulations provide a controlled environment to analyse how these conditions affect material stability and device performance, offering valuable data for designing more resilient materials (Colherinhas et al. 2011; Xiong et al. 2021). Computational approaches facilitate design optimization, such as modifying bandgaps, optimizing layer thicknesses, and using different interfacial materials to enhance both efficiency and stability. This allows the researchers to gain an increased degree of control over the parameters of the devices surpassing the limitations of experimental methods (Rehman et al. 2024; Njema et al. 2025).

Key Computational Methods

Below is the list of primary computational methods used for studying the stability of the PSCs.

Density Functional Theory (DFT)

Density functional theory (DFT) is one of the most widely used computational techniques for studying the stability of PSCs. As an ab initio method (derives from the fundamental physical principles without relying on empirical parameters), DFT allows researchers to calculate the electronic structure of materials by solving the quantum mechanical equations governing the behaviour of electrons. For the study that involves DFT, rather than using the wavefunctions for calculating the ground state energy, functional of electron density is used, hence the name DFT (Burke and Wagner 2012). The theory can be further refined by using different approximations for the energy term that includes the interaction between electrons, which is called exchange correlation energy. Two commonly used approximation are the local density approximation (LDA) which includes the effect of only local electron density and generalized gradient approximation (GGA) which included the spatial variation along with the electron's spin densities (Schleder et al. 2019). Each of these approximations further use different functionals, which differ in complexity and the computational power needed to calculate the energy. A functional uses the approximation to help in computing the value of energy with the help of electron density. The DFT calculations help in obtaining important information about the system or the material. One such information is the density of states. A plot of density of states of $CH_3NH_3PbI_3$ obtained by using DFT is given (Wang et al. 2013). Figure 8.4 shows plots made by results obtained with the help of DFT. Figure 8.4a shows the plots between density of states and their bandgap energy levels, whereas Fig. 8.4b shows

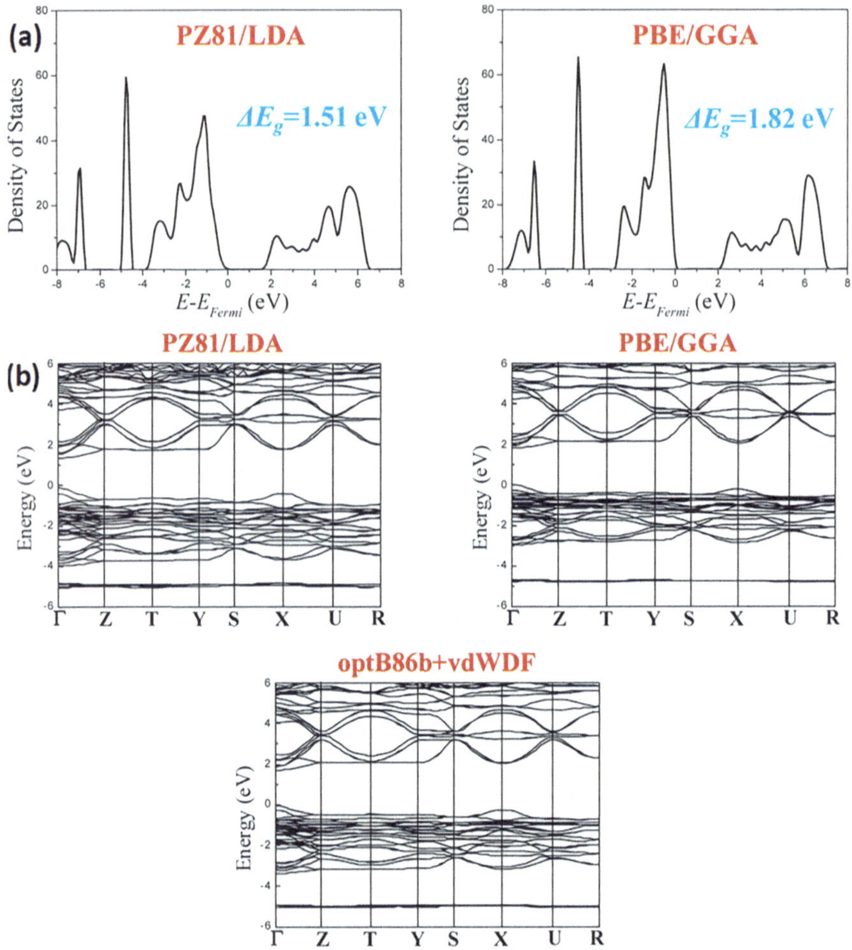

Fig. 8.4 Different diagrams made with the help of results gained from DFT. Panel (**a**) shows the graphs obtained between density of states and their bandgap energy using different functionals of orthorhombic $CH_3NH_3PbI_3$ crystals and Panel (**b**) shows calculated band structure along symmetry lines of first Brillouin zone using different functionals of orthorhombic $CH_3NH_3PbI_3$ crystals. (Reproduced with permission from Wang et al. 2013 Copyright (2014), Royal Society of Chemistry)

the plots between band structure along the symmetry lines of first Brillouin zone. This diagram gives us an example by showing two types of information we can gather by using a technique like DFT and proving how useful DFT can be (Wang et al. 2013).

This section gives some examples of how DFT has been used in the study of Perovskite material to observe the different phenomenon related to stability such as mechanical or thermodynamical stability of those materials. In regard to the perovskite material, Mosconi et al. investigated the structural and electronic properties of mixed $CH_3NH_3PbI_2X$ (where X = I, Br, Cl) using periodic DFT calculations (Periodic DFT involves periodic boundary conditions in atoms (Mazurek et al. 2020)). The $CH_3NH_3PbI_2Br$ perovskite had a larger bandgap than $CH_3NH_3PbI_3$. With the structural stability analysis, it was seen that $CH_3NH_3PbI_2Cl$ was less stable (formation energy being 0.16 eV) than $CH_3NH_3PbI_2Br$ (formation energy being 0.05 eV) and $CH_3NH_3PbI_3$. This explained why Br integrates more easily into the perovskite lattice than Cl. The study highlighted the role of hydrogen bonding between ammonium groups and halides (Mosconi et al. 2013). Pingak et al. used DFT for the investigation of $TlSnX_3$ (X = Cl, Br, I). The results showed that these perovskites are structurally and mechanically stable, with $TlSnBr_3$ and $TlSnI_3$ also being dynamically stable. The formation energy of these perovskite materials was calculated for information the chemical stability, bulk modulus for their mechanical stability, and the study of phonon dispersion for the dynamic stability. This study used GGA-PBE functional (Pingak et al. 2023). In another study, Yadav et al. conducted DFT calculations using GGA-PBE functional to see the effect of structural strain on the performance of PSCs and its properties. The strain of a device influences its stability. In the study, it was seen that compressive strain (the strain that is produced inside the perovskite crystal) reduced electronic bandgap, increased efficiency whereas the tensile strain increased the bandgap which led to the decrease in performance (Yadav et al. 2024). Pindolia et al. performed a study using DFT to investigate the structural, mechanical, optical, and electronic properties of $KSnI_3$ for use as an absorber material in PSCS. The thermodynamic stability was assessed through tolerance factor, octahedral factor, and formation energy, while elastic constants and mechanical properties such as Young's modulus and Poisson's ratio confirmed its mechanical stability and ductility which are important for thin-film fabrication. In this study, primarily from the stability point of view, thermodynamic and mechanical stability of the material was observed (Pindolia et al. 2023). Not only for the perovskite layer, but DFT and its results have also been used to find out more suitable materials to be used in the place of charge transport layers, electrodes, and adsorbents. Jouybar et al. used DFT on $ATiO_3$ perovskites (A = Li, Na, K, Rb, Cs) for this purpose itself where different calculated parameters such as magnetic moment, bandgap, and refractive index were used as an indicator for using different materials for different purposes (Jouybar et al. 2024).

These studies are examples to show that DFT can be used to measure the performance and stability of the PSCs. This is by calculating various parameters which in turn influence the performance the device. Other than this, DFT can also help in providing insights into how defects and ion migration can affect the material's performance and longevity (Guo et al. 2024; Hu et al. 2023). DFT, although being extremely useful, is still not perfect. It has some drawbacks that can make the results seem inaccurate. The functionals and their solutions are approximations and may not always be accurate in all cases. These approximations can also lead to errors which were not account for before (Burke 2012; Cohen et al. 2012).

Molecular Dynamics (MD)

Molecular dynamics (MD) simulations model the movement of atoms using classical physics, involving force computation and solving Newton's equations. Most computational effort goes into calculations of force, which use molecular mechanics force fields to approximate potential energy based on bonded (bond length, bond angle, dihedral angle) and nonbonded (electrostatic, van der Waals) interactions (Chow et al. 2012). MD simulations are used to study the thermal and structural stability of PSCs. This is the difference between MD simulations and ab initio methods. Ab initio methods are typically used for static calculations but MD simulations take into account the movement of atoms. MD simulations are extremely helpful in modelling the atom with respect to their thermodynamic properties. Atoms can be assigned velocities based on the Boltzmann distribution, and extrinsic thermodynamic variables like temperature, pressure, volume, and particle number can be controlled (Chroneos et al. 2013). MD simulations have been primarily used for biomolecular research, but in recent years (Hollingsworth and Dror 2018) this method has been used by researchers in studying the performance of PSCs as well.

Gutierrez-Sevillano et al. used MD as a computational method and the effect of moisture on the perovskite material was seen. This study involved MD simulations in the formation of the perovskite layer in pentane and water. Their observations led them to the knowledge that the degradation due to water was not due to its bipolar nature but because of additional chemical reactions that happen between the perovskite material and water (Gutierrez-Sevillano et al. 2015). Figure 8.5 shows an example of how the simulations in MD may look like. In the diagram along with the balls that represent the precursor, the stick-model represents the molecules of solvent. This figure shows that MD simulations may provide way to study how different species react with each other. For example, how they may aggregate in one case and stay dispersed in another case as seen in case of pentane and water, respectively, in the figure itself (Gutierrez-Sevillano et al. 2015).

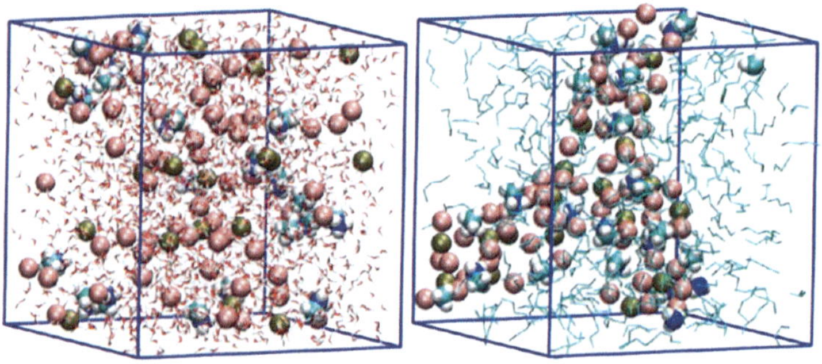

Fig. 8.5 Images of MD simulations of $CH_3NH_3PbI_3$ precursors in water (left) and in pentane (right). Pink spheres represent I, Tan spheres represent Pb, Cyan spheres represent C, Blue spheres represent N, and white spheres represent H. (Reproduced with permission from Gutierrez-Sevillano et al. 2015 Copyright (2015), Royal Society of Chemistry)

Pols et al. used MD simulations to study the perovskite material which was $CsPbI_3$. Reactive force field was used for the simulation. The study shed light on the effect of temperature on the phase stability of the perovskite material. Their findings showed that at lower temperatures the Cs cations shift away from their optimal positions, which may possibly lead to phase instability and conversion of perovskite phase to a non-perovskite phase. This study also showed that iodine vacancies contributed to degradation of the perovskite material by promoting Frenkel defect formation and causing the perovskite to break down into lead iodide (PbI_2). To improve stability, the study suggested using larger organic cations or smaller anions and minimizing iodine vacancies through iodine-rich synthesis or stabilizing additives (Pols et al. 2021). In another study by Caddeo et al. the degradation of perovskite material due to moisture was studied. By calibrating a simple force field for water and $CH_3NH_3PbI_3$, researchers studied how $CH_3NH_3PbI_3$ interacted with water in both vapour and liquid forms. This study demonstrated that moisture-induced degradation of the perovskite material can be controlled using an energy barrier. They found that at room temperature, water vapour absorption was energetically favourable and allowed water molecules to diffuse without harming the crystal structure. However, in liquid water, $CH_3NH_3PbI_3$ dissolved irreversibly due to a thermally activated reaction (Caddeo et al. 2017). Phung et al. used molecular dynamics along with experimental data of photoluminescence spectroscopy to understand the ionic defect migration primarily due to grain boundaries. Since ionic defects are known to be a cause of the degradation of Perovskite Solar Cells, the results of this study provide insight into long-term stability issues (Phung et al. 2020).

MD simulations are also an extremely useful tool for studying large-scale and long-term processes and for studying hybrid perovskites (Mattoni et al. 2016). Molecular dynamics simulations have proven to be a powerful tool for studying the stability of PSCs. These simulations have provided key insights into the effects of moisture, temperature, and structural defects on perovskite materials. By enabling a deeper understanding of degradation mechanisms, MD helps guide material engineering strategies to enhance the stability and performance of perovskite-based devices. Some drawbacks of MD simulations include the risk of wrong simulations if the force field used is inaccurate. This method uses high computational power and thus, for large systems, this can become a bottleneck (Prašnikar et al. 2024).

Ab Initio Molecular Dynamics (AIMD)

Ab initio molecular dynamics (AIMD) combines the principles of ab initio calculations with molecular dynamics to simulate the behaviour of materials under dynamic conditions. AIMD uses quantum Schrodinger equation and real physical potentials for simulations instead of Newton's equations and semiempirical potentials as for molecular dynamics (Paquet and Viktor 2018). In a way, this method is a hybrid between the previous two mentioned methods (DFT: which is an ab initio method and MD simulations) since it uses ab initio calculations as a basis for MD simulations. AIMD simulations have been used by researchers in studying the structural and electronic properties of different perovskite materials (Quarti et al. 2015).

In a study by Busipalli et al., DFT and AIMD simulations were performed on caesium lead iodide ($CsPbI_3$) Perovskite material to find out more about the stability of the PSCs in the presence of moisture. This study revealed that $CsPbI_3$ has better surface moisture stability than conventionally used $CH_3NH_3PbI_3$ for the PSCs (Busipalli et al. 2020). In research performed by Garrote-Márquez et al., the role of hydrogen bonds (HBs) in hybrid organic/inorganic halide perovskites, specifically $MAPbBr_3$, $FAPbI_3$, and their solid solution was studied. The findings suggested that HBs influenced the rotational dynamics of organic cations and, consequently, the structural and electronic properties of perovskites. It was predicted in this study HBs are expected to play a more prominent role near phase transitions at different temperatures (Garrote-Márquez et al. 2023). As seen in the previous sections 2D layering of perovskite materials can lead to an increased stability of the solar cells. So, naturally, computational studies can be performed over the performance of the solar cells that involve this method, for example, Mahal et al. used AIMD simulations to investigate the stability of 2D hybrid perovskites at the water interface, identifying potential degradation pathways. This study helped in increasing the understanding of two major classes of 2D perovskites, namely Ruddlesden–Popper (RP) phases and Dion–Jacobson (DJ) phases (Mahal et al. 2024). Kaiser et al. performed a study involving AIMD simulations to understand and compare the reaction of Tin-halide perovskite ($MASnI_3$ and $DMASnBr_3$) and lead-halide perovskite ($MAPbI_3$) with water. The study showed that $MASnI_3$ had lower stability against water as compared to $MAPbI_3$ even though, bulk $MAPbI_3$ absorbed more water. But this absorption of water can lead to degradation at a later stage. This study also showed that $DMASnBr_3$ exhibits impressive stability against water as it forms a protective amorphous layer of $SnBr_3$ complexes at its surface (Kaiser et al. 2022).

AIMD simulations is an excellent method that makes it easier for researchers to observe the material in different conditions of temperature and pressure (Quarti et al. 2015). By simulating the motion of atoms and ions at the atomic level, AIMD simulations help researchers understand how structural and chemical changes can impact the overall stability of the device. AIMD simulations serve as a powerful tool for understanding the fundamental processes governing perovskite solar cell performance and degradation.

Machine Learning (ML) and Data-Driven Approaches

Machine learning (ML) is a branch of artificial intelligence (AI) that enables computers to learn patterns from data and make decisions or predictions without being explicitly programmed. Instead of following predefined rules, ML models improve their performance over time by analysing and adapting to new data (Parikh et al. 2021). ML and data-driven approaches are emerging tools in the study of perovskite solar cell stability. These methods use large datasets from experiments and simulations to identify patterns and predict material properties, including stability-related behaviours. ML algorithms can be trained to predict how different material compositions, device architectures, or environmental conditions will affect the stability of PSCs (Mammeri et al. 2025). These approaches can also help in material screening, allowing researchers to identify promising candidates for stable perovskite formulations more efficiently (Li et al. 2019). By predicting stability trends and helping discover new compositions for perovskite materials, ML can significantly accelerate the development of more robust and durable materials for solar cells.

Mammeri et al. observed that probabilistic neural networks (give results with probability distribution) outperformed classic neural networks (give definite results). The study emphasizes the impact of data uncertainty and noise on model performance, highlighting the need for improved noise-handling techniques in predictive applications (Mammeri et al. 2025). Alsulami et al. used ML technique to analyse a comprehensive dataset from 2013 to 2020, focusing on how different materials affect the stability of PSCs. The analysis was done in two parts: first, by splitting the dataset based on cell architecture, and second, using a specific ML algorithm to identify which materials most positively and negatively influence stability. The findings showed that the p–i–n architecture was the most effective for achieving higher stability. The study suggested that optimizing materials for the transport layers of Perovskite Solar cells is crucial for improving their stability (Alsulami et al. 2024). Odabasi et al. also performed a study of the stability of PSCs using ML. This study also analysed the long-term stability of PSCs using data from past research. The impact of environmental conditions was examined through descriptive plots and association rule mining (finding associations between different data points). A decision tree was also developed to identify key patterns for fabricating stable cells. The results showed the stability benefits of 2D/3D perovskites and encapsulation. The results also showed that environmental factors such as high humidity, temperature, and illumination were detrimental to the device (Odabaşı and Yıldırım 2019). Mammeri et al. conducted a study using ML and discovered the most optimum materials for each layer of the solar cell. This study shows how versatile and useful ML can be, by even providing the flexibility of choosing individual layers of the device. It was also shown that for long-term stability, 2D-3D perovskite layers were a great option that always showed more stability (Mammeri et al. 2022).

So, it can be understood that machine learning and data-driven approaches are indeed an excellent method to find out trends and predict different characteristics and properties of PSCs. It is also important to understand that this method is not without flaws. It is very important to be careful in choosing the correct data for the training of programme. If the data that is fed into the system is flawed, then the output of the system will be full of errors, or it may be biased. Also, there is an issue of lack of explanation of the decision-making process (Barbierato and Gatti 2024).

Using SCAPS-1D Software

SCAPS (Solar Cell Capacitance Simulator) is a 1D solar cell simulation programme which is an extremely useful software to study the working of solar cells for various purposes. It is designed by the department of electronics and information systems at the University of Gent, Belgium (Burgelman et al. 2000). This software works on the principle of solving semiconductor equations like Poisson equation, continuity equations for holes and electrons and transport equation. This software uses Poisson's equation to determine the internal potential distribution inside the cell, continuity equations to measure the change in concentration of charge carriers, and transport equation provides insight in the movement of charge carriers within the cell (Sk et al. 2024; Mehmood et al. 2023).

Abadi et al. used SCAPS-1D software to analyse the impact of different parameters of ETL and HTL for the PSCs with the absorber layer of $(FAPbI_3)_{0.85}(MAPbBr_3)_{0.15}$. This study was conducted keeping stability in mind along with other factors. The choice of the absorber material was due to its high stability in real-life conditions. Different factors including bandgap and electron affinity of the transport layers were studied (Seyyed Abadi et al. 2019). In another study by Hachimi et al. SCAPS-1D was used to simulate $CsPbCl_{3-y}I_y$ ($y = 0$, 1, 2, and 3)-based solar cells. Different parameters like thickness and defect density of the perovskite layer, work function of the back contact, types of resistances, and temperature were investigated. In this study, it was seen that although the PCE of the solar cell based on $CsPbI_3$ was the highest ($y = 3$) as compared to $CsPbI_2Cl$, $CsPbCl_2I$, and $CsPbCl_3$, there was the limitation of thermal instability of $CsPbI_3$. This low stability could be improved by replacing iodine with chlorine and maintaining a balance (Hachimi et al. 2023). Devi et al. conducted a study using SCAPS-1D to examine the performance of $FAPbI_3$-based solar cells. Different parameters like thickness of absorber layer, ETL and HTL, defect density, type of resistance, and their impact were examined. This study conducted a numerical simulation using SCAPS-1D to evaluate the performance of FAPbI3-based PSCs with and without a defect passivation layer, CBz-PAI. The result of this study indicated WS_2 as ETL and CuI as HTL which proved to be important in enhancing the efficiency and stability of the device (Devi et al. 2024).

There are some limitations to SCAPS-1D as well. The input parameters such as defect densities, radiative recombination coefficient, and doping concentration should be realistic. There are also optical limitations which include improper accounting of scattering, interference, or reflection losses. Because of these issues, it is possible that the results may be better in theory but not in practice (Saidarsan et al. 2025).

Conclusion

The chapter addressed the topic of stability concerning the development of PSCs. The different ways by which PSCs can be degraded were discussed in detail. These included their reactions and their mechanisms. Several environmental factors were identified as contributing causes, including moisture, oxygen, and ultraviolet light. Additionally, the phenomenon of ion migration was examined, providing further understanding of the solar cell's operation. Alongside these factors, strategies for addressing them were also explored. These ways included compositional engineering, nanoengineering, encapsulation, interfacial modification and 2d/3d interface. After the discussion on the various approaches to enhancing the stability of PSCs, attention was turned to the methods for measuring and calculating stability. This segment placed particular emphasis on computational techniques, highlighting their numerous advantages. Additionally, the global applications of these computational methods by researchers were also explored.

References

Aftab S, Li X, Hussain S, Aslam M, Rajpar AH, Al-Sehemi AG (2023) Nanoscale enhancements in perovskite-based photovoltaics. Chem Eng J 480:148143. https://doi.org/10.1016/j.cej.2023.148143

Ahn N, Choi M (2023) Towards long-term stable perovskite solar cells: degradation mechanisms and stabilization techniques. Adv Sci 11(4):e2306110. https://doi.org/10.1002/advs.202306110

Ali J et al (2020) Interfacial and structural modifications in perovskite solar cells. Nanoscale 12(10):5719–5745. https://doi.org/10.1039/c9nr10788f

Alsulami BNN et al (2024) Application of large datasets to assess trends in the stability of perovskite photovoltaics through machine learning. J Mater Chem A 12(5):3122–3132. https://doi.org/10.1039/d3ta05966a

Aristidou N et al (2015) The role of oxygen in the degradation of methylammonium lead trihalide perovskite photoactive layers. Angew Chem Int Ed 54(28):8208–8212. https://doi.org/10.1002/anie.201503153

Azad SS, Keshavarzi R, Mirkhani V, Moghadam M, Tangestaninejad S, Mohammadpoor-Baltork I (2024) Stability enhancement of perovskite solar cells using multifunctional inorganic materials with UV protective, self cleaning, and high wear resistance properties. Sci Rep 14(1):6466. https://doi.org/10.1038/s41598-024-57133-8

Azzopardi B, Mutale J (2009) Life cycle analysis for future photovoltaic systems using hybrid solar cells. Renew Sust Energ Rev 14(3):1130–1134. https://doi.org/10.1016/j.rser.2009.10.016

Barbierato E, Gatti A (2024) The challenges of machine learning: a critical review. Electronics 13(2):416. https://doi.org/10.3390/electronics13020416

Bencherif H et al (2022) Performance enhancement of (FAPbI3)1-x(MAPbBr3)x perovskite solar cell with an optimized design. Micro Nanostruct 171:207403. https://doi.org/10.1016/j.micrna.2022.207403

Bi E, Song Z, Li C, Wu Z, Yan Y (2021) Mitigating ion migration in perovskite solar cells. Trends Chem 3(7):575–588. https://doi.org/10.1016/j.trechm.2021.04.004

Burgelman M, Nollet P, Degrave S (2000) Modelling polycrystalline semiconductor solar cells. Thin Solid Films 361–362:527–532. https://doi.org/10.1016/S0040-6090(99)00825-1

Burke K (2012) Perspective on density functional theory. J Chem Phys 136:150901. https://doi.org/10.48550/ARXIV.1201.3679

Burke K, Wagner LO (2012) DFT in a nutshell. Int J Quantum Chem 113(2):96–101. https://doi.org/10.1002/qua.24259

Busipalli DL, Lin K, Nachimuthu S, Jiang J (2020) Enhanced moisture stability of cesium lead iodide perovskite solar cells—a first-principles molecular dynamics study. Phys Chem Chem Phys 22(10):5693–5701. https://doi.org/10.1039/c9cp06341b

Caddeo C, Saba MI, Meloni S, Filippetti A, Mattoni A (2017) Collective molecular mechanisms in the CH3NH3PbI3 dissolution by liquid water. ACS Nano 11(9):9183–9190. https://doi.org/10.1021/acsnano.7b04116

Cao DH, Stoumpos CC, Farha OK, Hupp JT, Kanatzidis MG (2015) 2D homologous perovskites as light-absorbing materials for solar cell applications. J Am Chem Soc 137(24):7843–7850. https://doi.org/10.1021/jacs.5b03796

Cha M et al (2016) Enhancing perovskite solar cell performance by interface engineering using CH3NH3PbBr0.9I2.1 quantum dots. J Am Chem Soc 138(27):8581–8587. https://doi.org/10.1021/jacs.6b04519

Cha J et al (2022) Perovskite nanowires as defect passivators and charge transport networks for efficient and stable perovskite solar cells. Chem Eng J 451:138920. https://doi.org/10.1016/j.cej.2022.138920

Chen B, Yang M, Priya S, Zhu K (2016) Origin of J–V hysteresis in perovskite solar cells. J Phys Chem Lett 7(5):905–917. https://doi.org/10.1021/acs.jpclett.6b00215

Chen W et al (2018) Eliminating J-V hysteresis in perovskite solar cells via defect controlling. Org Electron 58:283–289. https://doi.org/10.1016/j.orgel.2018.04.017

Cho KT et al (2018) Selective growth of layered perovskites for stable and efficient photovoltaics. Energy Environ Sci 11(4):952–959. https://doi.org/10.1039/c7ee03513f

Chow E, Klepeis J, Rendleman C, Dror R, Shaw D (2012) 9.6 New technologies for molecular dynamics simulations. In: Comprehensive biophysics. Elsevier, Amsterdam, pp 86–104. https://doi.org/10.1016/b978-0-12-374920-8.00908-5

Christians JA, Herrera PAM, Kamat PV (2015) Transformation of the excited state and photovoltaic efficiency of CH3NH3PbI3 perovskite upon controlled exposure to humidified air. J Am Chem Soc 137(4):1530–1538. https://doi.org/10.1021/ja511132a

Chroneos A, Rushton M, Jiang C, Tsoukalas L (2013) Nuclear wasteform materials: atomistic simulation case studies. J Nucl Mater 441(1–3):29–39. https://doi.org/10.1016/j.jnucmat.2013.05.012

Chu Q et al (2023) Encapsulation: the path to commercialization of stable perovskite solar cells. Matter 6(11):3838–3863. https://doi.org/10.1016/j.matt.2023.08.016

Cohen AJ, Mori-Sánchez P, Yang W (2012) Challenges for density functional theory. Chem Rev 112(1):289–320. https://doi.org/10.1021/cr200107z

Colherinhas G, Fonseca TL, Fileti EE (2011) Theoretical analysis of the hydration of C60 in normal and supercritical conditions. Carbon 49(1):187–192. https://doi.org/10.1016/j.carbon.2010.09.002

Devi M, Kumar M, Singh DV (2024) Impact on green energy conversion and stability of PSC device with the insertion of bifunctional molecule as an interfacial layer. J Phys Chem Solids 194:112247. https://doi.org/10.1016/j.jpcs.2024.112247

Dong Q et al (2016) Encapsulation of perovskite solar cells for high humidity conditions. ChemSusChem 9(18):2597–2603. https://doi.org/10.1002/cssc.201600868

Eames C, Frost JM, Barnes PRF, O'Regan BC, Walsh A, Islam MS (2015) Ionic transport in hybrid lead iodide perovskite solar cells. Nat Commun 6(1):7497. https://doi.org/10.1038/ncomms8497

Emery Q et al (2022) Encapsulation and outdoor testing of perovskite solar cells: comparing industrially relevant process with a simplified lab procedure. ACS Appl Mater Interfaces 14(4):5159–5167. https://doi.org/10.1021/acsami.1c14720

Fatima Q, Haidry AA, Zhang H, Jery AE, Aldrdery M (2024) A critical review on advancement and challenges in using TiO2 as electron transport layer for perovskite solar cell. Mater Today Sustain 27:100857. https://doi.org/10.1016/j.mtsust.2024.100857

Frost JM, Butler KT, Brivio F, Hendon CH, Schilfgaarde M, Walsh A (2014) Atomistic origins of high-performance in hybrid halide perovskite solar cells. Nano Lett 14(5):2584–2590. https://doi.org/10.1021/nl500390f

Fu Q, Chen M, Li Q, Liu H, Wang R, Liu Y (2023) Selenophene-based 2D Ruddlesden-Popper perovskite solar cells with an efficiency exceeding 19%. J Am Chem Soc 145(39):21687–21695. https://doi.org/10.1021/jacs.3c08604

Garrote-Márquez A, Lodeiro L, Suresh R, Hernández NC, Grau-Crespo R, Menéndez-Proupin E (2023) Hydrogen bonds in lead halide perovskites: insights from ab initio molecular dynamics. J Phys Chem C 127(32):15901–15910. https://doi.org/10.1021/acs.jpcc.3c02376

González-Carrero S, Galian RE, Pérez-Prieto J (2015) Organometal halide perovskites: bulk low-dimension materials and nanoparticles. Part Part Syst Charact 32(7):709–720. https://doi.org/10.1002/ppsc.201400214

Grancini G et al (2017) One-year stable perovskite solar cells by 2D/3D interface engineering. Nat Commun 8(1):15684. https://doi.org/10.1038/ncomms15684

Guo M, Shen D, Li Y, Akram MA, Wei M (2020) TiO2 mesocrystals processed at low temperature as the electron-transport material in perovskite solar cells. ChemSusChem 13(19):5256–5263. https://doi.org/10.1002/cssc.202001486

Guo Z, Yuan M, Chen G, Liu F, Lu R, Yin W (2024) Understanding defects in perovskite solar cells through computation: current knowledge and future challenge. Adv Sci 11(20):e2305799. https://doi.org/10.1002/advs.202305799

Gutierrez-Sevillano JJ, Ahmad S, Calero S, Anta JA (2015) Molecular dynamics simulations of organohalide perovskite precursors: solvent effects in the formation of perovskite solar cells. Phys Chem Chem Phys 17(35):22770–22777. https://doi.org/10.1039/c5cp03220b

Habisreutinger SN, McMeekin DP, Snaith HJ, Nicholas RJ (2016) Research update: strategies for improving the stability of perovskite solar cells. APL Mater 4(9). https://doi.org/10.1063/1.4961210

Hachimi MA, Tarbi A, El Mrabet M, Erguig H, Chtouki T (2023) Performance and stability optimization of CsPbCl3-yIy (y = 0, 1, 2, and 3) lead-based perovskites solar cells using SCAPS-1D. J Phys Chem Solids 183:111651. https://doi.org/10.1016/j.jpcs.2023.111651

Hollingsworth SA, Dror RO (2018) Molecular dynamics simulation for all. Neuron 99(6):1129–1143. https://doi.org/10.1016/j.neuron.2018.08.011

Hu Y, Jin N, Sun L, Lin Z, Liu Y (2023) Theoretical study of Li-ion migration in perovskitetype AVO3 (A = Ca, La, Ce, and La0.75Ca0.25) with DFT+U methods. Mater Today Commun 35:106029. https://doi.org/10.1016/j.mtcomm.2023.106029

Ikeuba AI et al (2024) Review of computational methods used in the evaluation corrosion inhibition of metallic materials. Discov Chem Eng 4(1):28. https://doi.org/10.1007/s43938-024-00065-6

Im J et al (2015) Nanowire perovskite solar cell. Nano Lett 15(3):2120–2126. https://doi.org/10.1021/acs.nanolett.5b00046

Ito S, Tanaka S, Manabe K, Nishino H (2014) Effects of surface blocking layer of SB2S3 on nanocrystalline TIO2 for CH3NH3PBI3 perovskite solar cells. J Phys Chem C 118(30):16995–17000. https://doi.org/10.1021/jp500449z

Jeon NJ et al (2015) Compositional engineering of perovskite materials for high-performance solar cells. Nature 517(7535):476–480. https://doi.org/10.1038/nature14133

Jošt M et al (2020) Perovskite solar cells go outdoors: field testing and temperature effects on energy yield. Adv Energy Mater 10(25):2000454. https://doi.org/10.1002/aenm.202000454

Jouybar S, Naji L, Tafreshi SS, Leeuw NH (2024) A density functional theory study of the physicochemical properties of alkali metal titanate perovskites for solar cell applications. Molecules 29(14):3355. https://doi.org/10.3390/molecules29143355

Kaiser W, Ricciarelli D, Mosconi E, Alothman AA, Ambrosio F, Angelis F (2022) Stability of tin-versus lead-halide perovskites: ab initio molecular dynamics simulations of perovskite/water interfaces. J Phys Chem Lett 13(10):2321–2329. https://doi.org/10.1021/acs.jpclett.2c00273

Khalid M, Mallick TK (2023) Stability and performance enhancement of perovskite solar cells: a review. Energies 16(10):4031. https://doi.org/10.3390/en16104031

Khan I, Saeed K, Khan I (2017) Nanoparticles: properties, applications and toxicities. Arab J Chem 12(7):908–931. https://doi.org/10.1016/j.arabjc.2017.05.011

Kheralla A, Chetty N (2021) A review of experimental and computational attempts to remedy stability issues of perovskite solar cells. Heliyon 7(2):e06211. https://doi.org/10.1016/j.heliyon.2021.e06211

Kim H, Park N (2014) Parameters affecting I–V hysteresis of CH3NH3PBI3 perovskite solar cells: effects of perovskite crystal size and mesoporous TiO2 layer. J Phys Chem Lett 5(17):2927–2934. https://doi.org/10.1021/jz501392m

Kim H, Seo J, Park N (2016) Material and device stability in perovskite solar cells. ChemSusChem 9(18):2528–2540. https://doi.org/10.1002/cssc.201600915

Koh TM et al (2013) Formamidinium-containing metal-halide: an alternative material for near-IR absorption perovskite solar cells. J Phys Chem C 118(30):16458–16462. https://doi.org/10.1021/jp411112k

Lee YI et al (2017) A low-temperature thin-film encapsulation for enhanced stability of a highly efficient perovskite solar cell. Adv Energy Mater 8(9):1701928. https://doi.org/10.1002/aenm.201701928

Li Z, Yang M, Park J, Wei S, Berry JJ, Zhu K (2015) Stabilizing perovskite structures by tuning tolerance factor: formation of formamidinium and cesium lead iodide solid-state alloys. Chem Mater 28(1):284–292. https://doi.org/10.1021/acs.chemmater.5b04107

Li J, Pradhan B, Gaur S, Thomas J (2019) Predictions and strategies learned from machine learning to develop high-performing perovskite solar cells. Adv Energy Mater 9(46):1901891. https://doi.org/10.1002/aenm.201901891

Li H et al (2020) Interface modification for enhanced efficiency and stability perovskite solar cells. J Phys Chem C 124(24):12948–12955. https://doi.org/10.1021/acs.jpcc.0c02628

Li X et al (2024a) Dimensional diversity (0D, 1D, 2D, and 3D) in perovskite solar cells: exploring the potential of mixed-dimensional integrations. J Mater Chem A 12(8):4421–4440. https://doi.org/10.1039/d3ta06953b

Li W et al (2024b) Internal encapsulation enables efficient and stable perovskite solar cells. Adv Funct Mater 35(4):2414004. https://doi.org/10.1002/adfm.202414004

Lin Y et al (2018) Enhanced thermal stability in perovskite solar cells by assembling 2D/3D stacking structures. J Phys Chem Lett 9(3):654–658. https://doi.org/10.1021/acs.jpclett.7b02679

Lu H, Krishna A, Zakeeruddin SM, Grätzel M, Hagfeldt A (2020) Compositional and interface engineering of organic-inorganic lead halide perovskite solar cells. iScience 23(8):101359. https://doi.org/10.1016/j.isci.2020.101359

Mahal E, Manna SS, Das S, Pathak B (2024) Understanding moisture stability and degradation mechanisms in 2D hybrid perovskites: insights from ab initio molecular dynamics simulations. Energy Adv 3(8):1992–2001. https://doi.org/10.1039/d4ya00235k

Mahapatra A, Prochowicz D, Tavakoli MM, Trivedi S, Kumar P, Yadav P (2019) A review of aspects of additive engineering in perovskite solar cells. J Mater Chem A 8(1):27–54. https://doi.org/10.1039/c9ta07657c

Mammeri M, Dehimi L, Bencherif H, Pezzimenti F (2022) Paths towards high perovskite solar cells stability using machine learning techniques. Sol Energy 249:651–660. https://doi.org/10.1016/j.solener.2022.12.002

Mammeri M et al (2025) Stability forecasting of perovskite solar cells utilizing various machine learning and deep learning techniques. J Opt 54:930–938. https://doi.org/10.1007/s12596-024-01819-9

Mattoni A, Filippetti A, Caddeo C (2016) Modeling hybrid perovskites by molecular dynamics. J Phys Condens Matter 29(4):043001. https://doi.org/10.1088/1361-648x/29/4/043001

Mazurek AH, Szeleszczuk Ł, Pisklak DM (2020) Periodic DFT calculations—review of applications in the pharmaceutical sciences. Pharmaceutics 12(5):415. https://doi.org/10.3390/pharmaceutics12050415

McDonald C et al (2019) Nanostructured perovskite solar cells. Nanomaterials 9(10):1481. https://doi.org/10.3390/nano9101481

Mehmood S, Xia Y, Qu F, He M (2023) Investigating the performance of efficient and stable planer perovskite solar cell with an effective inorganic carrier transport layer using SCAPS-1D simulation. Energies 16(21):7438. https://doi.org/10.3390/en16217438

Mello Donegá C (2010) Synthesis and properties of colloidal heteronanocrystals. Chem Soc Rev 40(3):1512–1546. https://doi.org/10.1039/c0cs00055h

Metcalf I et al (2023) Synergy of 3D and 2D perovskites for durable, efficient solar cells and beyond. Chem Rev 123(15):9565–9652. https://doi.org/10.1021/acs.chemrev.3c00214

Mosconi E, Amat A, Nazeeruddin MK, Grätzel M, Angelis F (2013) First-principles modeling of mixed halide organometal perovskites for photovoltaic applications. J Phys Chem C 117(27):13902–13913. https://doi.org/10.1021/jp4048659

Niu G, Guo X, Wang L (2014) Review of recent progress in chemical stability of perovskite solar cells. J Mater Chem A 3(17):8970–8980. https://doi.org/10.1039/c4ta04994b

Njema GG, Kibet JK, Ngari SM (2025) Performance optimization of a novel perovskite solar cell with power conversion efficiency exceeding 37% based on methylammonium tin iodide. Next Energy 6:100182. https://doi.org/10.1016/j.nxener.2024.100182

Noman M, Shahzaib M, Jan ST, Shah SN, Khan AD (2023) 26.48% efficient and stable FAPbI3 perovskite solar cells employing SrCu2O2 as hole transport layer. RSC Adv 13(3):1892–1905. https://doi.org/10.1039/d2ra06535e

Nozik AJ (2010) Nanoscience and nanostructures for photovoltaics and solar fuels. Nano Lett 10(8):2735–2741. https://doi.org/10.1021/nl102122x

Odabaşı Ç, Yıldırım R (2019) Machine learning analysis on stability of perovskite solar cells. Sol Energy Mater Sol Cells 205:110284. https://doi.org/10.1016/j.solmat.2019.110284

Paquet E, Viktor HL (2018) Computational methods for ab initio molecular dynamics. Adv Chem 2018:1–14. https://doi.org/10.1155/2018/9839641

Parikh N et al (2021) Is machine learning redefining the perovskite solar cells? J Energy Chem 66:74–90. https://doi.org/10.1016/j.jechem.2021.07.020

Park YH et al (2017) Inorganic rubidium cation as an enhancer for photovoltaic performance and moisture stability of HC(NH2)2PBI3 perovskite solar cells. Adv Funct Mater 27(16):1605988. https://doi.org/10.1002/adfm.201605988

Park YJ et al (2023) Selective control of novel TiO2 nanorods: excellent building blocks for the electron transport layer of mesoscopic perovskite solar cells. ACS Appl Mater Interfaces 15(7):9447–9456. https://doi.org/10.1021/acsami.2c21731

Philippe B et al (2015) Chemical and electronic structure characterization of lead halide perovskites and stability behavior under different exposures—a photoelectron spectroscopy investigation. Chem Mater 27(5):1720–1731. https://doi.org/10.1021/acs.chemmater.5b00348

Phung N et al (2020) The role of grain boundaries on ionic defect migration in metal halide perovskites. Adv Energy Mater 10(20):1903735. https://doi.org/10.1002/aenm.201903735

Pindolia G, Shinde SM, Jha PK (2023) Non-leaded, KSnI3 based perovskite solar cell: a DFT study along with SCAPS simulation. Mater Chem Phys 297:127426. https://doi.org/10.1016/j.matchemphys.2023.127426

Pingak RK et al (2023) A DFT investigation of lead-free TlSnX3 (X = Cl, Br, or I) perovskites for potential applications in solar cells and thermoelectric devices. RSC Adv 13(48):33875–33886. https://doi.org/10.1039/d3ra06685a

Pols M, Vicent-Luna JM, Filot I, Duin ACT, Tao S (2021) Atomistic insights into the degradation of inorganic halide perovskite CSPBI3: a reactive force field molecular dynamics study. J Phys Chem Lett 12(23):5519–5525. https://doi.org/10.1021/acs.jpclett.1c01192

Prašnikar E, Ljubič M, Perdih A, Borišek J (2024) Machine learning heralding a new development phase in molecular dynamics simulations. Artif Intell Rev 57(4):102. https://doi.org/10.1007/s10462-024-10731-4

Quarti C, Mosconi E, Angelis F (2015) Structural and electronic properties of organo-halide hybrid perovskites from ab initio molecular dynamics. Phys Chem Chem Phys 17(14):9394–9409. https://doi.org/10.1039/c5cp00599j

Raj A, Kumar M, Anshul A (2022a) Recent progress in cesium-based lead-free halide double perovskite materials for photovoltaic applications. Phys Status Solidi A 219(22):2200425. https://doi.org/10.1002/pssa.202200425

Raj A et al (2022b) Comparative analysis of 'La' modified BiFeO3-based perovskite solar cell devices for high conversion efficiency. Ceram Int 49(1):1317–1327. https://doi.org/10.1016/j.ceramint.2022.09.112

Raj A et al (2023) Photovoltaic energy conversion in multiferroic perovskite absorber-based devices via experiment and theoretical calculations. Phys B Condens Matter 673:415504. https://doi.org/10.1016/j.physb.2023.415504

Raj A, Kumar M, Singh DV, Anshul A (2024) A critical review on transition metal dichalcogenides (TMDs): an efficiency booster for perovskite solar cells. FlatChem 44:100629. https://doi.org/10.1016/j.flatc.2024.100629

Rao MK, Sangeetha D, Selvakumar M, Sudhakar Y, Mahesha M (2021) Review on persistent challenges of perovskite solar cells' stability. Sol Energy 218:469–491. https://doi.org/10.1016/j.solener.2021.03.005

Rehman AU et al (2024) Performance optimization of FASnI3 based perovskite solar cell through SCAPS-1D simulation. Hybrid Adv 7:100301. https://doi.org/10.1016/j.hybadv.2024.100301

Rocks C, Svrcek V, Velusamy T, Macias-Montero M, Maguire P, Mariotti D (2018) Type-I alignment in MAPbI3 based solar devices with doped-silicon nanocrystals. Nano Energy 50:245–255. https://doi.org/10.1016/j.nanoen.2018.05.036

Saidarsan A, Guruprasad S, Malik A, Basumatary P, Ghosh DS (2025) A critical review of unrealistic results in SCAPS-1D simulations: causes, practical solutions and roadmap ahead. Sol Energy Mater Sol Cells 279:113230. https://doi.org/10.1016/j.solmat.2024.113230

Sakhatskyi K et al (2022) Assessing the drawbacks and benefits of ion migration in lead halide perovskites. ACS Energy Lett 7(10):3401–3414. https://doi.org/10.1021/acsenergylett.2c01663

Saliba M et al (2016) Cesium-containing triple cation perovskite solar cells: improved stability, reproducibility and high efficiency. Energy Environ Sci 9(6):1989–1997. https://doi.org/10.1039/c5ee03874j

Sandhu S, Park N (2024) Methodologies to improve the stability of high-efficiency perovskite solar cells. Acc Mater Res 5(12):1544–1557. https://doi.org/10.1021/accountsmr.4c00237

Schleder GR, Padilha ACM, Acosta CM, Costa M, Fazzio A (2019) From DFT to machine learning: recent approaches to materials science—a review. J Phys Mater 2(3):032001. https://doi.org/10.1088/2515-7639/ab084b

Seyyed Abadi NM, Banihashemi M, Kashaninia A (2019) Simulation and analysis of a perovskite solar cell with $(FAPbI_3)_{0.85}(MAPbBr_3)_{0.15}$ as absorber layer. In: 2019 27th Iranian conference on electrical engineering (ICEE), Yazd, Iran. IEEE, New York, pp 173–177. https://doi.org/10.1109/IranianCEE.2019.8786587

Sk M, Islam MT, Gourav (2024) Exploring the structural, electronic, optical, transport, and photovoltaic properties of Rb2LiGa(Br/I)6 using DFT and SCAPS-1D simulations. Sci Rep 14(1):24813. https://doi.org/10.1038/s41598-024-76593-6

Tiep NH, Ku Z, Fan HJ (2016) Recent advances in improving the stability of perovskite solar cells. Adv Energy Mater 6(3):1501420. https://doi.org/10.1002/aenm.201501420

Tsai H et al (2016) High-efficiency two-dimensional Ruddlesden–Popper perovskite solar cells. Nature 536(7616):312–316. https://doi.org/10.1038/nature18306

Valerio L et al (2019) Characterization and analysis of FAxCs(1−x) Pb(IyBr(1−y))3 perovskite solar cells with thickness controlled transport layers for performance optimization. AIP Adv 9(10):105006. https://doi.org/10.1063/1.5123400

Walsh A, Scanlon DO, Chen S, Gong XG, Wei S (2014) Self-regulation mechanism for charged point defects in hybrid halide perovskites. Angew Chem Int Ed 54(6):1791–1794. https://doi.org/10.1002/anie.201409740

Wang Y et al (2013) Density functional theory analysis of structural and electronic properties of orthorhombic perovskite CH3NH3PbI3. Phys Chem Chem Phys 16(4):1424–1429. https://doi.org/10.1039/c3cp54479f

Wang D, Wright M, Elumalai NK, Uddin A (2016) Stability of perovskite solar cells. Sol Energy Mater Sol Cells 147:255–275. https://doi.org/10.1016/j.solmat.2015.12.025

Wang H et al (2017) Adverse effects of excess residual PBI2 on photovoltaic performance, charge separation, and trap-state properties in mesoporous structured perovskite solar cells. Chem Eur J 23(16):3986–3992. https://doi.org/10.1002/chem.201605668

Wang F, Bai S, Tress W, Hagfeldt A, Gao F (2018) Defects engineering for high-performance perovskite solar cells. NPJ Flex Electron 2(1):22. https://doi.org/10.1038/s41528-018-0035-z

Wang Y et al (2020) Interfacial modification towards highly efficient and stable perovskite solar cells. Nanoscale 12(36):18563–18575. https://doi.org/10.1039/d0nr05136e

Wang J, Karimi S, Zeinalzad P, Zhang J, Gong Z (2024) Using machine learning and experimental study to correlate and predict accelerated aging with natural aging of GFRP composites in hygrothermal conditions. Constr Build Mater 438:137264. https://doi.org/10.1016/j.conbuildmat.2024.137264

Wu R-J, Wu K-T, Nian G-H, Chou W-Y, Cheng H-L (2022) Preparation and characterization of low-dimensional MAPbI3 perovskite nanowires with enhanced photoluminescence and photoresponsive properties by incorporating PEAI. J Phys Chem Solids 170:110931. https://doi.org/10.1016/j.jpcs.2022.110931

Xiong Q-L, Shimada T, Kitamura T, Li Z (2021) Shock response and defect evolution of copper single crystals at room and elevated temperatures. Model Simul Mater Sci Eng 29(4):045006. https://doi.org/10.1088/1361-651X/abea68

Xu D et al (2022) An internal encapsulating layer for efficient, stable, repairable and low-lead-leakage perovskite solar cells. Energy Environ Sci 15(9):3891–3900. https://doi.org/10.1039/d2ee01016j

Yadav AP, Bhojani AK, Bhargava K, Singh DK (2024) Impact of structural strain on performance of halide perovskite solar cells: a combined DFT and SCAPS-1D analysis. Adv Theory Simul 7(11):2400407. https://doi.org/10.1002/adts.202400407

Yang J et al (2023) Enhanced performance of CsPbIBr2 perovskite solar cell by modified zinc oxide nanorods array with [6,6]-phenyl C61 butyric acid. Chem Eur J 29(40):e202300566. https://doi.org/10.1002/chem.202300566

Zhang T, Hu C, Yang S (2019) Ion migration: a 'double-edged sword' for halide-perovskite-based electronic devices. Small Methods 4(5):1900552. https://doi.org/10.1002/smtd.201900552

Zhang X, Zheng P, Ma Y, Jiang Y, Li H (2022) Atomic-scale understanding of oxidation mechanisms of materials by computational approaches: a review. Mater Des 217:110605. https://doi.org/10.1016/j.matdes.2022.110605

Zhang W, Song Y, Zhang H, La A, Lu Y (2024) High efficiency and stability of perovskite solar cells prepared by alkali metal interfacial modification. Opt Express 32(10):17132. https://doi.org/10.1364/oe.522663

Zhou Y et al (2021) Improving thermal stability of perovskite solar cells by suppressing ion migration using copolymer grain encapsulation. Chem Mater 33(15):6120–6135. https://doi.org/10.1021/acs.chemmater.1c01675

Zhu C et al (2019) Strain engineering in perovskite solar cells and its impacts on carrier dynamics. Nat Commun 10(1):815. https://doi.org/10.1038/s41467-019-08507-4

Chapter 9
Computational Investigations of the Impact of 2D Materials on Perovskite Solar Cells

Akshay Kumar, Abhishek Raj, Garima Jain, Bhupendra Singh, and Manish Kumar

Abstract Perovskite solar cells are emerging as a transformative technology in photovoltaic community owing to their rapid growth in efficiency and low production cost. However, their commercial application is hindered due to lack of sustained stability primarily caused by interfacial defects and vulnerable metal electrodes. The incorporation of two-dimensional (2D) materials in perovskite solar cells have gained significant interest in recent times because of their excellent properties like high charge mobility, high conductivity, tunable bandgap, good optical transparency and their ability to form van der Waals heterojunction at the bottom interface. The role of 2D materials in perovskite solar cells as electron transport layer (ETL), hole transport layer (HTL), interfacial layer, and contact back electrodes has been discussed thoroughly in this chapter. The comparison of photovoltaic performance before and after incorporating 2D materials clearly indicates that 2D materials not only helps in increasing the efficiency but enhances the sustained stability also because of reduced nonradiative recombination, reduced defects and improved energy level alignment. In the end, some challenges in this field of study are recognized and future prospects are discussed.

Keywords Interfacial modifications · 2D materials · TMDs · Recombination · Energy alignment

---

A. Kumar · G. Jain (✉)
Department of Physics, DAV(PG) College Muzaffarnagar, Maa Shakumbhari University, Saharanpur, Uttar Pradesh, India
e-mail: garima@davcollegemzn.org

A. Raj
CSIR-National Environmental Engineering Research Institute (NEERI), Nagpur, India

Academy of Scientific and Innovative Research (AcSIR), Ghaziabad, India

B. Singh · M. Kumar (✉)
Experimental Research Laboratory, Department of Physics, ARSD College, University of Delhi, New Delhi, India
e-mail: mkumar2@arsd.du.ac.in

M. Kumar (ed.), *Computational Analysis of Perovskite Solar Cells*, Springer Tracts in Electrical and Electronics Engineering,
https://doi.org/10.1007/978-981-95-6521-4_9

Introduction

The global energy demand has increased at a rapid rate from last decade due to industrial development and population growth. Burning of fossil fuels like oils and coals is still the main source of energy supply whose effect on environmental degradation in the form of climate change, greenhouse effect is no longer hidden. As per a report from the International Energy Agency (IEA), by 2019, an astonishing 81% of the global primary energy supply came from fossil fuels (IEA 2019). The energy demand will just increase more and more with time so the shift from nonrenewable to renewable energy sources is the need of hour. Different types of renewable energy include solar, wind, hydroelectric, geothermal, and biomass will play a crucial role in this transition of energy from nonrenewable to renewable sources. These sources are not merely technical options, they signify a major shift in energy production and consumption, highlighting a deeper dedication to sustainability, resilience, and a more harmonious relationship with nature. The focus of various governments has shifted in implementing this transition which can be seen clearly in the target set in different conventions. To fulfil the 2030 agenda for sustainable development and meet the Paris agreement targets, renewable energy must account for at least 63% of the global energy system by 2040, ensuring a sustainable and low-carbon future (BloombergNEF 2024).

Solar energy, one of the most abundant and accessible form of renewable energy, has emerged as a cornerstone of the transition to a cleaner energy source. A solar cell is a device which directly transforms sunlight into electricity through the photovoltaic effect. Solar cells are a key component of solar panels, which can be deployed on residential rooftops, large-scale solar farms and even integrated into portable devices. At present silicon solar cells holds over 90% of solar market because of their high efficiency (25–26%) and module's stability which can last up to 20 years (Green et al. 2023). But the high installation cost (up to 75%) acts as a barrier in the wide use of PV technology (Myagmarsereejid et al. 2021). Among the other types of solar cells, perovskite solar cells (PSCs) lead the pack to replace conventional Si solar cells. PSCs have light absorbing material similar to the mineral perovskite ($CaTiO_3$). The general formula for perovskite material used in PSCs is ABX_3, where A is cation (e.g., methylammonium, formamidinium, cesium), B is metal cation (lead or tin), X is the halide anion. This unique structure endows perovskites with exceptional optoelectronic properties, like high absorption coefficient, tunable bandgap and high charge carrier diffusion length. An exponential growth in efficiency from 3.8 in 2010 to 26% (Kojima et al. 2009; National Laboratory of the Rockies n.d.) has been observed. The simple manufacturing process makes estimated cost to be half of Si solar cells (Meng et al. 2018a; Raj et al. 2023). However, there are some concerns which needs to be addressed like, thermal stability, photostability, moisture stability, large area fabrication and lead toxicity (Berhe et al. 2016). The lead-free halide double perovskite materials with the general formula $ABB'X_6$ have emerged as an alternative to traditional lead based perovskite (Raj et al. 2022). A typical PSCs is composed of several layers such as metal contact, electron transport layer (ETL), perovskite absorber and hole transport layer (HTL). The incompatibility among different materials possibly due to different crystal lattice size, energy levels mismatch, chemical reactivity, thermal expansion coefficients, can lead to more nonradiative recombination, degradation of perovskite and adjacent layers, high defect density and hysteresis in current-voltage curves. These effects lead to reduced efficiency and instability (Nazir et al. 2022). The longevity of PSCs is also limited by conventional metal electrodes, which easily degrades due to ion migration from perovskite layer (O'Sullivan et al. 2023). So in order to make more efficient and stable PSCs, improved interface quality and development of more stable electrodes are required.

From recent studies, it is observed that different categories of 2D materials, including graphene and its derivatives, transition metal dichalcogenides (TMDs), MXenes and black phosphorous (BP) are useful in enhancing the efficiency and stability of PSCs. 2D materials are highly suitable for application as ETL, HTL, interfacial layer and electrode material in PSCs (Xia et al. 2023; Raj et al. 2024). The defects at the interface are minimized as 2D materials forms van der Waals heterojunction through physical absorption with other materials (Liu et al. 2019). This interaction results in epitaxial growth of perovskite thin films, enhancing their crystallization and orientation (Bellani et al. 2021). Additionally, 2D materials assist in energy-level alignment at the perovskite/charge transport layer interface, ensuring efficient charge transport. As electrodes, 2D materials offer a less costly alternative to traditional metal electrodes and eliminate issues like metal ion diffusion into the perovskite layer, which can degrade device performance over time (Meng et al. 2018b).

For instance, WS_2 nanoflakes act as a lubricant between ETL and perovskite releasing interfacial strain which helps in achieving ultrahigh open-circuit voltage (V_{OC}) of 1.70 V and significantly improved stability (Zhou et al. 2020). Huang et al. have shown that an efficiency of 21.73% can be achieved by using TiS_2 as an ETL (Huang et al. 2019). Heo et al. achieved a PCE of 18.9% using TFSA-doped graphene electrode (Heo et al. 2018).

In this chapter the role of several 2D materials as different components of perovskite solar cell has been discussed thoroughly and their effect on the efficiency and stability is highlighted. The primary objective of the discussion is to explore various computational methods that have been reported to improve the efficiency of PSCs.

2D Materials

Two-dimensional (2D) materials are a class of materials characterized by their ultrathin layered structure, where the thickness is typically on atomic molecular scale. The properties shown by 2D materials are very different from their bulk counterpart (Miró et al. 2014). Since 2004, 2D materials have garnered significant attention from the research community, following Geim and Novoselov's successful isolation of single-layer graphene through micromechanical cleavage of graphite (Novoselov et al. 2004). These materials are characterized by strong covalent bonds within the plane and weak van der Waals forces between planes (Shen et al. 2024; Cai et al. 2018). The weak out-of-plane interaction makes it easier for exfoliation of 2D layers. Because of their unique physical and chemical properties like tunable electrical conductivity, tunable energy bands, mechanical strength, highly tailored optoelectronic properties, and solution processability, 2D materials are found to be useful in photovoltaic devices (Liu et al. 2020; Qin et al. 2021). Some of the 2D materials which have found a use in PSCs are discussed further in this chapter.

Graphene and Its Derivatives

Graphene is one of the most widely utilized 2D materials due to its high mobility, easily adjustable carrier concentration, and straightforward isolation. Typically, graphene is produced on a large scale using the chemical vapor deposition (CVD) method (Li et al. 2009). Graphene is a single layer of carbon atoms arranged in a hexagonal honeycomb structure with SP^2 hybridization (Xu et al. 2013). Graphene exhibits exceptional properties, including a thermal conductivity of 5×10^3 W m^{-1} K^{-1} at room temperature (Balandin et al. 2008), carrier mobility reaching up to $\sim 2 \times 10^5$ cm^2 V^{-1} S^{-1} (Bolotin et al. 2008), optical transparency of 97.7% across infrared to UV spectrum (Nair et al. 2008) impressive surface to mass ratio of approximately 2630 m^2 g^{-1} (Stoller et al. 2008), high Young modulus of ~1 TPa (Lee et al. 2008), and good mechanical flexibility (Gwon et al. 2011). The engineering of graphene's electronic structure has advanced quickly, leading to the development of semiconducting modifications with an exceptionally small bandgap, thereby opening up new possibilities for photovoltaic and electronic applications (Miró et al. 2014). Recently, researchers discovered that the properties of graphene and its derivatives, such as the bandgap, conductivity, and dispersibility, can be accurately adjusted by controlling the functional groups and the number of layers (Kim et al. 2014; Morozov et al. 2008). Graphene oxide (GO), a key derivative of graphene, can be obtained through oxidation using Hummer's method, which involves treating graphene with strong acids and oxidizing agents (Hummers and Offeman 1958). This process introduces various functional groups such as epoxide, carbonyl, and hydroxyl groups onto the surface and edges of the graphene, making it hydrophilic. As a result, GO disperses well in aqueous solutions and can be produced in large quantities with high yield, facilitating the creation of uniform, large-area films, which is beneficial for commercial applications. Unlike, graphene, GO has poor electrical conductivity. However, it can be chemically reduced to form reduced graphene oxide (rGO), which exhibits enhanced conductivity and transparency (Loh et al. 2010; Eda et al. 2009). Although the reduction process does not fully restore the graphene structure and leaves some defects (Gomez-Navarro et al. 2010).

Besides GO and rGO, other graphene derivatives, such as graphene quantum dots (GQDs) and graphene nanoribbons, have also garnered significant attention and are being applied across various research fields.

Transition Metal Dichalcogenides

Transition metal dichalcogenides (TMDs) are a class of layered 2D materials composed of transition metals (like tungsten, molybdenum, or titanium) and chalcogen elements (such as sulfur, selenium, or tellurium). TMDs typically have the general chemical formula MX_2, where "M" represents a transition metal and "X" represents a chalcogen (Choi et al. 2017). The structure is composed of metal atom layers sandwiched between two chalcogen atom layers. These layers are bound by weak van der Waals forces, enabling them to be exfoliated into thin, monolayer sheets. When reduced to a monolayer (a single atomic layer thick), TMDs exhibit many exciting properties that differ significantly from their bulk counterparts. Unlike their bulk counterparts, 2D TMDs have a direct band gap, which is beneficial for layer-dependent electronics. These materials display a range of properties, from semiconducting to superconducting, while also offering excellent chemical and mechanical stability (Fu et al. 2021). These characteristics make TMDs promising candidates for photovoltaic applications.

MXenes

MXenes are a family of two-dimensional (2D) transition metal carbides, nitrides, or carbonitrides that have garnered significant attention in recent years due to their unique properties and versatile applications (VahidMohammadi et al. 2021). MXenes have the general formula $M_{n-1}X_nT_x$, where: M is a transition metal (such as titanium, tantalum, or molybdenum), X represents carbon or nitrogen, T represents surface terminations (like –OH, –O, or –F), which are typically introduced during the synthesis process (Saeed et al. 2022). MXenes are derived from the selective etching of aluminum (in the case of carbides) or other group A elements belonging to group 13 or 14 from the corresponding $M_{n+1}AX_n$ (MAX) phases (Palei et al. 2023). MXenes possess excellent electrical conductivity, due to the surface terminations (like –OH and –O), they tend to be hydrophilic and exhibit high mechanical strength and flexibility (Qamar et al. 2022). The properties of MXenes can be tuned by varying the transition metal, the surface terminations, or the synthesis conditions, allowing for a wide range of applications. Some of the widely used MXenes in PSCs are $Ti_3C_2T_X$, Ti_2CT_X, Mo_2CT_X, V_2CT_X, and Nb_2CT_X (Yang et al. 2023).

Black Phosphorous

Black phosphorus (BP) is a unique allotrope of phosphorus that has attracted considerable interest in recent years because of its exceptional properties and promising applications in electronics, optoelectronics, and energy storage. It is the most stable form of phosphorus at room temperature and pressure, and it has a distinctive layered structure, which makes it a fascinating 2D material, similar to graphene and transition metal dichalcogenides (TMDs). In its bulk form, black phosphorus consists of layers of phosphorus atoms arranged in stacked layers, bonded by weak van der Waals interactions, which facilitates straightforward exfoliation into thin layers or even monolayers (Wu et al. 2021). BP has a remarkable feature of tunable bandgap which can vary with thickness from 2 to 0.3 eV. Furthermore, bandgap of a 10 nm-thick black phosphorus can be tuned from ~300 to below 50 meV using a moderate displacement field up to 1.1 V nm^{-1} (Deng et al. 2017). The carrier mobility of BP is very high allowing the fast charge transport (Zhang et al. 2015). Furthermore, BP exhibits outstanding mechanical properties which includes high tensile strength and ductility, positioning it as an emerging material for flexible electronics and stretchable devices (Zhu et al. 2020).

Impact of 2D Materials as ETL

Electron transport layer (ETL) is a key layer in device structure of PSCs. It is a thin layer of material placed between the perovskite absorber layer and the electrode. The main function of the ETL is to efficiently transport electrons from the perovskite layer to the electron-collecting electrode while also preventing the recombination of charge carriers (electrons and holes). The design and selection of the ETL are essential for optimizing the efficiency, stability, and overall performance of the PSC. An ideal ETL should have high electron mobility, suitable band alignment with perovskite material and electrode, high transparency, low defect density, chemical stability, and mechanical flexibility. In n-i-p configuration of PSC, metal oxides such as titanium oxide (TiO_2), tin oxide (SnO_2), and zinc oxide are typically used as ETL while Fullerene and PCBM are typically used in inverted p-i-n structure of PSC (Rai et al. 2020). TiO_2 typically requires high-temperature processing (around 400–500 °C) to achieve the desired crystallinity and electrical properties which limits the potential for large scale and flexible fabrication of PSCs. SnO_2 is also a popular choice as ETL because of wide bandgap, high carrier mobility, good chemical stability, and requirement of low annealing temperature (<180 °C). However, SnO_2 often contains oxygen vacancies and other defects which can act as trap states that lead to limited PV performance and hysteresis. Also the conduction band of SnO_2 is much lower than perovskite which leads to voltage loss. 2D materials have demonstrated the ability to improve device performance, either by doping or by substituting traditional ETLs (Alzaid 2021). The effect of incorporation of various 2D materials like graphene and TMDs in ETL has been discussed below.

Gagandeep et al. conducted a detailed computational study on the integration of graphene as both the ETL and HTL in lead-free tin-based perovskite solar cells (TPSCs). Using AFORS-HET simulation software, they designed a device structure of ITO/n-graphene/$CH_3NH_3SnI_3$/p-Graphene/Ag as shown in Fig. 9.1a, where nitrogen-doped graphene (n-Graphene) served as the ETL and boron-doped graphene (p-Graphene) served as the HTL. The effectiveness of graphene as both ETL and HTL lies in its tunable work function. The nitrogen doping of graphene adjusts its work function to approximately 4.4 eV, aligning well with the conduction band minimum (CBM) of $CH_3NH_3SnI_3$, thereby enabling efficient electron extraction at the ETL interface. Conversely, boron-doped graphene modifies the work function to about 5.1 eV, aligning closely with the valence band maximum (VBM) of $CH_3NH_3SnI_3$, which facilitates efficient hole extraction at the HTL interface. This precise energy level alignment as shown in Fig. 9.1b minimizes energy barriers for charge transfer and suppresses interfacial recombination, significantly enhancing device performance. The simulation results demonstrated that this dual graphene-layered configuration achieved a PCE of 10.67%, with a J_{SC} of 19.93 mA/cm^2, V_{OC} of 687.1 mV, and the FF of 77.92%. The *J-V* curve and quantum efficiency curve for optimized parameters are shown in Fig. 9.1c, d, respectively. Additionally, incorporating a textured perovskite layer to enhance light trapping further boosted the PCE to 13.28%. These improvements are attributed to the superior charge carrier mobility and excellent interfacial contact provided by the graphene layers (Gagandeep et al. 2019).

Numerous studies have demonstrated that molybdenum disulfide (MoS_2) can also serve effectively as an ETL. MoS_2 is composed of a molybdenum (Mo) atom

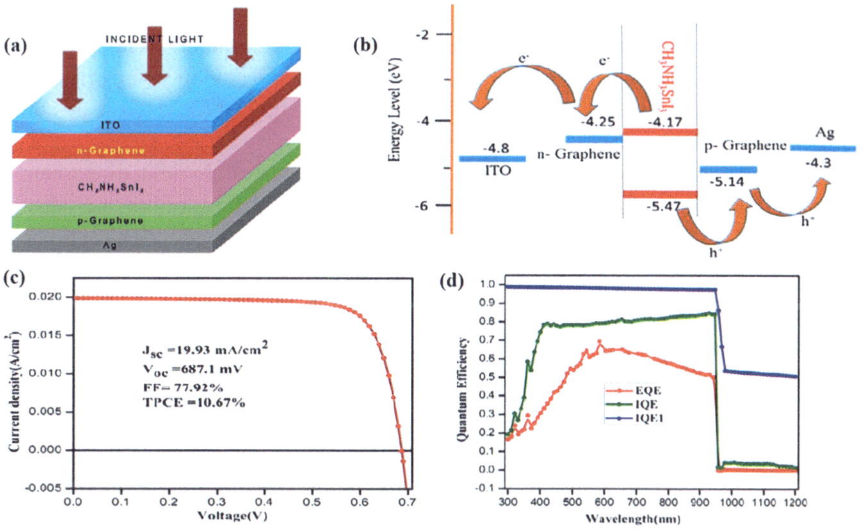

Fig. 9.1 (**a**) Simulated device structure and (**b**) energy level diagram of PSC with n-graphene, p-graphene as ETL and HTL, respectively, (**c**) *J-V* curve, (**d**) quantum efficiency vs. wavelength for optimized parameters. (Reproduced with permission from Gagandeep et al. 2019)

positioned between two sulfur (S) atoms, creating a layered structure. Bulk MoS_2 has an indirect bandgap of 1.23 eV while monolayer MoS_2 has a direct band gap with energy gap up to 1.90 eV (Rao et al. 2014). The use of thin layer MoS_2 as an ETL was first employed by Singh et al. (2019). They synthesized MoS_2 thin film over fluorine-doped tin oxide (FTO)-coated glass plate by microwave irradiation method. The suitable work function, good transmission in visible region, and high electrical conductivity make MoS_2 a good ETL candidate. Jani et al. demonstrated MoS_2 as a highly effective ETL in enhancing the performance of formamidinium lead triiodide ($FAPbI_3$)-based PSCs, outperforming the conventionally used titanium dioxide (TiO_2) in several key aspects. They did computational simulations using SCAPS-1D, which revealed that MoS_2, with its superior electron mobility of 100 cm^2 V^{-1} S^{-1} compared to TiO_2's 0.05 cm^2 V^{-1} S^{-1} leads to improved charge extraction and reduced recombination losses, resulting in enhanced performance metrics. The MoS_2-based cell (cell 2) achieved a significant efficiency improvement of 16.8%, with a higher J_{SC} of 22.5 mA/cm^2 and V_{OC} of 1.02 V, compared to the TiO_2-based cell (cell 1), which recorded only 8% efficiency. These outcomes are linked to the capability of MoS_2 to facilitate better electron diffusion and suppress reverse recombination at the $MoS_2/FAPbI_3$ interface, as depicted in the recombination rate-depth (RR-x) profiles. The quantum efficiency-wavelength (QE-λ) analysis also highlights MoS_2's superiority, especially in the visible spectrum, although minor absorption losses in the 730–830 nm range are attributed to the narrower bandgap of MoS_2. Furthermore, when absorber defect densities increased from 10^{15} to 10^{17} cm^{-3}, the MoS_2-based cell exhibited a significantly lower performance degradation of only 28% in efficiency, compared to a drastic 77% reduction for the TiO_2-based cell. This resilience is attributed to efficient electron extraction capabilities of MoS_2 that mitigate the impact of high defect densities in the absorber layer. Additionally, MoS_2 enhances the reproducibility of PSCs under variable carrier mobility conditions, as evidenced by its lower standard deviation in efficiency (0.07%) compared to TiO_2-based cells (1.21%), making it ideal for large-scale commercialization (Jani and Bhargava 2022).

Tungsten disulfide (WS_2) is another important TMD material with a direct bandgap of ~2 eV. Sobayel et al. computationally investigated the use of tungsten disulfide (WS_2) as the ETL in PSCs through numerical simulations using SCAPS-1D. Their results highlight that WS_2 with its high electron mobility and favorable band alignment offers a viable alternative to traditional TiO_2. WS_2 exhibits high electron mobility, good conductivity (10^{-3} Ω^{-1} cm^{-1}), and an ideal n-type semiconducting nature, which facilitate efficient electron transport and reduced recombination losses. Moreover, WS_2 is stable, nontoxic, and environmentally friendly, making it a sustainable choice for PSCs. The band alignment of WS_2 with the perovskite absorber layer ensures minimal interfacial energy loss, contributing to higher V_{oc} and improved charge extraction. The optimized PSC configuration (FTO/WS_2/$CH_3NH_3PbI_{3-x}Cl_x$/Spiro-OMeTAD/Ni) achieved a PCE of 25.70%, with V_{OC} of 1.056 V, J_{SC} of 25.48 mA/cm^2, and a fill factor (FF) of 88.54%. However, the performance is highly dependent on defect densities, with amphoteric defects in the perovskite layer ($CH_3NH_3PbI_3$) significantly reducing efficiency when exceeding

10^{15} cm^{-3}. Interface defects at the WS_2/perovskite and perovskite/HTL layers also impact performance, with defect densities above 10^{17} and 10^{16} cm^{-3}, respectively, leading to decrease in efficiency. Apart from this, the study highlights that PSCs with WS_2 ETL perform best at temperatures between 10 and 40 °C, with efficiency declining at higher temperatures. These findings suggest that WS_2 is a promising ETL material for PSCs, particularly in tropical climates, and encourage further research into transition metal dichalcogenides for solar cell applications (Sobayel et al. 2019).

Impact of 2D Materials as HTL

Hole transport layer (HTL) plays an important role in the photovoltaic performance and stability of a PSCs. The main function of a HTL is to collect and transport the holes generated in perovskite layer to electrodes hence minimizing the recombination in perovskite layer. An effective HTL serves as a shield, safeguarding the perovskite layer from environmental elements like moisture and oxygen, which helps improve the device's operational stability. An ideal HTL should possess high hole mobility and conductivity to facilitate efficient charge transport. The valence band maximum (VBM) of HTL should be slightly higher than that of the perovskite layer to allow seamless hole transfer while conduction band minimum should be sufficiently higher than that of perovskite to block electron conduction. The HTL should also exhibit a low density of hole traps to reduce recombination losses. Uniform and complete surface coverage is crucial to ensure effective charge extraction and prevent leakage currents (Ameen et al. 2015). Organic material 2,2′,7,7′-tetrakis (*N*,*N*-di-methoxyphenyl-amine)-9,9′-spirobifluorene (spiro-OMeTAD) is commonly used HTL in PSCs as it fulfils almost all requirements of HTL. However, it requires chemically aggressive additives like lithium-based and antimony-based salts, 4-*tert*-butylpyridine (TBP) to enhance the conductivity and improve the stability under environmental stress (e.g., moisture and oxygen) hence making its synthesis complex and expensive (Hawash et al. 2018). Another commonly used HTL is poly[bis(4-phenyl)(2,4,6-trimethylphenyl)amine] (PTAA) which offers more hole mobility and stability without the dopant requirements (Bi et al. 2023). But its high cost and uniform deposition challenge limits the application on a larger scale. Other promising candidates for HTL are poly(3,4-ethylenedioxythiophene) polystyrene sulfonate (PEDOT:PSS), inorganic HTLs like NiO, CuSCN, and MoO_x but all of them requires some sort of modification to meet the required criterion. Various 2D materials have demonstrated their effectiveness as HTL because of their high hole mobility, large surface area, and mechanical stability. These properties not only enhance the key photovoltaic parameters but also improves the stability.

Sun et al. were among the first group to use GO as HTL in inverted planar $CH_3NH_3PbI_{3-x}Cl_x$ perovskite. They achieved a PCE of up to 12.4% using GO as compared to PCE of 9.26% using PEDOT:PSS. GO worked efficiently as an HTM due to its aligned energy levels and work function of ~4.9 eV. Additionally, grazing incidence X-ray diffraction (GIXRD) confirms that perovskite film grown on GO, exhibited highly ordered crystal structures with preference for alignment along (110) crystal plane (Wu et al. 2014). The limited conductivity of graphene oxide (GO) restricts its use as a sole hole transport layer (HTL). Kumar et al. conducted a comprehensive computational investigation on the use of graphene quantum dots (GQDs) and their edge-functionalized derivatives (GQDs-3Xs, where X = CH_3, COOH, $COOCH_3$, NH_2, NMe_2, OH, OMe, SH, and SMe) as efficient hole transport materials (HTMs) for PSCs. Using density functional theory (DFT) and time-dependent DFT (TD-DFT), they analyzed the electronic structure, charge transport dynamics, and optical properties of both pristine and functionalized GQDs. For a material to function effectively as an HTM in PSCs, its HOMO level must be positioned above the valence band (VB) of the perovskite absorber, aligned at or below the Fermi level of the cathode, and its LUMO level must sit higher than the conduction band (CB) of the perovskite. In this study, $MAPbI_3$ perovskite, a widely used absorber material in PSCs, was selected with VB at −5.43 eV and CB at −3.93 eV. Gold (Au) was used as the cathode, possessing a fermi level at −5.1 eV. The DFT results revealed that HOMO levels of GQD-3COOH (−5.30 eV) and GQD-3$COOCH_3$ (−5.23 eV) are appropriately aligned between the VB of $MAPbI_3$ and the Fermi level of the Au cathode. This optimal alignment facilitates efficient hole extraction while minimizing charge recombination, making these functionalized GQDs highly promising as HTMs. Beyond optimal energy alignment, solubility is another critical factor influencing thin-film formation and overall device performance. Both GQD-3COOH and GQD-3$COOCH_3$ demonstrated excellent solubility, facilitating the formation of uniform and high-quality films essential for efficient charge transport. The introduction of functional groups on these GQDs significantly reduced hole reorganization energy, resulting in enhanced hole mobility. Quantitatively, the hole mobility of GQDs reached 2.45 $cm^2\ V^{-1}\ s^{-1}$, a substantial improvement over the traditional HTM Spiro-OMeTAD, which typically exhibits mobility between 2×10^{-4} and 6×10^{-3} $cm^2\ V^{-1}\ s^{-1}$. Devices incorporating GQD-3COOH and GQD-3$COOCH_3$ achieved open-circuit voltages of 1.07 and 1.00 V, respectively, outperforming devices using Spiro-OMeTAD. This performance boost is attributed to superior energy-level alignment, which facilitates efficient hole extraction and reduces charge recombination. Additionally, GQDs exhibited strong optical absorption in the visible range and lower exciton binding energy, enhancing light-harvesting and charge separation. The integration of GQD-COOH and GQD-$COOCH_3$ as HTMs leads to the improvement in PCE of PSCs and offers a more stable and an economical alternative to traditional metal-based electrodes. These results underscore the immense potential of functionalized GQDs in advancing PSC performance through improved charge transport, enhanced stability, and efficient energy-level alignment (Kumar et al. 2023).

Key properties like tunable bandgap, high carrier mobility, and optical transparency makes TMDs a good option to be used as HTL. MoS_2 is becoming a promising material for use as either an HTL or a buffer layer due to its solution-processability, affordability, and high carrier mobility. Kohnehpoushi et al. explored the use of

MoS_2 as a two-dimensional (2D) hole-transport material (HTM) for PSCs, emphasizing its potential in enhancing efficiency, cost-effectiveness, and durability. They conducted a computational investigation using SCAPS-1D simulations to analyze the influence of MoS_2's thickness and associated device configurations on PSC performance. The alignment of the work function between MoS_2 and the perovskite facilitated efficient hole extraction and minimized recombination losses, as shown in the energy band diagrams (Fig. 9.2a). Their findings revealed that incorporating a single layer of MoS_2 (0.67 nm) as the HTM significantly improved the V_{OC} to 0.93 V and the PCE to 20.43%, compared to a PCE of 10.41% for devices without an HTM. However, they observed that increasing the MoS_2 thickness beyond 1.34 nm adversely affected device performance due to increased resistivity and reduced light transmission as shown in Fig. 9.2b. The *J-V* curve for different thickness of MoS_2 is shown in Fig. 9.2c (Kohnehpoushi et al. 2018).

Impact of 2D Materials as Interfacial Layer

Interfacial engineering plays a critical role in optimizing the photovoltaic performance and long-term stability of PSCs. Since PSCs rely on multiple layers such as the perovskite absorber, ETL, HTL, and electrodes, efficient charge transfer at these

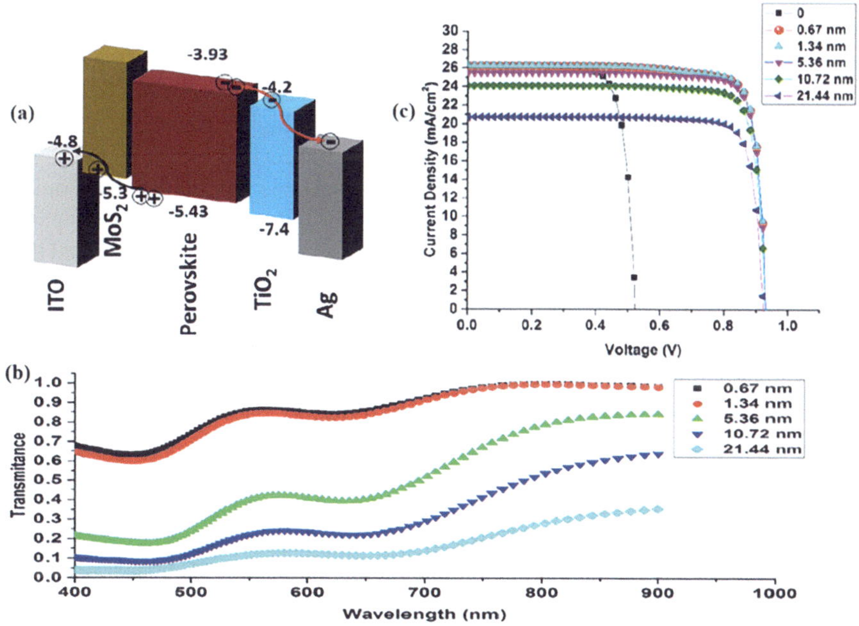

Fig. 9.2 (**a**) Energy level diagram of device having MoS_2 as HTL. (**b**) *J-V* curve. (**c**) Transmittance spectra of device with varied thickness of MoS_2 layer. (Reproduced with permission from Kohnehpoushi et al. 2018)

interfaces is essential to minimize recombination losses and improve device efficiency. Poor interfacial contact leads to energy level mismatches, charge accumulation, and increased nonradiative recombination, all of which contribute to reduced PCE and long-term instability. Additionally, interfacial defects or trap states at the perovskite/transport layer junction can create recombination centers, leading to voltage loss and hysteresis in PSCs. Therefore, tailoring the interfaces using engineered materials can enhance charge extraction, optimize energy band alignment, and improve the overall operational stability of PSCs. The incorporation of 2D materials, such as graphene, MoS_2, WS_2, and hexagonal boron nitride (h-BN), has significantly enhanced interfacial properties, leading to improved charge extraction, minimized recombination losses, and better energy band alignment. These materials exhibit high electron mobility, which enhances charge transport while reducing accumulation at the interface, resulting in higher V_{OC} and FF. Additionally, their atomically thin structure and smooth surface effectively passivate defects, suppressing nonradiative recombination and improving the overall photovoltaic performance. Many 2D materials also act as protective barriers against moisture, oxygen, and ion migration, addressing PSC stability issues and extending operational lifetimes. Furthermore, their solution-processable nature allows for scalable and cost-effective fabrication, making them highly attractive for commercial solar cell applications. By integrating 2D materials into PSC architectures, researchers have achieved significant efficiency gains, demonstrating their potential in advancing next-generation, highly stable, and high-performance PSCs.

Wang et al. were among the first group who integrated nanocomposite of graphene and TiO_2 nanoparticles as electron transport layer in meso-super structured PSCs. They discovered that the nanoflakes of graphene has significantly improved charge transport, reduced recombination losses, and enabled low-temperature fabrication. Unlike traditional TiO_2 layers that require high-temperature sintering (~500 °C), the graphene-TiO_2 composite allows device fabrication at ≤150 °C, making it compatible with flexible substrates and scalable production. The combination of TiO_2 and graphene delivered a champion PCE of up to 15.6% as compared to 10% PCE by using low temperature processed TiO_2 only. This study showed that nanocomposites of graphene and TiO_2 hold significant promise as ETLs for highly efficient PSCs processed at low temperatures (Wang et al. 2014). Rehman et al. achieved a champion PCE of 31.126% by using a lead-free alternative ($KGeCl_3$) as the absorber and incorporating graphene layer between absorber and Cu_2O (HTL). Graphene layer improves the performance of PSCs by enhancing charge carrier dynamics, reducing recombination losses, and optimizing energy band alignment. The high charge carrier mobility and exceptional electrical conductivity of graphene enable efficient charge separation and transport at the interface. The energy band alignment between the graphene layer and the adjacent $KGeCl_3$ and Cu_2O layers reduce potential barriers, facilitating smoother charge transfer and enhancing charge extraction efficiency, as shown in Fig. 9.3a. Additionally, the formation of a space charge region (SCR) at the graphene/$KGeCl_3$ interface further enhances charge separation and transport by minimizing energy dissipation and creating an electric field that drives charge carriers towards their respective electrodes as shown in

Fig. 9.3b. A computational optimization using SCAPS-1D simulations identified the ideal graphene thickness of 80 nm, which balances coverage and conductivity. This optimization minimized recombination losses while ensuring effective charge transport, resulting in a substantial improvement in the power conversion efficiency (PCE) from 29.03% (without graphene) to 31.13% (with graphene), as illustrated in Fig. 9.3c. Key photovoltaic parameters also improved, with V_{OC} reaching 0.921 V, J_{SC} 39.926 mA/cm^2, and FF rising to 84.62% (Rehman et al. 2025).

Meskini et al. conducted a computational investigation into the incorporation of Ti_3C_2 MXene as an interlayer between SnO_2 (ETL) and perovskite absorber, utilizing SCAPS-1D simulations to evaluate its impact on device performance. Initially, the PSC with SnO_2-Ti_3C_2 (1.0 wt%) as the ETL achieved a PCE of 18.32%. Through optimization of the device configuration, the simulated PCE was enhanced to 27.81%, demonstrating the significant potential of Ti_3C_2 to improve PSC performance. The computational analysis revealed that the improved efficiency was a result of enhanced energy-level alignment between the ETL and perovskite layer as shown in Fig. 9.3c, d, which reduced interfacial recombination losses and facilitated efficient charge extraction. The study identifies 1.0 wt% Ti_3C_2 as the optimal concentration for SnO_2 ETLs, achieving maximum PCE due to enhanced interfacial properties. As shown in Fig. 9.3e, this concentration provides the longest carrier lifetime in the absorber layer, reducing recombination and improving charge transport. The ideality factor decreases while the open-circuit voltage increases at 1.0 wt% MXene, striking a balance that minimizes recombination and maximizes efficiency. Higher concentrations, however, introduce adverse effects such as increased interfacial resistance, underscoring 1.0 wt% as the ideal addition for superior PSC performance. By optimizing the thickness of the SnO_2-Ti_3C_2 ETL to 10 nm and the perovskite absorber layer to 700 nm, achieving a balance between efficient light absorption and effective charge transport, increase in the PCE to 27.81%, V_{OC} to 1.051 V, J_{SC} to 36.21 mA/cm^2, and FF to 73.07% was achieved (Meskini and Asgharizadeh 2024) (Fig. 9.3f).

Impact of 2D Materials in Perovskite Layer

The perovskite layer is a key component of PSCs, serving as the primary light-absorbing and electron-hole pairs generating material. Its remarkable properties, including strong light absorption capability, extended carrier diffusion length, and low exciton binding energy, enable efficient conversion of light into electrical energy. These factors have a scope of improvement by improving the crystallization dynamics, surface morphology, and increasing the size of grain of perovskite layer which will directly influence the performance of PSCs (Nie et al. 2015). In recent years, numerous preparation techniques have been established to achieve perovskite films of high quality. Integrating 2D materials within the perovskite layer has demonstrated significant improvements in film crystallization, leading to enhanced overall performance of PSCs.

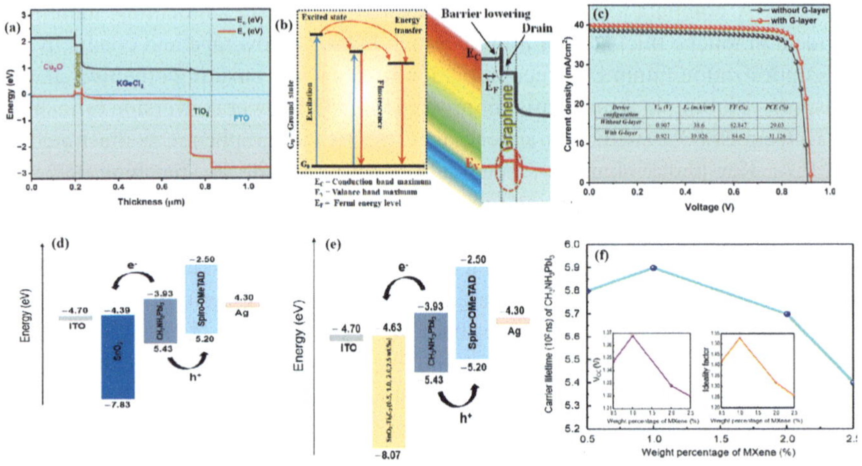

Fig. 9.3 (**a**) Energy band diagram of device using $KGeCl_3$ as the absorber and graphene as an interfacial layer. (**b**) Impact of graphene as an interfacial layer on charge transport and band alignment in PSC. (**c**) *J-V* curve with and without graphene layer. (Reproduced with permission from Rehman et al. 2025). (**d**) Energy level diagram without MXene. (**e**) Energy level diagram with MXene. (**f**) Carrier lifetime in absorber as a function of MXene weight percentage. (Reproduced with permission from Meskini and Asgharizadeh 2024)

For the first time Hadadian et al. introduced N-doped reduced graphene oxide (N-rGO) into perovskite layer. N-rGO was incorporated into the perovskite layer by dispersing it in the precursor solution containing formamidinium (FA), methylammonium (MA) cations, and halides before spin coating. This incorporation slows down the nucleation and crystallization process through interactions between the FA cations and nitrogen sites on N-rGO. This delayed crystallization allows the formation of fewer but larger crystals, increasing grain size from approximately 196 nm in pristine films to 490 nm in N-RGO-incorporated films, with some grains exceeding 800 nm. The increased grain dimensions minimize the presence of grain boundaries and enhance the ability of the material to capture light effectively. The surface passivation by N-rGO minimizes recombination losses, as evidenced by slower photoluminescence (PL) decay leads to higher V_{OC}. These enhancements collectively increase the PCE from 17.3% to 18.7% (Hadadian et al. 2016). 2D materials can also enhance the stability of PSCs by preventing the degradation of perovskite molecules caused by environmental moisture, ionic diffusion, thermal and photonic instability under solar irradiation. Mahmoudi et al. integrated silver nanoparticle-anchored rGO (Ag-rGO) with $MAPbI_{3-x}Cl_x$ perovskite. The graphene sheets in Ag-rGO act as a physical barrier, suppressing ion migration, particularly iodide ions having smaller diameter (0.412 nm), by blocking their diffusion pathways due to their small lattice constant (0.246 nm). Ag-rGO interlayer effectively prevents the iodide ion migration and the formation of degradation products like PbI_2 during thermal aging. Compared to pristine PSCs, which experienced a 39% reduction in PCE after 90 h of thermal aging at 90 °C, Ag-rGO-based PSCs retained over 94% of their initial PCE (Mahmoudi et al. 2019).

Singh et al. studied the impact of incorporating hafnium disulfide (HfS_2) into $CH_3NH_3PbI_3$ absorber using first-principles DFT calculations. HfS_2, a 2D TMD, has gained recognition as a highly promising material for photovoltaic applications due to its outstanding electronic and optical characteristics. Unlike conventional bulk semiconductors, monolayer HfS_2 exhibits a tunable indirect-to-direct bandgap transition, ultrahigh carrier mobility, and remarkable chemical stability, establishing it as an attractive candidate for use in next-generation solar cells. Its layered structure and weak van der Waals interactions facilitate easy integration with other 2D materials to form heterostructures, enabling enhanced light absorption and charge separation. The introduction of the HfS_2 monolayer to form $CH_3NH_3PbI_3/HfS_2$ van der Waals (vdW) heterostructure, effectively tunes the bandgap of the heterostructure, reducing transition energies and enhancing visible-light sensitivity, which is crucial for improved photovoltaic performance. The optimized heterostructure exhibits a high optical absorption coefficient in the visible spectrum, reaching values on the order of 10^5 cm^{-1} (Fig. 9.4a), indicating its strong capability for light harvesting. Additionally, the charge transfer mechanism in the $CH_3NH_3PbI_3/HfS_2$ interface, as shown in Fig. 9.4b, c, reveals a spontaneous electron transfer from HfS_2 to $CH_3NH_3PbI_3$, driven by a built-in electric field at the interface, which enhances the separation of photogenerated carriers and reduces recombination losses. The band alignment analysis (Fig. 9.4d) confirms a Type-II staggered heterostructure, facilitating efficient charge transfer and further minimizing recombination. The spectroscopic limited maximum efficiency (SLME) analysis (Fig. 9.4e) estimates a peak

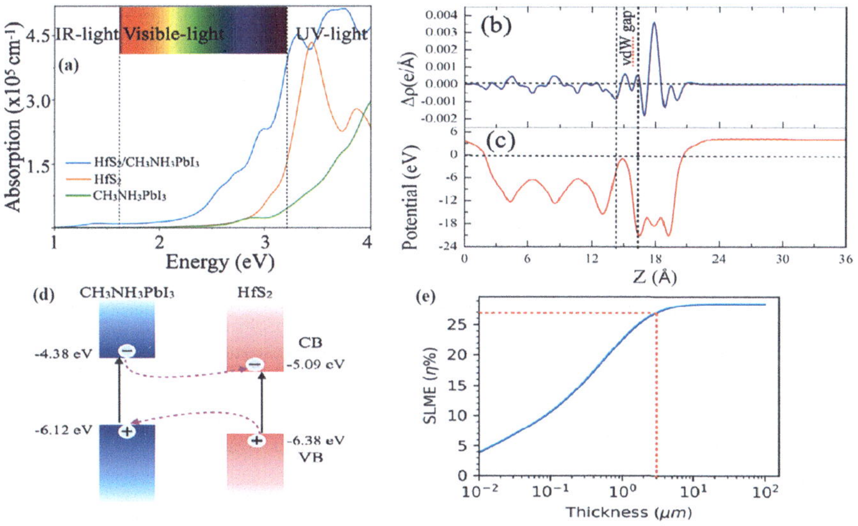

Fig. 9.4 (**a**) Optical absorption coefficient spectra comparing $CH_3NH_3PbI_3/HfS_2$ with pristine $CH_3NH_3PbI_3$. (**b**) Planar charge density difference profile showing charge redistribution at the $CH_3NH_3PbI_3/HfS_2$ interface. (**c**) Electrostatic potential profile illustrating the built-in electric field across the $CH_3NH_3PbI_3/HfS_2$ heterostructure. (**d**) Band alignment diagram of the $CH_3NH_3PbI_3/HfS_2$ heterostructure. (**e**) Spectroscopic limited maximum efficiency (SLME) as a function of thickness for the $CH_3NH_3PbI_3/HfS_2$ heterostructure. (Reproduced with permission from Singh and Ahuja 2022)

efficiency of 28.45% for the $CH_3NH_3PbI_3/HfS_2$ heterostructure, outperforming several existing layered photovoltaic materials. These findings indicate that the integration of 2D HfS_2 into the perovskite absorber layer significantly enhances charge separation, optical absorption, and overall device efficiency, establishing it as a promising candidate for next-generation photovoltaic applications (Singh and Ahuja 2022).

$Ti_3C_2T_X$ was first incorporated into $MAPbI_3$ perovskite layer by Ma et al. The incorporation of 2D $Ti_3C_2T_x$ enhanced both performance and stability. When 0.03 wt% of $Ti_3C_2T_X$ was added to the perovskite precursor, the PCE increased from 15.54% to 17.41%. This improvement is attributed to the MXene's ability to retard the nucleation process, resulting in larger perovskite grain sizes (increased from 150 to 358 nm), as observed through SEM images. The larger grains reduced grain boundaries, which improved carrier transport and minimized recombination losses. Furthermore, the high electrical conductivity and mobility of MXene facilitated efficient charge transfer, reducing the charge transfer resistance. The stability was further improved due to the MXene's surface passivation, which suppressed defect formation and reduced hysteresis (Guo et al. 2018). Agresti et al. demonstrated that the incorporation of $Ti_3C_2T_X$ MXene into the $MAPbI_3$ absorber layer reduces its electron affinity from 3.9 to 3.55 eV without altering the bandgap. This modification resulted in a 26% increment in the PCE (Agresti et al. 2019). The performance of PSCs is significantly influenced by parameters such as conduction band offset (CBO), valence band offset (VBO), doping concentrations, electron affinity, and layer thickness at the absorber layer interfaces. Optimizing CBO and VBO ensures efficient charge extraction, reduces recombination losses, and facilitates smooth charge transport by aligning energy levels between layers. Further optimization was conducted by Asgharizadeh et al., who replaced Spiro-OMeTAD with CuSCN as the HTL, adjusted the absorber layer thickness, and fine-tuned the CBO and VBO values. Using SCAPS-1D simulation software, they investigated the role of CBO and VBO in device performance. The optimal CBO of −0.3 eV, depicted in Fig. 9.5a of the study, significantly increased the open-circuit voltage without affecting the short-circuit current density. Similarly, a VBO value of 0.02 eV, minimized interfacial recombination. These adjustments created favorable energy alignments as shown in Fig. 9.5b, reducing recombination rates and enhancing charge carrier dynamics. Consequently, the PCE increased from 16.14% to 22.80% as shown in *J-V* curve and *P-V* curve in Fig. 9.5c (Asgharizadeh et al. 2023).

Impact of 2D Materials as Electrodes

A typical PSC structure have transparent electrode and back electrode. The function of transparent electrode is to transmit the light toward perovskite absorber layer and to collect and transport photogenerated electrons and holes to the external circuit. An ideal transparent electrode should have high optical transparency,

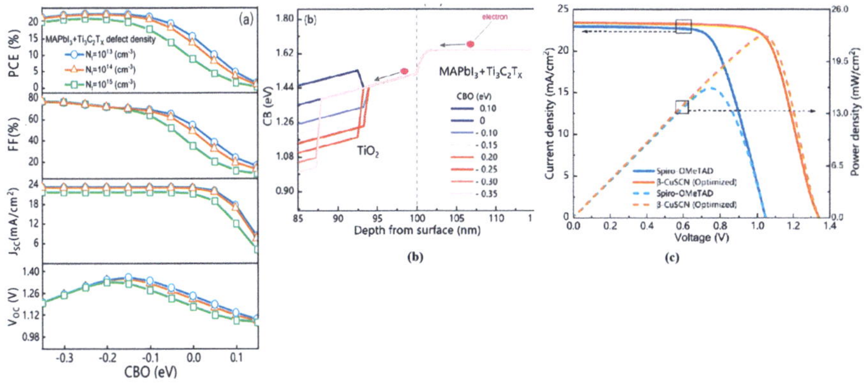

Fig. 9.5 (**a**) Variation of cell parameters with CBO. (**b**) Band alignment at ETL/absorber interface for different CBO. (**c**) *J-V* curve represented by solid lines and P-V curve by dashed lines. (Reproduced with permission from Asgharizadeh et al. 2023)

work function aligned with subsequent charge transport layer, and chemical and mechanical stability. Additionally, it should be cost-effective to enable large-scale production and practical application (Raj et al. 2023). Fluorine-doped tin oxide (FTO) and indium-doped tin oxide (ITO) are the commonly used transparent electrode. FTO is cost-effective and thermally stable but has higher sheet resistance and lower transparency. ITO offers superior transparency and lower sheet resistance but is brittle, expensive, and degrades under harsh conditions. The main function of back electrode is to collect and conduct charge carriers from the charge transport layer to the external circuit and to reflect unabsorbed light back into the perovskite layer. Ag and Au are widely used back electrodes in PSCs, each with distinct advantages and drawbacks. Ag is more affordable and highly conductive but is prone to corrosion, halide migration, and oxidation, which can degrade device stability. Au, while chemically inert and highly stable, is extremely expensive and requires complex vacuum deposition methods, making it less suitable for large-scale applications (You et al. 2019). These limitations highlight the need for cost-effective and stable alternatives for transparent and back electrodes. 2D materials, especially graphene, have attracted considerable interest in recent years due to their outstanding electrical conductivity, high transparency, extensive surface area, and robust mechanical stability. The effects of using various 2D materials as electrodes are discussed below.

One of the studies indicate that solar cells fabricated on ITO substrates with an initial PCE of 12.2% retained only 50% of their efficiency after enduring 1000 bending cycles at a 4 mm bending radius (Kim et al. 2015). In comparison, 2D graphene demonstrates outstanding optical and electrical properties, along with excellent mechanical stability and high chemical inertness. Additionally, its abundant carbon availability makes it an ideal choice for manufacturing flexible and transparent electrodes. Luo et al. constructed a flexible all carbon-electrode device where graphene was used as transparent electrode and cross-stacking carbon nanotubes (CSCNTs) as back electrode as shown in Fig. 9.6a and energy levels in Fig. 9.6b. The device obtained a PCE of 11.9% with Spiro-OMeTAD while 8.4% without it. The bending stability of this device at different bending radii and bending cycle was compared with device control device using ITO and Au electrodes having structure FEN/ITO/TiO_2/PCBM/$MAPbI_3$/spiro-OMeTAD/Au Fig. 9.6d, e. The control device lost its 13% of PCE after 200 bending cycles at a bending curvature of 4 mm while all-carbon-based devices showed no significant degradation. At a bending curvature of 2.2 mm, all-carbon devices maintained 85% of their initial efficiency, while control device failed completely due to conductivity loss. Over 2000 bending cycles, all-carbon PSCs retained 84% of their initial efficiency, whereas ITO-based devices degraded to just 13%. In photostability tests under continuous illumination for 1014 h, all-carbon devices retained 92% of their initial efficiency, while ITO-based devices dropped to 75% Fig. 9.6c. The enhanced stability of all-carbon devices is attributed to the chemical inertness of carbon materials, their excellent conductivity, and resistance to deformation and ion migration, whereas ITO electrode tend to crack and lose conductivity under stress. Also the metal electrodes are prone to diffusion into the perovskite layer, causing irreversible damage. The carbon electrodes remain stable, making them ideal for flexible and long-lasting PSCs (Luo et al. 2018). However, the device with graphene-based electrodes usually have lower PCE and FF value due to lower optical transparency which can be improved by using lesser number of graphene layers but it will increase the sheet resistance. Work function alignment is also not ideal which can be improved by doping of certain agent. Doping graphene electrodes with bis(trifluoromethanesulfonyl)amide (TFSA) significantly improved their properties for use in the PSCs. Heo et al. used TFSA-doped graphene TCE in inverted $FAPbI_{3-x}Cl_x$ based perovskite solar cells as shown in Fig. 9.6f. TFSA doping reduced the sheet resistance of graphene from 650 Ω/cm^2 to as low as 108 Ω/cm^2 while maintaining high optical transparency (>96%). The enhanced electrical conductivity and optimized work function (−4.90 eV at 20 mM TFSA) as shown in Fig. 9.6g improved charge carrier extraction and reduced recombination losses in the PSCs. Devices with TFSA-doped graphene achieved a PCE of 18.9% compared to 12.8% for pristine graphene (Heo et al. 2018). Graphene can also be used as back electrodes and replace the precious metals (e.g., Au, Ag) because of its excellent optoelectronics properties and low cost. Its combination of high electrical conductivity and excellent transparency makes it ideal for use in semi-transparent PSCs, which hold significant promise for tandem solar cell applications.

Apart from graphene, TMDs like NbS_2 has also been explored as back electrode. The study of Shao et al. highlights the potential of 2H-NbS_2 as a back electrode in perovskite solar cells (PSCs), offering a scalable and cost-efficient substitute to noble metals like gold (Au). NbS_2 electrodes exhibit high electrical conductivity (8.7×10^3 S/cm) and a work function of 5.20 eV, which aligns well with the spiro-OMeTAD hole transport layer ensuring efficient hole extraction and reduced energy

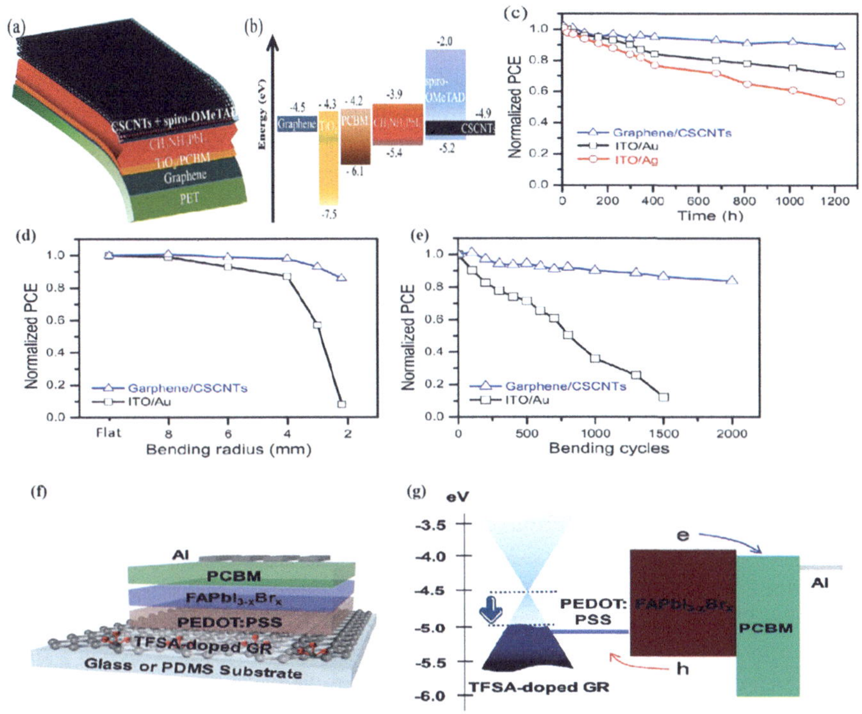

Fig. 9.6 (**a**) Device structure. (**b**) Energy level diagram of all-carbon electrode-based PSCs. Normalized stability with respect to (**c**) time, (**d**) bending radius, and (**e**) bending cycles for all-graphene electrode-based and standard electrode-based PSCs. (Reproduced with permission from Luo et al. 2018). (**f**) Device structure. (**g**) Energy level diagram of TFSA-doped graphene electrode used in mixed halide perovskite device. (Reproduced with permission from Heo et al. 2018)

losses. PSCs incorporating NbS_2 electrodes achieved a PCE of 8.3% (Shao et al. 2018).

Safie et al. investigated the feasibility of utilizing a reduced graphene oxide-molybdenum disulfide (rGO-MoS_2) nanocomposite as a top electrode in inverted planar perovskite solar cells (PSCs), through SCAPS-1D simulations. The device architecture modeled in SCAPS-1D consists of ITO/CuSCN/$MAPbI_3$/PCBM/rGO-MoS_2, where the rGO-MoS_2 composite replaces traditional metal electrodes. Computationally derived electrical characteristics indicate a significant improvement in photovoltaic performance of device, with a power conversion efficiency of 19.68%, V_{OC} of 1.16 V, a J_{SC} of 25.26 mA/cm^2, and FF of 67.15%. The SCAPS-1D simulations validate that the enhanced performance is primarily due to the lower sheet resistance (10.7 Ω/sq) of the 5:5 rGO-MoS_2 composition, which ensures better charge extraction and reduced recombination losses. Further computational analysis of band alignment and energy level matching confirms that the rGO-MoS_2

composite provides an improved interface for charge transport, with its energy level closely aligned with the perovskite absorber and electron transport layers. Additionally, optical absorption spectra reveal that the composite possesses a narrow bandgap of 1.25 eV, enhancing light absorption and charge carrier generation. This computationally validated result supports the experimental finding that a higher rGO content contributes to better charge transfer properties (Safie et al. 2022).

Conclusion

The application of 2D materials in PSCs has brought about significant advancements in addressing critical challenges like efficiency, stability, and scalability. Materials such as MXenes, transition metal dichalcogenides (TMDs), and graphene derivatives have shown remarkable versatility, offering unique properties that make them suitable for various roles in PSCs, including as electrodes, interfacial layers, and transport layers. MXenes contribute to improving grain size and passivating defects, resulting in enhanced charge transport and device stability. Similarly, TMDs such as MoS_2 and WS_2 support better perovskite crystallization and minimize recombination losses, ensuring improved energy efficiency and operational durability. Graphene, with its excellent conductivity and transparency, has proven effective for charge collection while enabling the development of flexible and semi-transparent PSCs. Black phosphorus, on the other hand, brings a tunable bandgap and high carrier mobility, which strengthens its role in defect passivation and hole extraction. Moving forward, efforts should focus on refining fabrication methods, reducing production costs, and ensuring compatibility among device layers. These advancements will unlock the full potential of 2D materials, leading to the development of more efficient and durable PSCs that can meet future energy demands.

References

Agresti A, Pazniak A, Pescetelli S et al (2019) Titanium-carbide MXenes for work function and interface engineering in perovskite solar cells. Nat Mater 18:1228–1234. https://doi.org/10.1038/s41563-019-0478-1

Alzaid M (2021) Recent progress in the role of two-dimensional materials as an efficient charge transport layer in perovskite solar cells. Int J Energy Res 45(12):12598–12613. https://doi.org/10.1002/er.6672

Ameen S, Rub MA, Kosa SA, Alamry KA et al (2015) Perovskite solar cells: influence of hole transporting materials on PCE. ChemSusChem 9(10). https://doi.org/10.1002/cssc.201501228

Asgharizadeh S, Azadi SK, Zeynali M (2023) DFT computational approach to interfacial recombination in MXene-assisted perovskite solar cells with β-CuSCN HTL. J Phys Chem C 127(41):20511–20522. https://doi.org/10.1021/acs.jpcc.3c04666

Balandin AA, Ghosh S, Bao W, Calizo I, Teweldebrhan D, Miao F, Lau CN (2008) Superior thermal conductivity of single-layer graphene. Nano Lett 8(3):902–907. https://doi.org/10.1021/nl0731872

Bellani S, Bartolotta A, Agresti A, Calogero G, Grancini G et al (2021) Solution-processed two-dimensional materials for next-generation photovoltaics. Chem Soc Rev 50(22):11870–11965. https://doi.org/10.1039/d1cs00106j

Berhe TA, Su W-N, Chen C-H, Pan C-J, Cheng J-H, Chen H-M, Tsai M-C, Chen L-Y, Dubale AA, Hwang B-J (2016) Organometal halide perovskite solar cells: degradation and stability. Energy Environ Sci 9(2):323–356. https://doi.org/10.1039/C5EE02733K

Bi H, Fujiwara Y, Kapil G, Tavgeniene D et al (2023) Perovskite solar cells consisting of PTAA modified with monomolecular layer and application to all-perovskite tandem solar cells with efficiency over 25%. Adv Funct Mater 33:2300089. https://doi.org/10.1002/adfm.202300089

BloombergNEF (2024) New energy outlook 2024: executive summary. Bloomberg New Energy Finance, London. https://assets.bbhub.io/professional/sites/24/847353_NEO24_ExecSum.pdf

Bolotin KI, Sikes KJ, Jiang Z, Klima M, Fudenberg G, Hone J, Kim P, Stormer HL (2008) Ultrahigh electron mobility in suspended graphene. Solid State Commun 146(9–10):351–355. https://doi.org/10.1016/j.ssc.2008.02.024

Cai X, Luo Y, Liu B, Cheng HM (2018) Preparation of 2D material dispersions and their applications. Chem Soc Rev 47(16):6224–6266. https://doi.org/10.1039/C8CS00254A

Choi W, Choudhary N, Han GH, Park J, Akinwande D et al (2017) Recent development of two-dimensional transition metal dichalcogenides and their applications. Mater Today 20(3):116–130. https://doi.org/10.1016/j.mattod.2016.10.002

Deng B, Tran V, Xie Y et al (2017) Efficient electrical control of thin-film black phosphorus bandgap. Nat Commun 8:14474. https://doi.org/10.1038/ncomms14474

Eda G, Lin Y-Y, Miller S, Chen C-W, Chhowalla M (2009) Insulator to semimetal transition in graphene oxide. J Phys Chem C 113(35):15768–15771. https://doi.org/10.1021/jp9051402

Fu Q, Han J, Wang X, Xu P, Yao T et al (2021) 2D transition metal dichalcogenides: design, modulation, and challenges in electrocatalysis. Adv Mater 33(21):2170045. https://doi.org/10.1002/adma.202170045

Gagandeep, Singh M, Kumar R, Singh V (2019) Graphene as charge transport layers in lead-free tin-based perovskite solar cells. Mater Res Express 6(11):115611. https://doi.org/10.1088/2053-1591/ab4b02

Gomez-Navarro C, Meyer JC, Sundaram RS, Chuvilin A, Kurasch S, Burghard M, Kern K, Kaiser U (2010) Atomic structure of reduced graphene oxide. Nano Lett 10(4):1144–1148. https://doi.org/10.1021/nl9031617

Green MA, Dunlop ED, Yoshita M, Kopidakis N, Bothe K (2023) Solar cell efficiency tables (version 62). Prog Photovolt Res Appl 31(7):651–663. https://doi.org/10.1002/pip.3726

Guo Z, Gao L, Xu Z, Teo S, Zhang C, Kamata Y, Hayase S, Ma T (2018) High electrical conductivity 2D MXene serves as additive of perovskite for efficient solar cells. Small 14:e1802738. https://doi.org/10.1002/smll.201802738

Gwon H, Kim H-S, Lee KU, Seo D-H, Park YC, Lee Y-S, Ahn BT, Kang K (2011) Flexible energy storage devices based on graphene paper. Energy Environ Sci 4(4):1277–1283. https://doi.org/10.1039/C0EE00627D

Hadadian M, Correa-Baena J-P, Goharshadi EK, Ummadisingu A, Seo J-Y, Luo J, Gholipour S, Zakeeruddin SM, Saliba M, Abate A, Grätzel M, Hagfeldt A (2016) Enhancing efficiency of perovskite solar cells via N-doped graphene: crystal modification and surface passivation. Adv Mater 28(37):8681–8686. https://doi.org/10.1002/adma.201602785

Hawash Z, Ono LK, Qi YB (2018) Recent advances in spiro-MeOTAD hole transport material and its applications in organic-inorganic halide perovskite solar cells. Adv Mater Interfaces 5:1700623. https://doi.org/10.1002/admi.201700623

Heo JH, Shin DH, Song DH, Kim DH, Lee SJ et al (2018) Super-flexible bis(trifuoromethanesulfonyl)-amide doped graphene transparent conductive electrodes for photo-stable perovskite solar cells. J Mater Chem A 6:8251–8258. https://doi.org/10.1039/c8ta02672f

Huang P, Chen Q, Zhang K, Yuan L, Zhou Y, Song B, Li Y (2019) 21.7% efficiency achieved in planar n–i–p perovskite solar cells via interface engineering with water-soluble 2D TiS2. J Mater Chem A 7(11):6213–6219. https://doi.org/10.1039/C8TA11841H

Hummers WS, Offeman RE (1958) Preparation of graphitic oxide. J Am Chem Soc 80(6):1339. https://doi.org/10.1021/ja01539a017

IEA (2019) World energy outlook 2019. International Energy Agency, Paris. https://www.iea.org/reports/world-energy-outlook-2019

Jani R, Bhargava K (2022) Efficiency and reproducibility enhancement in perovskite solar cell with MoS_2 as electron transport layer: a computational finding. IEEE Trans Electron Dev 69(8):4349–4354. https://doi.org/10.1109/TED.2022.3181533

Kim H, Kim HH, Jang JI, Lee SK, Lee G-W et al (2014) Doping graphene with an atomically thin two-dimensional molecular layer. Adv Mater 26(48):8141–8146. https://doi.org/10.1002/adma.201403196

Kim BJ, Kim DH, Lee Y-Y, Shin H-W, Han GS et al (2015) Highly efficient and bending durable perovskite solar cells: toward a wearable power source. Energy Environ Sci 8(4):916–921. https://doi.org/10.1039/c4ee02441a

Kohnehpoushi S, Nazari P, Nejand BA, Eskandari M (2018) MoS2: a two-dimensional hole-transporting material for high-efficiency, low-cost perovskite solar cells. Nanotechnology 29(20):205201. https://doi.org/10.1088/1361-6528/aab1d4

Kojima A, Teshima K, Shirai Y, Miyasaka T (2009) Organometal halide perovskites as visible-light sensitizers for photovoltaic cells. J Am Chem Soc 131:6050–6051. https://doi.org/10.1021/ja809598r

Kumar A, Sayyed MI, Punina D, Naranjo E, Jácome E, Abdulameer MK, Albazoni HJ, Shariatinia Z (2023) Graphene quantum dots (GQD) and edge-functionalized GQDs as hole transport materials in perovskite solar cells for producing renewable energy: a DFT and TD-DFT study. RSC Adv 13(42):29163–29173. https://doi.org/10.1039/D3RA05438A

Lee C, Wei X, Kysar JW, Hone J (2008) Measurement of the elastic properties and intrinsic strength of monolayer graphene. Science 321(5887):385–388. https://doi.org/10.1126/science.1157996

Li X, Cai W, An J, Kim S, Nah J, Yang D et al (2009) Large-area synthesis of high-quality and uniform graphene films on copper foils. Science 324(5932):1312–1314. https://doi.org/10.1126/science.1171245

Liu Y, Huang Y, Duan X (2019) Van der Waals integration before and beyond two-dimensional materials. Nature 567(7748):323–333. https://doi.org/10.1038/s41586-019-1013-x

Liu C, Chen H, Wang S, Liu Q, Jiang Y-G et al (2020) Two-dimensional materials for next-generation computing technologies. Nat Nanotechnol 15(6):545–557. https://doi.org/10.1038/s41565-020-0724-3

Loh KP, Bao Q, Eda G, Chhowalla M (2010) Graphene oxide as a chemically tunable platform for optical applications. Nat Chem 2(12):1015–1024. https://doi.org/10.1038/nchem.907

Luo Q et al (2018) All-carbon-electrode-based endurable flexible perovskite solar cells. Adv Funct Mater 28(11):1706777. https://doi.org/10.1002/adfm.201706777

Mahmoudi T, Wang Y, Hahn Y-B (2019) Stability enhancement in perovskite solar cells with perovskite/silver–graphene composites in the active layer. ACS Energy Lett 4(1):235–241. https://doi.org/10.1021/acsenergylett.8b02201

Meng L, You J, Yang Y (2018a) Addressing the stability issue of perovskite solar cells for commercial applications. Nat Commun 9(1):5265. https://doi.org/10.1038/s41467-018-07255-1

Meng X, Zhou J, Hou J, Tao X, Cheung SH et al (2018b) Versatility of carbon enables all-carbon-based perovskite solar cells to achieve high efficiency and high stability. Adv Mater 30:e1706975. https://doi.org/10.1002/adma.201706975
Meskini M, Asgharizadeh S (2024) Performance simulation of the perovskite solar cells with Ti_3C_2 MXene in the SnO_2 electron transport layer. Sci Rep 14(1):5723. https://doi.org/10.1038/s41598-024-56461-z
Miró P, Audiffred M, Heine T (2014) An atlas of two-dimensional materials. Chem Soc Rev 43(18):6537–6554. https://doi.org/10.1039/c4cs00102h
Morozov SV, Novoselov KS, Katsnelson MI, Schedin F, Elias DC et al (2008) Giant intrinsic carrier mobilities in graphene and its bilayer. Phys Rev Lett 100(1):016602. https://doi.org/10.1103/PhysRevLett.100.016602
Myagmarsereejid P, Ingram M, Batmunkh M, Zhong YL (2021) Doping strategies in Sb_2S_3 thin films for solar cells. Small 17:2100241. https://doi.org/10.1002/smll.202100241
Nair RR, Blake P, Grigorenko AN, Novoselov KS, Booth TJ, Stauber T, Peres NMR, Geim AK (2008) Fine structure constant defines visual transparency of graphene. Science 320(5881):1308. https://doi.org/10.1126/science.1156965
National Laboratory of the Rockies (n.d.) <best research cell efficiencies.PDF>. https://www.nrel.gov/pv/assets/pdfs/best-research-cell-efciencies.pdf
Nazir G, Lee SY, Lee JH, Rehman A, Lee JK et al (2022) Stabilization of perovskite solar cells: recent developments and future perspectives. Adv Mater 34:e2204380. https://doi.org/10.1002/adma.202204380
Nie W, Tsai H, Asadpour R, Blancon J-C, Neukirch A, Gupta G, Crochet J, Chhowalla M, Tretia S, Alam M, Wang H-L, Mohite A (2015) Solar cells: high-efficiency solution-processed perovskite solar cells with millimeter-scale grains. Science 347(6221):522–525. https://doi.org/10.1126/science.aaa0472
Novoselov KS, Geim AK, Morozov SV, Jiang DE, Zhang Y, Dubonos SV et al (2004) Electric field effect in atomically thin carbon films. Science 306(5696):666–669. https://doi.org/10.1126/science.1102896
O'Sullivan J, Wright M, Niu X, Miller P, Wilshaw PR et al (2023) Towards a graphene transparent conducting electrode for perovskite/silicon tandem solar cells. Prog Photovolt Res Appl 31(11):1478–1492. https://doi.org/10.1002/pip.3739
Palei S, Murali G, Kim C-H, In I, Lee S-Y et al (2023) A review on interface engineering of MXenes for perovskite solar cells. Nano-Micro Lett 15:123. https://doi.org/10.1007/s40820-023-01083-9
Qamar S, Fatima K, Ullah N, Akhter Z, Waseem A et al (2022) Recent progress in use of MXene in perovskite solar cells: for interfacial modification, work-function tuning, and additive engineering. Nanoscale 14(36):13018–13039. https://doi.org/10.1039/d2nr02799b
Qin Z, Chen Y, Zhu K, Zhao Y (2021) Two-dimensional materials for perovskite solar cells with enhanced efficiency and stability. ACS Mater Lett 3(9):1402–1416. https://doi.org/10.1021/acsmaterialslett.1c00327
Rai N, Rai S, Singh PK et al (2020) Analysis of various ETL materials for an efficient perovskite solar cell by numerical simulation. J Mater Sci Mater Electron 31:16269–16280. https://doi.org/10.1007/s10854-020-04175-z
Raj A, Kumar M, Anshul A (2022) Recent progress in cesium-based lead-free halide double perovskite materials for photovoltaic applications. Phys Status Solidi A 219(22):2200425. https://doi.org/10.1002/pssa.202200425
Raj A, Kumar M, Anshul A (2023) Topical advances in fabrication technologies of perovskite solar cell heterostructures: performance and future perspective. Mater Lett 340:134171. https://doi.org/10.1016/j.matlet.2023.134171
Raj A, Kumar M, Singh DV, Anshul A (2024) A critical review on transition metal dichalcogenides (TMDs): an efficiency booster for perovskite solar cells. FlatChem 44:100629. https://doi.org/10.1016/j.flatc.2024.100629

Rao CNR, Maitra U, Waghmare UV (2014) Extraordinary attributes of 2-dimensional MoS_2 nanosheets. Chem Phys Lett 609:172–183. https://doi.org/10.1016/j.cplett.2014.06.003

Rehman UU, Almufarij RS, Abd-Elwahed AR, Sahar KU, Hussain E, Ashfaq A, Mahmood K, Wang C-M (2025) Improving efficiency of germanium-based perovskite solar cells with graphene interface layer: a strategy to minimize charge recombination. J Phys Chem Solids 198:112487. https://doi.org/10.1016/j.jpcs.2024.112487

Saeed MA, Shahzad A, Rasool K, Mateen F, Oh J-M et al (2022) 2D MXene: a potential candidate for photovoltaic cells? A critical review. Adv Sci 9(21):2104743. https://doi.org/10.1002/advs.202104743

Safie NE, Sairi MNFM, Azam MA, Takasaki A (2022) Development and analysis of rGO-MoS_2 nanocomposite as a top electrode for the application of inverted planar perovskite solar cells via SCAPS-1D device simulation. J Mater Res 37(20):3372–3383. https://doi.org/10.1557/s43578-022-00652-9

Shao F, Tian Z, Qin P, Bu K, Zhao W, Xu L, Wang D, Huang F (2018) 2H-NbS_2 film as a novel counter electrode for meso-structured perovskite solar cells. Sci Rep 8:7033. https://doi.org/10.1038/s41598-018-25449-x

Shen X, Lin X, Peng Y et al (2024) Two-dimensional materials for highly efficient and stable perovskite solar cells. Nano Micro Lett 16:201. https://doi.org/10.1007/s40820-024-01417-1

Singh A, Ahuja G (2022) Two-dimensional perovskite-HfS_2 van der Waals heterostructure as an absorber material for photovoltaic applications: a first-principles study. J Mater Sci 57(12):8904–8915. https://doi.org/10.1021/acsaem.1c03796

Singh R, Giri A, Pal M, Thiyagarajan K, Kwak J, Lee J-J, Jeong U, Cho K (2019) Perovskite solar cells with an MoS_2 electron transport layer. J Mater Chem A 7:7151–7158. https://doi.org/10.1039/C8TA12254G

Sobayel K, Akhtaruzzaman M, Rahman KS, Ferdaous MT, Al-Mutairi ZA, Alharbi HF, Alharthi NH, Karim MR, Hasmady S, Amin N (2019) A comprehensive defect study of tungsten disulfide (WS_2) as electron transport layer in perovskite solar cells by numerical simulation. Results Phys 12:1097–1103. https://doi.org/10.1016/j.rinp.2018.12.049

Stoller MD, Park S, Zhu Y, An J, Ruoff RS (2008) Graphene-based ultracapacitors. Nano Lett 8(10):3498–3502. https://doi.org/10.1021/nl802558y

VahidMohammadi A, Rosen J, Gogotsi Y (2021) The world of two-dimensional carbides and nitrides (MXenes). Science 372(6547):eabf1581. https://doi.org/10.1126/science.abf1581

Wang JT-W, Ball JM, Barea EM, Abate A, Alexander-Webber JA, Huang J, Saliba M, Mora-Sero I, Bisquert J, Snaith HJ, Nicholas RJ (2014) Low-temperature processed electron collection layers of graphene/TiO2 nanocomposites in thin film perovskite solar cells. Nano Lett 14(2):724–730. https://doi.org/10.1021/nl403997a

Wu Z, Bai S, Xiang J, Yuan Z, Yang Y, Cui W, Gao X, Liu Z, Jin Y, Sun B (2014) Efficient planar heterojunction perovskite solar cells employing graphene oxide as hole conductor. Nanoscale 6(18):10505–10510. https://doi.org/10.1039/c4nr03181d

Wu Z, Lyu Y, Zhang Y, Ding R, Zheng B et al (2021) Large scale growth of few-layer two-dimensional black phosphorus. Nat Mater 20(9):1203–1209. https://doi.org/10.1038/s41563-021-01001-7

Xia J, Sohail M, Nazeeruddin MK (2023) Efficient and stable perovskite solar cells by tailoring of interfaces. Adv Mater 35:2211324. https://doi.org/10.1002/adma.202211324

Xu M, Liang T, Shi M, Chen H (2013) Graphene-like two-dimensional materials. Chem Rev 113(5):3766–3798. https://doi.org/10.1021/cr300263a

Yang L, Li P, Ma J, Zhang X, Wang X-F et al (2023) MXenes for perovskite solar cells: progress and prospects. J Energy Chem 81:443–461. https://doi.org/10.1016/j.jechem.2023.02.041

You P, Tang G, Yan F (2019) Two-dimensional materials in perovskite solar cells. Mater Today Energy 11:128–158. https://doi.org/10.1016/j.mtener.2018.11.006

Zhang S, Yan Z, Li Y, Chen Z, Zeng H (2015) Atomically thin arsenene and antimonene: semimetal–semiconductor and indirect–direct band-gap transitions. Angew Chem Int Ed 54(11):3112–3115. https://doi.org/10.1002/anie.201411246

Zhou Q, Duan J, Yang X, Duan Y, Tang Q (2020) Interfacial strain release from the WS2/CsPbBr3 van der Waals heterostructure for 1.7 V voltage all-inorganic perovskite solar cells. Angew Chem 132(49):22181–22185. https://doi.org/10.1002/ange.202010252

Zhu J, Xiao G, Zuo X (2020) Two-dimensional black phosphorus: an emerging anode material for lithium-ion batteries. Nano-Micro Lett 12:120. https://doi.org/10.1007/s40820-020-00453-x

Chapter 10
Interfacial Engineering in Perovskite Solar Cells by the Simulation Approaches

Puneet Sehgal and Himani Dua Sehgal

Abstract The exceptional power conversion efficiency (PCE), affordability, and adaptability of perovskite solar cells (PSCs) have made them a game-changer in the photovoltaics industry. The main component, hybrid organic-inorganic halide perovskites, has special qualities like extended charge-carrier diffusion lengths, variable bandgaps, high absorption coefficients, and low-temperature solution processability. With PCEs surpassing 25% within 10 years of development, these characteristics have allowed PSCs to make impressive strides and compete with more well-established technologies like silicon solar cells. The rapid growth of photovoltaic (PV) technologies has been driven by the pressing need for clean, renewable energy sources. Perovskites, a class of materials with the general formula ABX_3, are at the forefront of next-generation solar technology because of their distinctive optoelectronic features. Their quick increase in efficiency—from early reports of 3.8% in 2009 to over 25% today—highlights their revolutionary potential in the photovoltaics industry. The term "perovskite" originates from a naturally occurring mineral, calcium titanate ($CaTiO_3$), discovered in 1839 by Gustav Rose. However, the perovskite materials used in solar cells are typically hybrid organic-inorganic lead halides, such as methylammonium lead iodide ($CH_3NH_3PbI_3$). Research into perovskites as light absorbers began in 2009 when Kojima et al. first demonstrated their application in dye-sensitized solar cells. Since then, PSCs have witnessed unprecedented advancements, with a steep trajectory in efficiency and extensive exploration of their properties.

Keywords Perovskite solar cells (PSCs) · Charge-carrier diffusion lengths · Variable bandgaps · High absorption coefficients

---

P. Sehgal (✉)
Atma Ram Sanatan Dharma College, University of Delhi, Delhi, India

H. D. Sehgal
Shaheed Rajguru College of Applied Sciences for Women, University of Delhi, Delhi, India

M. Kumar (ed.), *Computational Analysis of Perovskite Solar Cells*, Springer Tracts in Electrical and Electronics Engineering,
https://doi.org/10.1007/978-981-95-6521-4_10

Introduction

Perovskite photovoltaic cells have advanced quickly in recent years, greatly increasing their capacity to compete with commercial solar cells such as Si-, CdTe-, and CIGS that are now on the market (Kojima et al. 2009; Kim et al. 2012; Lee et al. 2012a; Burschka et al. 2013; Chen et al. 2015; Zhou et al. 2014; Chen et al. 2017; Jung et al. 2019; Jiang et al. 2019; Wang et al. 2019; Vahedi et al. 2022). Despite their potential, PSCs have environmental safety, scalability, and stability limitations. Perovskite materials can deteriorate due to environmental variables such as oxygen, moisture, and heat stress, impacting device performance. The unique semiconducting properties of Perovskite materials help in the formation of photovoltaic devices like n-i-p (regular) or p-i-n (inverted) structures (Jiang et al. 2016a). Techniques including encapsulation, surface passivation, and compositional engineering are used to improve stability. Additionally, scalable fabrication techniques like roll-to-roll printing and blade coating are being investigated to facilitate commercial deployment. Researchers are looking at lead-free substitutes and recycling techniques to reduce environmental hazards in response to worries about the use of lead in perovskites. Among other solar cells, PSCs based on organic-inorganic hybrid perovskites have become popular due to their rapidly rising power conversion efficiencies (PCEs), ease of processing, and relatively low production costs (Wu et al. 2017). An important turning point for the excitonic dye-sensitizer solar cells (DSSC) was reached in 2009 when the first use of organic-inorganic hybrid perovskites was reported, with a conversion efficiency of 3.9% (Park 2016; Hashemian et al. 2022). A lot of work has been done in the past 5 years to increase PSC longevity and performance. The stability and performance of PSCs can be increased by interface engineering their various layers. For example, P4VP polymer material interface engineering has been instrumental in boosting PSC performance by raising the fill factor to more than 82%. Its enhanced energy level alignment and decreased fault density caused its power conversion efficiency to rise from 17.46% to 20.02% (Zhang et al. 2020). Several other effective strategies have been proposed, including doping of various elements, controlled interfaces with different modifications, modifications of various additives, different engineering solvents, designing the cell configurations as per the application where they are required, and optimizing the design of the absorber layer, hole transport material (HTM), and electron transport material (ETM) are proposed to meet the requirements as whenever they are needed. Enhancing PSC performance was the aim of these strategies (Wei et al. 2018; Liu et al. 2021a, b; Yelzhanova et al. 2022; Xiang et al. 2022; Zhao et al. 2019; Senguttuvan et al. 2024), and they are proved to be beneficial under different applications. Phase change materials (PCMs) were incorporated into photovoltaic panels to lower panel temperature and boost electrical outputs and efficiency (Khanafer et al. 2024). The ability of charge carriers to flow across PSC interfaces is very important to the operation of optoelectronic devices. If the charge carriers encounter a larger barrier or interface density of states, the charge collection or recombination will take place. Buffer layers (BLs) have been widely used to address this problem by enhancing carrier injection to increase the conductivity, lowering trap-states, and minimizing contact resistance in various solar cell designs (Shrotriya et al. 2006). One critical element that improves PSC performance is the deposition of BL between the absorber and HTM layers and Al_2O_3 (aluminium oxide) is the buffer material most frequently employed for this application (Guarnera et al. 2015). In solid-state solar cells, additional materials such SiO_2, ZrO_2, ZnO, glycerol-doped materials and C_{60} are also used as BL, as observed in the past (Pathak et al. 2014; Li et al. 2014; Chang et al. 2016; Jia et al. 2016; Li et al. 2015; Liu et al. 2015). To prevent shunting pathways, it was proposed in a different study to add Al_2O_3 as an insulated mesoporous layer between the absorber and the Spiro-OMeTAD layers. With these devices operating for more than 350 hours, the researchers saw a notable improvement in their long-term stability (Liao et al. 2022). After 350 h of operation, the top-performing cell's initial efficiency (η) of 13% only dropped by 5% (Liao et al. 2022). Moreover, in an experimental work, Al_2O_3 was used at $ZnO/CH_3NH_3PbI_3$ interface with the optimized efficiency of 15.55% (Si et al. 2016). The incorporation of SiO_2 at PEDOT:PSS/$CH_3NH_3PbI_3$ interface presented a PCE of 16% (Xu et al. 2022). According to the literature review, PSCs containing SiO_2 and Al_2O_3 conventional BLs have a lower PCE than Cd Te BL. Additionally, it was investigated that spikes produced with SiO_2 and Al_2O_3 layers and that their VBMs were deeper than CdTe BL. As a result, our BL was better suited to the absorber layer than their band offsets. As a result, Cd Te was thought to be more effective as BL than these conventional materials (Touafek et al. 2020). Additionally, the performance and stability of inverted PSCs were enhanced by the use of bathocuproine (BCP) as BL (Touafek et al. 2020). When BCP was used as a hole-blocking layer between the metal electrode and ETM layer, PSC efficiency rose (Touafek et al. 2020). Additionally, the BCP alterations at the absorber/ETM and absorber/HTM interfaces enhanced PSC stability and efficiency, which enhanced crystallinity and decreased defects (Yakusheva et al. 2022; Chen et al. 2021; Seshaiah and Kim 2024). Between the absorber layers and CuSCN HTM, the BL of CsBr was employed. Additionally, it might passivate the surface flaws of $CsPbI_2Br$ (Ramos et al. 2015). Cadmium telluride continues to be a leading contender in the race to develop affordable photovoltaic systems for terrestrial applications as compared to other BLs. This material has a straight optical bandgap of 1.5 eV, which closely matches the solar spectrum, makes it ideal for photovoltaic energy conversion. Furthermore, the absorption coefficient of CdTe is extremely high. Numerous methods can be used to produce this chemically stable molecule (Nykyruy et al. 2019; Ferekides et al. 2004). It has notable hole mobility and is classified as a p-type semiconductor (Xiao et al. 2018). The long-term stability of PSCs employing CdTe can be impacted by several factors. One serious risk that might cause the breakdown of metal-halide perovskite absorbers and the deterioration of charge transport layers is moisture. Enhancing moisture resistance and effective encapsulation are generally crucial tactics to raise the long-term durability of perovskite solar cells based on CdTe. Additionally, the hydrophobic properties of CdTe shield the absorber from corrosion and moisture (Jiang et al. 2016b). Additionally, CdTe has high UV resistance that minimizes UV-induced degradation in PSCs, enhancing their stability (Gao et al. 2012). An unique, environmentally useful, and extremely sustainable lead-free perovskite structure based on tin (Sn) is demonstrated in one case study (Aslam et al. 2022; Aslam et al. 2023).

Main Properties of Perovskite Solar Cell (PSC) Materials

The extraordinary performance of perovskite solar cells (PSCs) is attributed to the exceptional optoelectronic properties of perovskite materials. Perovskite solar cells (PSCs) have certain properties like high absorption coefficient, i.e., perovskites can absorb sunlight effectively with thin active layers, reducing material usage and cost, by altering the composition (e.g., substituting halide ions), researchers can adjust the bandgap to optimize light absorption thus matching the solar spectrum hence electrons, holes can travel long distances without recombination, ensuring efficient charge extraction; perovskites possess a high tolerance to defects, which helps to maintain high efficiency even with material imperfections and perovskite layers can be fabricated using low-cost, scalable techniques such as spin-coating and inkjet printing. The basic parameters of PSC are electron affinity, energy band gap, electron mobility, etc. (Han et al. 2025; Jha et al. 2023; Haddout et al. 2019; Liu et al. 2016; Qasim et al. 2022; Ezealigo et al. 2019; Boschloo et al. 2006; Tinedert et al. 2020; Madelung 2004; Zhao et al. 2016; Sunny et al. 2021; Zhang et al. 2009; Omarova et al. 2022; El Jouad et al. 2015). By suitably altering the absorbers, ETM, and HTM layers, electron affinity (χ), defect density (N_t), and dopant concentration, the photovoltaic performance can be somewhat improved (Siddique and Rahman 2024).

Device Architecture of Perovskite Solar Cells (PSCs)

Several simulation tools, including APS-1D, AFORS-HET, PC-1D, and SCAPS-1D, are commonly utilized for the theoretical modeling and analysis of solar cell performance (Mojiri et al. 2013; Stangl et al. 2010; Kowsar et al. 2019). One of the most widely adopted among these is SCAPS-1D (Solar Cell Capacitance Simulator), a one-dimensional simulation program developed by the Department of Electronics and Information Systems (ELIS) at the University of Ghent, Belgium. While many of these tools share foundational simulation principles, they differ in aspects such as computational speed, available features, graphical interface, user experience, and overall functionality. SCAPS-1D has been widely applied in research involving perovskite solar cells (PSCs). However, the simulator has certain built-in assumptions that may limit its accuracy in real-world scenarios. A notable limitation is its tendency to idealize conditions—often neglecting real environmental influences such as atmospheric pressure variations or long-term material degradation (Stangl et al. 2010; Kowsar et al. 2019).

The underlying physics of SCAPS-1D is based on a set of fundamental semiconductor equations, which include Poisson's equation, the continuity equations for electrons and holes, and the charge transport equations for both carrier types. Additionally, it incorporates models for the optical absorption coefficient, enabling a more detailed analysis of photon interaction and carrier generation within the device structure (Moiz et al. 2021; Verschraegen and Burgelman 2007; Burgelman et al. 2004; Xiao et al. 2016). Typical PSCs are composed of multiple functional layers: a transparent conducting electrode (TCE) such as indium tin oxide (ITO), which allows incident light to pass through; an electron transport layer (ETL), often using titanium dioxide (TiO_2), which facilitates efficient electron extraction while blocking hole movement; a perovskite absorber layer, which serves as the primary site for light absorption and electron-hole pair generation; a hole transport layer (HTL) made from materials like Spiro-OMeTAD or nickel oxide (NiO); and a metallic back electrode, typically silver or gold, for charge collection.

Figure 10.1a provides a schematic representation of the structure and optical design of a perovskite solar cell incorporating photonic crystal (PC) elements. Figure 10.1b features a scanning electron microscope (SEM) image that shows the cross-sectional view of a PC-enhanced cell, fabricated through alternate layering of SiO_2 nanoparticles and TiO_2 over a glass substrate. Meanwhile, Fig. 10.1c displays the visual variations (color-coded from blue to red) in devices incorporating different photonic crystal configurations (Zhang et al. 2015).

PSC architectures are generally divided into two categories: planar and mesoporous. The planar type consists of a smooth, continuous perovskite layer, as shown in Fig. 10.2. In contrast, the mesoporous structure includes a porous scaffold, often made of mesoporous TiO_2, which aids in improving charge transport and collection. Figure 10.2a outlines a generic hybrid perovskite solar cell configuration, including optional layers that may enhance performance but are not strictly required. Figure 10.2b depicts the corresponding energy level diagram (in eV) for key materials such as $CH_3NH_3PbI_3$ and TiO_2 (Han et al. 2025).

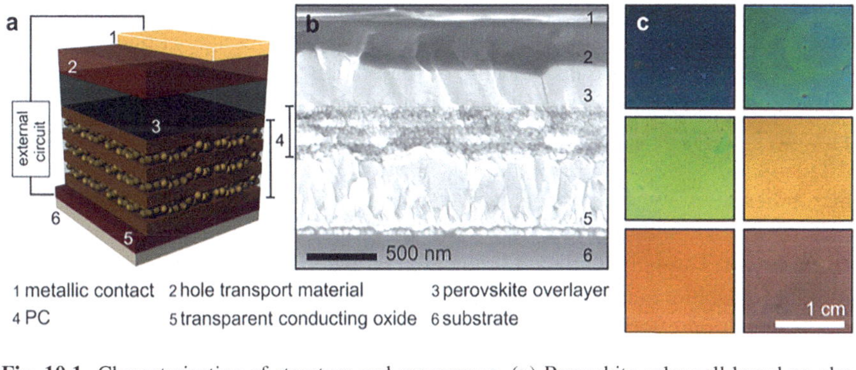

Fig. 10.1 Characterization of structure and appearance. (**a**) Perovskite solar cell based on photonic crystals (PCs). (**b**) Cross-section of a PC-based solar cell captured by a secondary electron scanning electron microscope. (**c**) Devices that integrate various PCs display colors (ranging from blue to red) (Zhang et al. 2015)

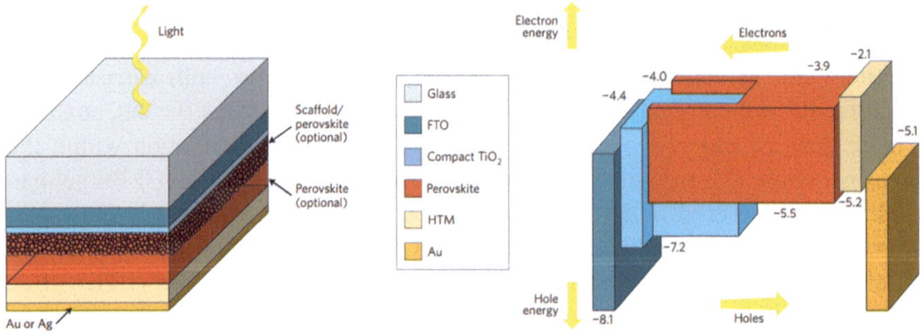

Fig. 10.2 (**a**) The organic-inorganic halide solar cell's structure corresponds to a simple planar thin film cell, and (**b**) displays the vacuum energy levels (in eV) for corresponding materials ($CH_3NH_3PbI_3$ perovskite, conducting TiO_2 scaffold) (Han et al. 2025)

Advantages, Challenges, and Opportunities of Perovskite Solar Cells (PSCs)

Compared to traditional silicon solar cells, PSCs offer several advantages like cost-effectiveness, i.e., perovskites can be synthesized and processed using low-temperature, energy-efficient methods. PSCs have achieved efficiencies comparable to silicon cells, with the potential for tandem integration to surpass the Shockley-Queisser limit. Thin-film PSCs enable lightweight and flexible devices suitable for portable and wearable applications. Techniques like roll-to-roll printing allow for large-scale production at low costs. Despite their potential, PSCs face critical challenges. They are sensitive to environmental factors such as moisture, oxygen, and heat, leading to degradation over time. Improving device stability is crucial for commercial viability. The presence of lead in most perovskite formulations raises environmental and health concerns. Lead-free alternatives, such as tin-based perovskites, are actively being explored. Transitioning from laboratory-scale fabrication to large-scale production involves overcoming challenges such as uniformity and defect control. Efforts to address these challenges include compositional engineering, encapsulation techniques, and the development of advanced device architectures. Additionally, tandem solar cells combining perovskites with silicon or other materials hold promise for breaking efficiency records. The field of PSCs is rapidly advancing, with ongoing research focused on improving stability, efficiency, and environmental safety. As these challenges are addressed, PSCs have the potential to revolutionize the solar energy industry, contributing to a sustainable and carbon-neutral future. Through continued innovation in materials science, device engineering, and scalable manufacturing, PSCs are poised to play a pivotal role in global energy transition efforts. Thus, PSCs represent a groundbreaking advancement in photovoltaic technology, combining exceptional performance with the promise of low-cost, sustainable energy generation. As the field matures, the potential for widespread adoption and integration into diverse applications remains immense.

Simulation approaches have become indispensable in addressing these challenges, enabling researchers to model physical and chemical processes at interfaces, optimize materials and device architectures, and predict the impact of various interfacial engineering strategies. This chapter provides an in-depth discussion of simulation methodologies and their application to interfacial engineering in PSCs.

Fundamentals of Interfaces in Perovskite Solar Cells (PSCs)

Main Interfaces in Perovskite Solar Cells (PSCs)

In perovskite solar cells (PSCs), the interfaces between different functional layers are vital in dictating overall device efficiency, long-term stability, and scalability. Optimal device performance—particularly efficient charge extraction, suppressed recombination losses, and operational durability—is highly dependent on the quality and characteristics of these interfacial regions. A thorough understanding and strategic optimization of these interfaces are therefore essential for the development of high-performance, stable PSCs.

PSCs feature multiple internal interfaces, each playing a specific role in charge separation, transport, and recombination dynamics. Key interfaces include:

- The electron transport layer (ETL)/perovskite interface.
- The perovskite/hole transport layer (HTL) interface.
- Front and back electrode interfaces.
- Grain boundaries within the perovskite absorber layer.

Among these, the ETL/perovskite interface is particularly critical for facilitating the extraction and transport of photogenerated electrons while simultaneously blocking hole movement. Several parameters impact the efficiency of this interface, with energy band alignment being a primary factor. For efficient electron extraction, the conduction band edge of the ETL must be equal to or lower than that of the perovskite. Any misalignment can result in energy losses and reduced device efficiency. Additionally, non-radiative recombination due to trap states at the interface can further compromise performance. To address these issues, various surface passivation techniques have been employed—such as incorporating interfacial buffer layers or applying self-assembled monolayers (SAMs)—to minimize interfacial defects and improve charge transfer. Common ETL materials include titanium dioxide (TiO_2), tin oxide (SnO_2), zinc oxide (ZnO), and fullerene derivatives. While TiO_2 has been widely adopted due to its favorable electronic properties, its susceptibility to photocatalytic degradation under UV exposure has driven interest in alternatives like SnO_2, which offer improved chemical and UV stability. Figure 10.3 illustrates the energy band alignment at the ETL/perovskite interface, highlighting the importance of matching conduction bands for efficient electron transfer (Yang et al. 2019).

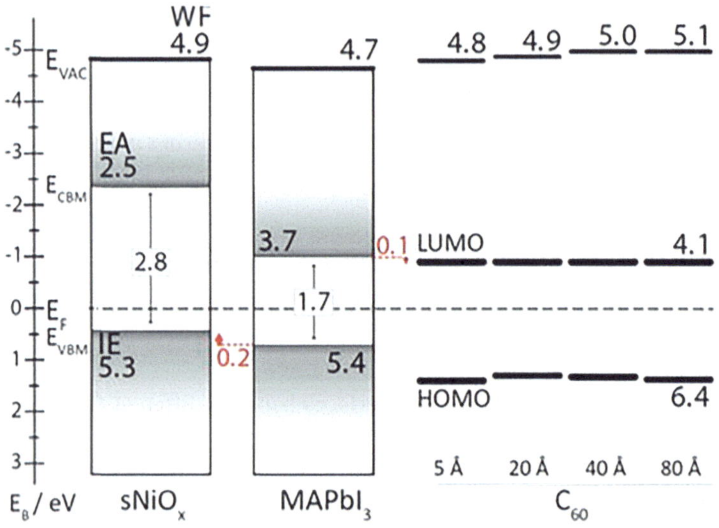

Fig. 10.3 Shows the energy band diagram of ETL/Perovskite Interface (Yang et al. 2019)

The perovskite/hole transport layer (HTL) interface plays a critical role in enabling efficient hole extraction and transport, while simultaneously acting as a barrier to prevent electron backflow. The characteristics of this interface significantly affect key device parameters such as the open-circuit voltage (Voc) and fill factor (FF). For optimal hole transfer, there must be a close alignment between the valence band levels of the perovskite and the HTL. Misalignment at this interface can lead to increased resistance losses and enhanced carrier recombination, both of which degrade device performance. To minimize recombination and enhance interfacial stability, surface passivation strategies are commonly employed. These include the use of Lewis base molecules, ammonium halides, and two-dimensional (2D) perovskite interlayers, which help in mitigating interfacial defects and stabilizing the interface.

HTL materials can be either organic—such as Spiro-OMeTAD and poly(triarylamine) (PTAA)—or inorganic, including materials like nickel oxide (NiO) and copper thiocyanate (CuSCN). While organic HTLs often offer excellent energy level alignment and are easy to process, they typically require chemical dopants to enhance conductivity, which can adversely impact long-term stability. In contrast, inorganic HTLs are generally more robust and stable but may demand advanced interfacial engineering to achieve efficient charge transfer and minimize energy losses. Figure 10.4 presents the energy band diagram of the HTL/perovskite interface, illustrating the importance of proper band alignment for efficient hole extraction (Yang et al. 2019).

In perovskite solar cells (PSCs), the electrode interfaces—comprising the front transparent conducting electrode and the rear metal contact—are essential for efficient charge collection and transport. To minimize resistive losses and enhance

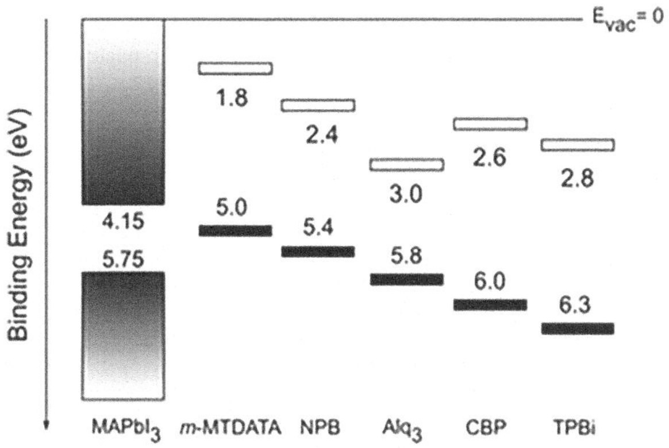

Fig. 10.4 Shows the energy band diagram of HTL/Perovskite Interface (Yang et al. 2019)

device stability, these interfaces must be carefully engineered. For the front contact, materials such as indium tin oxide (ITO) and fluorine-doped tin oxide (FTO) are commonly employed due to their excellent electrical conductivity and optical transparency. To improve interfacial compatibility with the electron transport layer (ETL), additional buffer layers like compact TiO_2 or SnO_2 are often incorporated. The back electrode typically consists of metals such as aluminum (Al), silver (Ag), or gold (Au), chosen for their ability to form good electrical contact with the hole transport layer (HTL) while maintaining a minimal energy barrier. However, the high cost and potential instability of noble metals have encouraged the exploration of low-cost, carbon-based electrodes, which offer promising scalability and chemical stability for large-area devices.

One major challenge at the electrode interfaces is the potential for chemical interaction between metal contacts and halide ions from the perovskite layer, which can lead to corrosion and performance degradation. To address this, protective barrier layers such as molybdenum oxide (MoOx) or aluminum oxide (Al_2O_3) are often introduced to suppress direct contact and enhance long-term durability. These interfacial strategies are illustrated in Fig. 10.5. In the control device, a NiOx nanoparticle layer is employed as the HTL, paired with a blade-coated carbon electrode as the anode. To enhance interfacial contact between the perovskite layer and the HTL, a spin-coated polythiophene (P3HT) layer is introduced beneath the NiOx layer (Fig. 10.5a). P3HT is chosen for its versatile functionality as a hole transport material in PSCs, although other conductive polymers are also widely utilized due to their high specific capacitance and ease of processing (Lu et al. 2023).

Thus, Fig. 10.5a shows the schematic of the fabrication process for PSCs featuring a bilayer HTL and carbon electrode, Figs. 10.5b, d, f illustrates the SEM images of pristine $MAPbI_3$, $MAPbI_3$ with NiOx, and $MAPbI_3$ with P3HT/NiOx, respectively while Figs. 10.5c, e, g depicts the surface morphology of the corresponding films after blade coating with a carbon paste solvent (cyclohexanone and phenolic resin) and Fig. 10.5(h) shows the X-ray diffraction (XRD) patterns of perovskite films with different surface modifications (Lu et al. 2023).

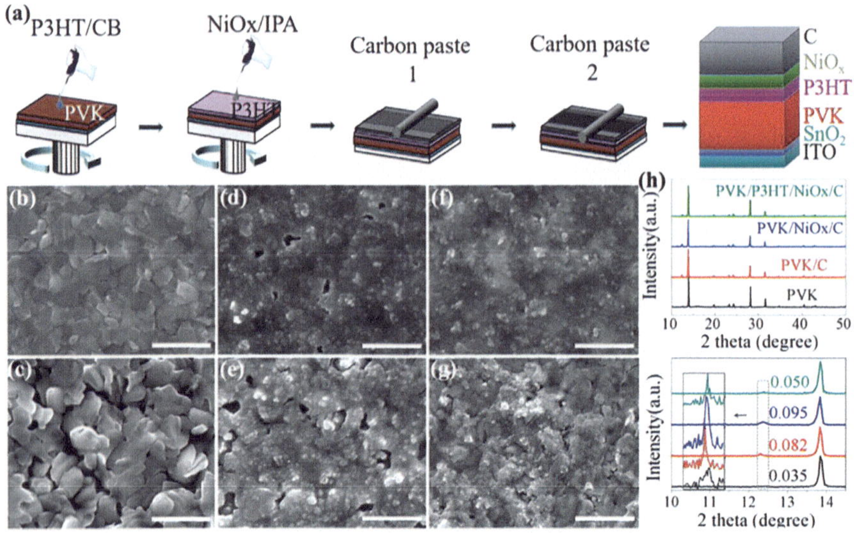

Fig. 10.5 (**a**) Schematic illustration of the bilayer HTL carbon electrode perovskite solar cell. (**b**, **d**, **f**): SEM pictures of $MAPbI_3$ and $MAPbI_3$ spin-coated with NiOx and P3HT/NiOx. (**c**, **e**, **g**): The matching surfaces of the three films mentioned above following blade coating with carbon paste solvents (phenolic resin and cyclohexanone). (**h**) Perovskite film XRD pattern with various surface coatings (Lu et al. 2023)

Grain boundaries within polycrystalline perovskite films serve as internal organic interfaces and play a crucial role in governing charge transport and recombination dynamics. These boundaries often contain defect states that act as non-radiative recombination centers, especially along the grain edges, which negatively impact carrier lifetimes and overall device efficiency. To mitigate the adverse effects of these defect-rich regions, surface passivation techniques are commonly employed. This includes the incorporation of organic additives or alkali metals, which help reduce the defect density and suppress recombination. However, ion migration along grain boundaries remains a concern, as it can influence both the stability and the hysteresis behavior of the device. Non-radiative recombination at these interfacial sites results in a loss of open-circuit voltage (Voc) and reduced power conversion efficiency (PCE). Additionally, ion movement and chemical degradation triggered by environmental factors—such as heat, moisture, and ultraviolet (UV) exposure—further compromise long-term device reliability.

One of the notable consequences of poor interfacial control is hysteresis in the current–voltage (I–V) curves, which manifests as inconsistent performance depending on the scan direction. This is often attributed to ion migration and imperfect interfacial contact, both of which must be addressed to ensure stable and reproducible device operation.

Strategies for Interface Optimization and Advanced Characterization Techniques in Perovskite Solar Cells (PSCs)

To overcome these challenges, several approaches have been developed in the past such as surface passivation in which defect states and recombination at interfaces are decreased by incorporating passivation layers or molecules (such as alkali halides and phenyl-C61-butyric acid methyl ester, or PCBM). Also, interfacial Layers are developed where energy level alignment and stability are enhanced by the introduction of ultrathin layers, such as self-assembled monolayers (SAMs) or transition metal oxides (e.g., MoOx, WO_3). A glass/FTO substrate, a SnO_2 electron transport layer, a $CsPbI_2Br$ perovskite layer, and a carbon layer constitute the all-inorganic PSC device structure (Figure 10.6a). Spin-coating 4-ADPA in ethyl acetate at varying concentrations creates an interlayer on top of the perovskite layer (Figure 10.6b). Figure 10.6c illustrates the 4-ADPA passivation procedure. After applying 4-ADPA and thermally annealing the perovskite film with a hot air gun, it is thermally annealed again (Sun et al. 2023). Similarly, chemical engineering is done in which interfacial characteristics can be improved by altering the composition of the perovskite or nearby layers (for example, by partially substituting A, B, or X-site ions). Grain size and grain boundary defects can be improved by optimizing the perovskite layer's development process (for example, by solvent engineering or thermal annealing). By shielding delicate interfaces from external influences, protective encapsulation layers improve device stability. Understanding interfaces requires detailed characterization. Techniques such as X-ray Photoelectron

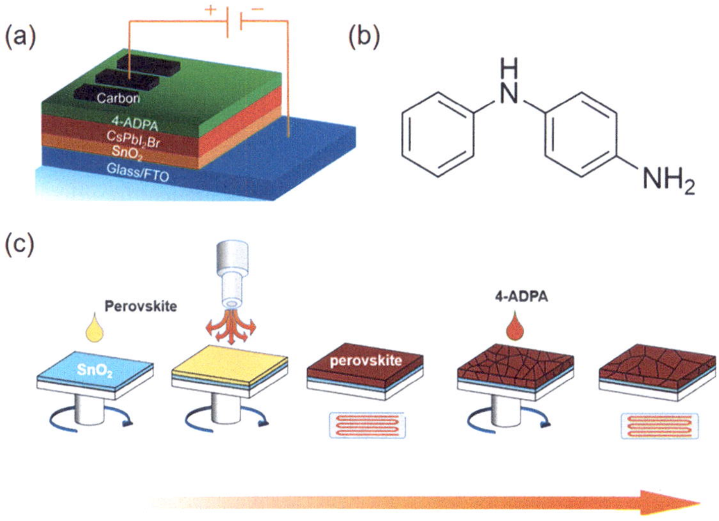

Fig. 10.6 (**a**) The schematic diagram of 4-ADPA modified all-inorganic PSC device structure, (**b**) the molecular structure of 4-ADPA, and (**c**) schematic diagram of 4-ADPA passivation process (Sun et al. 2023)

Spectroscopy (XPS) and Time-of-Flight Secondary Ion Mass Spectrometry (ToF-SIMS) in which we analyze chemical composition and interfacial reactions, Photoluminescence (PL) Mapping which evaluates defect density and recombination behaviour, Transient Photovoltage (TPV) and Transient Photocurrent (TPC) which provide insights into charge dynamics and recombination at interfaces and Atomic Force Microscopy (AFM) and Kelvin Probe Force Microscopy (KPFM) to study surface morphology and work function variations. Thus, various techniques have been developed in the past and used under specific conditions. Hence, solving interfacial issues is essential to PSCs' continued development. Novel strategies for achieving stable and effective PSCs include tandem device topologies, quantum-dot-based interfacial layers, and machine learning for interface tuning. A better knowledge of interfacial dynamics will accelerate PSCs' transition from lab study to large-scale implementation. By resolving these interfacial issues, PSCs can realize their full potential as cutting-edge renewable energy technology and make a substantial contribution to global sustainability goals.

Simulation Approaches for Interfacial Engineering in Perovskite Solar Cells (PSCs)

Simulation approaches play a vital role in advancing interfacial engineering for perovskite solar cells (PSCs). They provide insights into the physical, chemical, and electronic phenomena occurring at interfaces, enabling the design and optimization of materials and structures. By leveraging computational tools, researchers can predict interfacial properties, understand degradation mechanisms, and propose strategies for improving device performance and stability. Various simulation approaches used in interfacial engineering, such as **Density Functional Theory (DFT),** are widely employed to investigate the atomic-scale properties of materials at interfaces. It provides insights into electronic structure, band alignment, defect states, and chemical interactions. DFT is used to determine the energy levels of the perovskite, ETL, and HTL to guarantee correct band alignment for effective charge extraction. DFT assists in determining passivation techniques to reduce trap states by simulating point defects and vacancies. It assesses the binding energy between layers (such as perovskite/HTL or ETL/perovskite) and offers information on the stability of the interface. Potential interfacial materials, including surface passivation or novel transport layers, are screened using DFT. DFT has an advantage of high accuracy for small systems, especially in predicting electronic and chemical properties and challenges of Computationally intensive, limiting its application to small models (e.g., <1000 atoms). Another simulation technique of **Molecular Dynamics (MD) Simulations** models the dynamic behaviour of atoms and molecules over time, enabling the study of interfacial interactions and structural evolution. MD models shed light on adhesion and structural integrity by revealing how layers develop and interact at the atomic level. One important aspect affecting stability and hysteresis in PSCs is ion transport pathways across interfaces, which are studied using MD. To forecast long-term stability, it simulates how encapsulation materials and perovskites interact. It offers insights on interfacial processes that vary with time, requires parameterization of force fields, and is restricted to limited spatial domains and short timeframes (nanoseconds). **Drift-diffusion modeling** solves carrier transport equations to simulate charge generation, transport, recombination, and extraction within PSCs. This approach evaluates how interfacial properties influence charge mobility and extraction. It identifies recombination losses at interfaces and proposes ways to minimize them. Drift-diffusion models help in optimizing layer thickness, doping levels, and material selection for better interfacial performance. It enables macroscopic device-level simulations, directly connecting interfacial properties to photovoltaic performance, and relies on empirical input parameters, which may require experimental validation. **Ab Initio Molecular Dynamics (AIMD)** combines quantum mechanics (e.g., DFT) with molecular dynamics to study time-dependent processes at the atomic scale. AIMD simulates the interaction of water, oxygen, or other environmental factors with interfaces to understand degradation pathways. It models the behaviour of interfaces under thermal stress, providing insights into phase stability. AIMD predicts interfacial reactions, such as halide migration or metal electrode corrosion. It has High accuracy for dynamic interfacial phenomena at the atomic level. It is computationally expensive, restricting its use to small systems and short timescales. **Finite Element Modeling (FEM)** is used to simulate mechanical, thermal, and electrical properties across interfaces and throughout the entire device. FEM evaluates the mechanical stress at interfaces caused by thermal expansion mismatch between layers. It predicts heat dissipation across interfaces, which is critical for device stability. FEM models large-area devices, identifying potential interfacial issues in scaled-up fabrication. It is suitable for large-scale and multi-physics problems and requires material-specific input parameters, which may not always be available. **Kinetic Monte Carlo (KMC) Simulations** model charge carrier dynamics and recombination processes at interfaces by simulating probabilistic events. KMC tracks individual carrier movements to study how interfacial defects influence transport and recombination. It quantifies the effect of trap densities at interfaces on device performance. KMC models long-term stability by simulating ion migration and degradation processes over extended timescales. It provides insights into stochastic processes like recombination and ion migration and is also computationally intensive for large systems or long timescales. **Hybrid Multiscale Simulations** approaches combine multiple simulation techniques, such as DFT with MD or drift-diffusion with FEM, to bridge different spatial and temporal scales. Hybrid simulations link atomic-scale phenomena (e.g., defect chemistry) with device-scale performance (e.g., efficiency and stability). By integrating quantum, atomistic, and macroscopic models, hybrid simulations provide a holistic understanding of interfaces. It captures multi-scale phenomena, providing a complete picture of interfacial behaviour and complex implementation and high computational requirements. **Machine Learning and Data-Driven** approaches large datasets from simulations and experiments to predict interfacial properties and optimize materials. ML algorithms predict new interfacial materials with desired properties. Data-driven models identify the most critical interfacial parameters affecting device performance. ML helps map the relationship between interfacial defects and stability. It has rapid predictions and the ability to handle large datasets and requires high-quality training data, which may be limited for specific systems.

Future Directions to Use Simulation Approaches in Perovskite Solar Cells (PSCs)

Various future aspects should be noted while using these simulation methods as per the requirements like integrated frameworks in which we combine various simulation approaches into unified frameworks for a comprehensive analysis of interfacial phenomena. Real-time simulations, which involve the development of algorithms capable of simulating dynamic interfacial processes in real time. Stronger integration of simulations with experimental studies to validate predictions and refine models. High-Performance Computing (HPC), i.e. leveraging HPC resources for large-scale simulations of complex interfacial systems. Simulation approaches are indispensable for understanding and optimizing interfaces in PSCs. Techniques like DFT, MD, drift-diffusion modeling, and machine learning enable researchers to tackle interfacial challenges systematically. By providing predictive insights and guiding experimental efforts, simulations accelerate the development of efficient, stable, and scalable PSCs, paving the way for their commercialization and widespread adoption in renewable energy technologies.

Various Case Studies to Understand Simulation Approaches in Perovskite Solar Cells (PSCs)

First Case Study

In the first case study, we studied the Band Alignment at the ETL/Perovskite Interface Using Density Functional Theory (DFT) to optimize band alignment between the electron transport layer (ETL) and the perovskite layer for improved charge extraction. We used the simulation Technique of Density functional theory (DFT) with hybrid functional corrections (e.g., HSE06) with the use of various materials like ETLs: TiO_2, SnO_2, and Perovskite: Methylammonium lead iodide ($MAPbI_3$). and various parameters are studied like Bandgap, conduction band minima (CBM), valence band maxima (VBM), and many more. It was found that the energy levels of TiO_2 showed a conduction band offset (CBO) of ~0.3 eV with $MAPbI_3$, and SnO_2 exhibited a smaller CBO (~0.1 eV), favouring efficient charge extraction. The effect of Surface Modification showed that fluorination of SnO_2

reduced surface trap states, further improving alignment (National Renewable Energy Laboratory (NREL) n.d.; Lee et al. 2012b; Yin et al. 2016). Figure 10.3 shows the Energy level diagram of ETL/perovskite interfaces, and Table 10.1 shows the band alignment parameters for different ETL materials (National Renewable Energy Laboratory (NREL) n.d.; Lee et al. 2012b; Yin et al. 2016).

Second Case Study

In the second case study, the ion migration and interfacial stability using molecular dynamics (MD) were studied to study the impact of ion migration at the perovskite/HTL interface and its effect on stability. The Simulation Technique of Molecular dynamics (MD) using force fields parameterized for iodide ions (I^-) was used. Various materials are used, like Perovskite: $MAPbI_3$ and HTLs: Spiro-OMeTAD and NiO. Various parameters are studied in simulation, like electric field, temperature, and ion diffusion coefficients. It was found that Iodide ions preferentially accumulated at the Spiro-OMeTAD interface, forming defect clusters and NiO interfaces that showed reduced ion migration due to stronger bonding. This could be due to the passivation effect, i.e., the application of 2D perovskite passivation layers significantly reduced ion migration. Figure 10.4 shows the MD snapshot of iodide ion distribution at the perovskite/HTL interface, and Table 10.2 shows the Ion migration rates and interface stability metrics (National Renewable Energy Laboratory (NREL) n.d.; Lee et al. 2012b; Yin et al. 2016).

Third Case Study

In another case study, recombination Loss Analysis Using Drift-Diffusion Modeling was studied to quantify recombination losses at the perovskite/transport layer interface and propose mitigation strategies. The technique of Drift-diffusion modeling

Table 10.1 Band alignment parameters for different ETL materials (National Renewable Energy Laboratory (NREL) n.d.; Lee et al. 2012b; Yin et al. 2016)

Material	Bandgap (eV)	CBM (eV)	VBM (eV)	CBO with $MAPbI_3$ (eV)
TiO_2	3.2	−4.0	−7.2	0.3
SnO_2	3.6	−4.2	−7.8	0.1

Table 10.2 Ion migration rates and interface stability metrics (National Renewable Energy Laboratory (NREL) n.d.; Lee et al. 2012b; Yin et al. 2016)

Interface	Ion migration rate (cm^2/s)	Defect density (cm^{-3})	Stability (Hours)
$MAPbI_3$/Spiro-OMeTAD	1.2×10^{-7}	5.4×10^{1}·	500
$MAPbI_3$/NiO	4.6×10^{-8}	1.8×10^{1}·	800

Table 10.3 Recombination rates and device performance metrics (National Renewable Energy Laboratory (NREL) n.d.; Lee et al. 2012b; Yin et al. 2016)

Interface	Recombination rate (cm^{-2}/s)	V_{oc} (V)	PCE (%)
Unpassivated	1.2×10^{-3}	0.95	18.5
SAM-passivated	3.6×10^{-3}	1.15	23.5

Table 10.4 Stability metrics under moisture exposure (National Renewable Energy Laboratory (NREL) n.d.; Lee et al. 2012b; Yin et al. 2016)

Material	Water adsorption energy (eV)	Efficiency retention (%)
$MAPbI_3$	−0.8	65
2-D perovskite	−0.2	90

using TCAD tools was used. Various parameters like Defect density, recombination rate, and charge mobility were studied. The findings were recombination Loss, i.e., passivated interfaces exhibited recombination rates >10^{-3} cm^2/s and self-assembled monolayers (SAMs) reduced recombination by ~70%. Thus, SAM-treated interfaces achieved a V_{oc} improvement of 0.2 V and a 5% increase in PCE. Table 10.3 shows the recombination rates and device performance metrics (National Renewable Energy Laboratory (NREL) n.d.; Lee et al. 2012b; Yin et al. 2016).

Fourth Case Study

In the last case Study, degradation Mechanisms Under Moisture Exposure Using AIMD were studied to study moisture-induced interfacial degradation in PSCs. Various simulation technique was used, like Ab initio molecular dynamics (AIMD) for water interaction with perovskite surfaces with materials $MAPbI_3$ and 2D perovskites. It was found that water molecules disrupted Pb–I bonds, leading to iodide release and material decomposition, and 2D perovskites exhibited hydrophobic properties, repelling water molecules. The devices with hydrophobic interfacial layers retained 90% of their efficiency after 1000 hours of moisture exposure. Table 10.4 shows the Stability metrics under moisture exposure (National Renewable Energy Laboratory (NREL) n.d.; Lee et al. 2012b; Yin et al. 2016).

Simulation tools provide unparalleled insights into interfacial interactions in PSCs. Researchers can use techniques like drift-diffusion modeling, DFT, MD, and AIMD to enhance material properties, reduce recombination losses, and increase device stability. The case studies demonstrate how computational techniques enable focused interfacial engineering, paving the way for the commercialization of incredibly stable and efficient PSCs. Simulation approaches continue to evolve, enabling Real-time modeling of operational interfaces. Integration with machine learning for predictive analysis and development of universal design principles for interfacial engineering.

Conclusion

Interfacial engineering is important for improving PSC performance and reliability. Simulation techniques provide a robust framework for understanding and optimizing interfacial phenomena, paving the way for the next generation of high-efficiency, stable solar cells. From this perspective, we have highlighted the role that interfaces play in minimizing nonradiative recombination and achieving the S-Q efficiency limit. Specific transport and interface design techniques have been suggested based on the carrier dynamics in PSCs. High electrical/hole conductivity for carrier transport, high electron/hole mobility, and appropriate energy levels for carrier extraction should all be processed by a good ETL/HTL. Additionally, carrier recombination is suppressed by interfacial alteration. Even though PSCs are developing quickly, stability, wide-area scalability, and flexibility place greater demands on the interface qualities for successful commercialization. Environmentally stable counterparts could be used to replace these moisture, heat, and light-sensitive materials. For example, SnO_2 could be used as an ETL in place of TiO_2 because of its sensitivity to UV light, and doping-free organic material could be used as an HTL instead of Spiro-OMeTAD with hygroscopic Li-TFSI doping. Transport layers for flexible and large-area PSCs should be made at low temperatures (< 150 °C) that is consistent with routine industrial production. It was possible to prepare ZnO as a few metal oxides at low temperatures. ZnO's surface needs to be chemically altered since it interacts with organometal halide perovskites. Large-scale roll-to-roll preparation is suitable for organic materials with flexibility and solution-processing capabilities. Furthermore, controlling the nucleation and development of perovskites is challenging in scalable preparation. Effective methods for ETL or HTL include constructing two-dimensional ETL or HTL to regulate the crystallization kinetics of perovskite and adding surface-organic groups to customize the surface energy of ETL or HTL. We predict that a thorough analysis of PSC interfaces will yield a PCE that is nearly equal to the theoretical value within 0.1 cm^2 and maintain 90% PCE as the area grows to about 800 cm^2, which is comparable to silicon-based solar cells with the same area.

References

Aslam S, Farooqi AS, Rahman MYA, Samsuri SAM (2022) Titanium-based vacancy-ordered double halide family in perovskite solar cells. Phys Status Solidi A 219(14):2100671. https://doi.org/10.1002/pssa.202100671

Aslam S, Bakhsh S, Yusof Y, Rahman MYA, Ibrahim AR, Samsuri SAM (2023) Structural and optical properties of vacancy-ordered double halide perovskites, Cs_2TiI_6 films. Mater Sci Eng B 296:116645. https://doi.org/10.1016/j.mseb.2023.116645

Boschloo G, Edvinsson T, Hagfeldt A (2006) Dye-sensitized nanostructured ZnO electrodes for solar cell applications. In: Nanostructured materials for solar energy conversion. Elsevier, Amsterdam, pp 227–254

Burgelman M, Verschraegen J, Degrave S, Nollet P (2004) Modeling thin-film PV devices. Prog Photovolt Res Appl 12(2–3):143–153. https://doi.org/10.1002/pip.526

Burschka J, Pellet N, Moon S-J, Humphry-Baker R, Gao P, Nazeeruddin MK, Grätzel M (2013) Sequential deposition as a route to high-performance perovskite-sensitized solar cells. Nature 499(7458):316–319. https://doi.org/10.1038/nature12340

Chang C-Y, Huang W-K, Wu J-L, Chang Y-C, Lee K-T, Chen C-T (2016) Room temperature solution-processed n-doped zirconium oxide cathode buffer layer for efficient and stable organic and hybrid perovskite solar cells. Chem Mater 28(2):242–251. https://doi.org/10.1021/acs.chemmater.5b04137

Chen W, Wu Y, Yue Y, Liu J, Zhang W, Yang X, Chen H, Bi E, Ashraful I, Grätzel M (2015) Efficient and stable large-area perovskite solar cells with inorganic charge extraction layers. Science 350(6263):944–948. https://doi.org/10.1126/science.aad1015

Chen H, Ye F, Tang W, He J, Yin M, Wang Y, Xie F, Bi E, Yang X, Grätzel M (2017) A solvent- and vacuum-free route to large-area perovskite films for efficient solar modules. Nature 550(7674):92–95. https://doi.org/10.1038/nature23877

Chen R, Long B, Wang S, Liu Y, Bai J, Huang S, Li H, Chen X (2021) Efficient and stable perovskite solar cells using bathocuproine bilateral-modified perovskite layers. ACS Appl Mater Interfaces 13(21):24747–24755. https://doi.org/10.1021/acsami.1c04223

El Jouad Z, Morsli M, Louarn G, Cattin L, Addou M, Bernède JC (2015) Improving the efficiency of sub-phthalocyanine-based planar organic solar cells through the use of MoO_3/CuI double anode buffer layer. Sol Energy Mater Sol Cells 141:429–435. https://doi.org/10.1016/j.solmat.2015.06.016

Ezealigo BN, Nwanya AC, Simo A, Osuji RU, Bucher R, Maaza M, Ezema FI (2019) Optical and electrochemical capacitive properties of copper (I) iodide thin film deposited by SILAR method. Arab J Chem 12(8):5380–5391. https://doi.org/10.1016/j.arabjc.2016.11.014

Ferekides C, Balasubramanian U, Mamazza R, Viswanathan V, Zhao H, Morel D (2004) CdTe thin film solar cells: device and technology issues. Sol Energy 77(6):823–830. https://doi.org/10.1016/j.solener.2004.06.007

Gao X, Wu J, Wei X, He C, Wang X, Guo G, Pu Q (2012) Facile one-step photochemical synthesis of water-soluble CdTe(S) nanocrystals with high quantum yields. J Mater Chem 22(13):6367–6373. https://doi.org/10.1039/c2jm16279e

Guarnera S, Abate A, Zhang W, Foster JM, Richardson G, Petrozza A, Snaith HJ (2015) Improving the long-term stability of perovskite solar cells with a porous Al_2O_3 buffer layer. J Phys Chem Lett 6(3):432–437. https://doi.org/10.1021/jz5027924

Haddout A, Raidou A, Fahoume M, Elharfaoui N, Lharch M (2019) Influence of CZTS layer parameters on cell performance of kesterite thin-film solar cells. In: Proceedings of the 1st international conference on electronic engineering and renewable energy: ICEERE 2018, 15-17 April 2018, Saidia, Morocco. Springer, Cham, pp 640–646. https://doi.org/10.1007/978-3-319-97816-1_59

Han J, Park K, Tan S, Vaynzof Y, Xue J, Diau EW-G, Bawendi MG, Lee J-W, Jeon I (2025) Perovskite solar cells. Nat Rev Methods Primers 5(1):3. https://doi.org/10.1038/s43586-024-00373-9

Hashemian M, Jafarmadar S, Khalilarya S, Faraji M (2022) Energy harvesting feasibility from photovoltaic/thermal (PV/T) hybrid system with Ag/Cr_2O_3-glycerol nanofluid optical filter. Renew Energy 198:426–439. https://doi.org/10.1016/j.renene.2022.07.015

Jha PK, Chourasia NK, Srivastava A, Sharma AK, Kumar R, Sharma S, Kumar M, Chourasia RK (2023) Study of eco-friendly organic–inorganic heterostructure $CH_3NH_3SnI_3$ perovskite solar cell via SCAPS simulation. J Electron Mater 52:4321–4329. https://doi.org/10.1007/s11664-023-09758-x

Jia X, Zhang L, Luo Q, Lu H, Li X, Xie Z, Yang Y, Li Y-Q, Liu X, Ma C-Q (2016) Power conversion efficiency and device stability improvement of inverted perovskite solar cells by using a ZnO: PFN composite cathode buffer layer. ACS Appl Mater Interfaces 8(29):18410–18417. https://doi.org/10.1021/acsami.6b05145

Jiang Q, Zhang L, Wang H, Yang X, Meng J, Liu H, Yin Z, Wu J, Zhang X, You J (2016a) Enhanced electron extraction using SnO_2 for high-efficiency planar structure HC $(NH_2)_2PbI_3$-based perovskite solar cells. Nat Energy 2:1–7. https://doi.org/10.1038/nenergy.2016.18

Jiang Z, Matras-Postolek K, Yang P (2016b) Hydrophobic CdSe and CdTe quantum dots: Shell coating, shape control, and self-assembly. RSC Adv 6(27):25656–25661. https://doi.org/10.1039/C6RA00388A

Jiang Q, Zhao Y, Zhang X, Yang X, Chen Y, Chu Z, Ye Q, Li X, Yin Z, You J (2019) Surface passivation of perovskite film for efficient solar cells. Nat Photonics 13(7):460–466. https://doi.org/10.1038/s41566-019-0398-2

Jung EH, Jeon NJ, Park EY, Moon CS, Shin TJ, Yang T-Y, Noh JH, Seo J (2019) Efficient, stable and scalable perovskite solar cells using poly (3-hexylthiophene). Nature 567(7749):511–515. https://doi.org/10.1038/s41586-019-1036-3

Khanafer K, Al-Masri A, Marafie A, Vafai K (2024) Thermal performance of solar photovoltaic panel in hot climatic regions: applicability and optimization analysis of PCM materials. Numer Heat Transf A Appl 85(16):1612–1632. https://doi.org/10.1080/10407782.2024.1234567

Kim H-S, Lee C-R, Im J-H, Lee K-B, Moehl T, Marchioro A, Moon S-J, Humphry-Baker R, Yum J-H, Moser JE (2012) Lead iodide perovskite sensitized all-solid-state submicron thin film mesoscopic solar cell with efficiency exceeding 9%. Sci Rep 2:591. https://doi.org/10.1038/srep00591

Kojima A, Teshima K, Shirai Y, Miyasaka T (2009) Organometal halide perovskites as visible-light sensitizers for photovoltaic cells. J Am Chem Soc 131(17):6050–6051. https://doi.org/10.1021/ja809598r

Kowsar A, Billah M, Dey S, Debnath SC, Yeakin S, Farhad SFU (2019) Comparative study on solar cell simulators. In: 2019 2nd international conference on innovation in engineering and technology (ICIET). IEEE, New York, pp 1–6. https://doi.org/10.1109/ICIET48527.2019.9290649

Lee MM, Teuscher J, Miyasaka T, Murakami TN, Snaith HJ (2012a) Efficient hybrid solar cells based on meso-superstructured organometal halide perovskites. Science 338(6107):643–647. https://doi.org/10.1126/science.1228604

Lee MM, Teuscher J, Miyasaka T, Murakami TN, Snaith HJ (2012b) Efficient hybrid solar cells based on meso-superstructure organometal halide perovskites. Science 338(6107):643–647. https://doi.org/10.1126/science.1228604

Li W, Dong H, Wang L, Li N, Guo X, Li J, Qiu Y (2014) Montmorillonite as bifunctional buffer layer material for hybrid perovskite solar cells with protection from corrosion and retarding recombination. J Mater Chem A 2(38):13587–13592. https://doi.org/10.1039/C4TA02250F

Li J-F, Zhao C, Zhang H, Tong J-F, Zhang P, Yang C-Y, Xia Y-J, Fan D-W (2015) Improving the performance of perovskite solar cells with glycerol-doped PEDOT:PSS buffer layer. Chin Phys B 25(2):028402. https://doi.org/10.1088/1674-1056/25/2/028402

Liao Y, Tian N, Wang J, Yao D, Zheng G, Zhou B, Yang Y, Long F (2022) Performance enhancement of evaporated $CsPbI_2Br$ perovskite solar cells with a CuSCN hole transport layer via a caesium bromide buffer layer. ACS Appl Energy Mater 5(9):9542–9548. https://doi.org/10.1021/acsaem.2c00987

Liu X, Yu H, Yan L, Dong Q, Wan Q, Zhou Y, Song B, Li Y (2015) Triple cathode buffer layers composed of PCBM, C_{60}, and LiF for high-performance planar perovskite solar cells. ACS Appl Mater Interfaces 7(11):6230–6237. https://doi.org/10.1021/am508515x

Liu J, Zhang Y, Liu C, Peng M, Yu A, Kou J, Liu W, Zhai J (2016) Piezo phototronic effect enhanced UV photodetector based on CuI/ZnO double-shell grown on flexible copper microwire. Nanoscale Res Lett 11:1–7. https://doi.org/10.1186/s11671-016-1315-7

Liu P, Chen Y, Xiang H, Yang X, Wang W, Ran R, Zhou W, Shao Z (2021a) Benefitting from synergistic effect of anion and cation in antimony acetate for stable $CH_3NH_3PbI_3$-based perovskite solar cell with efficiency beyond 21%. Small 17(37):2102186. https://doi.org/10.1002/smll.202102186

Liu P, Han N, Wang W, Ran R, Zhou W, Shao Z (2021b) High-quality Ruddlesden–Popper perovskite film formation for high-performance perovskite solar cells. Adv Mater 33(20):2002582. https://doi.org/10.1002/adma.202002582

Lu X, Li Y, Dong C, Gao Y, Yue G, Liu K, Tan F (2023) Efficient carbon electrode perovskite solar cells with robust buffer interfaces. J Mater Res Technol 24:8162–8170. https://doi.org/10.1016/j.jmrt.2023.05.038

Madelung O (2004) Semiconductors: data handbook. Springer Science & Business Media, Cham. https://doi.org/10.1007/978-3-662-06418-4

Moiz SA, Alahmadi ANM, Aljohani AJ (2021) Design of a novel lead-free perovskite solar cell for 17.83% efficiency. IEEE Access 9:54254–54263. https://doi.org/10.1109/ACCESS.2021.3071016

Mojiri A, Taylor R, Thomsen E, Rosengarten G (2013) Spectral beam splitting for efficient conversion of solar energy—a review. Renew Sust Energ Rev 28:654–663. https://doi.org/10.1016/j.rser.2013.08.052

National Renewable Energy Laboratory (NREL) (n.d.) Best research-cell efficiencies. https://www.nrel.gov/pv/cell-efficiency.html. Accessed 7 Feb 2025

Nykyruy L, Yavorskyi R, Zapukhlyak Z, Wisz G, Potera P (2019) Evaluation of CdS/CdTe thin film solar cells: SCAPS thickness simulation and analysis of optical properties. Opt Mater 92:319–329. https://doi.org/10.1016/j.optmat.2019.04.025

Omarova Z, Yerezhep D, Aldiyarov A, Tokmoldin N (2022) In silico investigation of the impact of hole-transport layers on the performance of $CH_3NH_3SnI_3$ perovskite photovoltaic cells. Crystals 12(6):699. https://doi.org/10.3390/cryst12060699

Park N-G (2016) Methodologies for high efficiency perovskite solar cells. Nano Convergence 3:1–13. https://doi.org/10.1186/s40580-016-0079-8

Pathak SK, Abate A, Leijtens T, Hollman DJ, Teuscher J, Pazos L, Docampo P, Steiner U, Snaith HJ (2014) Towards long-term photostability of solid-state dye-sensitized solar cells. Adv Energy Mater 4(10):1301667. https://doi.org/10.1002/aenm.201301667

Qasim I, Ahmad O, Abdin ZU, Rashid A, Nasir MF, Malik MI, Rashid M, Hasnain SM (2022) Design and numerical investigations of eco-friendly, non-toxic ($Au/CuSCN/CH_3NH_3SnI_3/CdTe/ZnO/ITO$) perovskite solar cell and module. Sol Energy 237:52–61. https://doi.org/10.1016/j.solener.2022.03.007

Ramos FJ, Ince M, Urbani M, Abate A, Grätzel M, Ahmad S, Torres T, Nazeeruddin MK (2015) Non-aggregated Zn(II) octa(2,6-diphenylphenoxy) phthalocyanine as a hole transporting material for efficient perovskite solar cells. Dalton Trans 44(26):10847–10851. https://doi.org/10.1039/C5DT00592A

Senguttuvan N, Ravichandran K, Prakasam A (2024) Exploring the effect of Ce doping on structural, magnetic, solar cell and optical characteristics of nano $ZnMn_2O_4$. J Supercond Nov Magn 37:1–8. https://doi.org/10.1007/s10948-023-06734-9

Seshaiah KV, Kim JH (2024) Nature of defects and their passivation engineering for advancements in perovskite solar cells. Chem Eng J 492:152370. https://doi.org/10.1016/j.cej.2024.152370

Shrotriya V, Li G, Yao Y, Chu C-W, Yang Y (2006) Transition metal oxides as the buffer layer for polymer photovoltaic cells. Appl Phys Lett 88(7):073508. https://doi.org/10.1063/1.2172216

Si H, Liao Q, Zhang Z, Li Y, Yang X, Zhang G, Kang Z, Zhang Y (2016) An innovative design of perovskite solar cells with Al_2O_3 inserting at ZnO/perovskite interface for improving the performance and stability. Nano Energy 22:223–231. https://doi.org/10.1016/j.nanoen.2016.02.048

Siddique MAF, Rahman ASMS (2024) The SCAPS-1D simulation of non-toxic $KGeCl_3$ perovskite from DFT-derived properties. Mater Sci Eng B 303:117268. https://doi.org/10.1016/j.mseb.2023.117268

Stangl R, Leendertz C, Haschke J (2010) Numerical simulation of solar cells and solar cell characterization methods: the open-source on-demand program AFORS-HET. Sol Energy 14(4):319–352

Sun Q, Wang T, Zhou C, Zhang C, Shao Y, Liu X, Chen X (2023) Interfacial modification to improve all-inorganic perovskite solar cells by a multifunctional 4-aminodiphenylamine layer. J Alloys Compd 960:170629. https://doi.org/10.1016/j.jallcom.2023.170629

Sunny A, Rahman S, Khatun M, Ahmed SRA (2021) Numerical study of high-performance HTL-free $CH_3NH_3SnI_3$-based perovskite solar cell by SCAPS-1D. AIP Adv 11(3):035308. https://doi.org/10.1063/5.0039125

Tinedert I, Pezzimenti F, Megherbi M, Saadoune A (2020) Design and simulation of a high-efficiency CdS/CdTe solar cell. Optik 208:164112. https://doi.org/10.1016/j.ijleo.2019.164112

Touafek N, Dridi C, Mahamdi R (2020) Bathocuproine buffer layer effect on the performance of inverted perovskite solar cells. J Technol Innov Renew Energy 9(1):1–6

Vahedi B, Golab E, Sadr AN, Vafai K (2022) Thermal, thermodynamic and exergoeconomic investigation of a parabolic trough collector utilizing nanofluids. Appl Therm Eng 206:118117. https://doi.org/10.1016/j.applthermaleng.2022.118117

Verschraegen J, Burgelman M (2007) Numerical modeling of intra-band tunneling for heterojunction solar cells in SCAPS. Thin Solid Films 515(15):6276–6279. https://doi.org/10.1016/j.tsf.2006.12.091

Wang Y, Liu X, Zhou Z, Ru P, Chen H, Yang X, Han L (2019) Reliable measurement of perovskite solar cells. Adv Mater 31(15):1803231. https://doi.org/10.1002/adma.201803231

Wei D, Ma F, Wang R, Dou S, Cui P, Huang H, Ji J, Jia E, Jia X, Sajid S (2018) Ion migration inhibition by the cation–π interaction in perovskite materials for efficient and stable perovskite solar cells. Adv Mater 30(9):1707583. https://doi.org/10.1002/adma.201707583

Wu Y, Xie F, Chen H, Yang X, Su H, Cai M, Zhou Z, Noda T, Han L (2017) Thermally stable $MAPbI_3$ perovskite solar cells with efficiency of 19.19% and area over 1 cm^2 achieved by additive engineering. Adv Mater 29(17):1701073. https://doi.org/10.1002/adma.201701073

Xiang H, Liu P, Ran R, Wang W, Zhou W, Shao Z (2022) Two-dimensional Dion–Jacobson halide perovskites as new-generation light absorbers for perovskite solar cells. Renew Sust Energ Rev 166:112614. https://doi.org/10.1016/j.rser.2022.112614

Xiao Z, Yuan Y, Wang Q, Shao Y, Bai Y, Deng Y, Dong Q, Hu M, Bi C, Huang J (2016) Thin-film semiconductor perspective of organometal trihalide perovskite materials for high-efficiency solar cells. Mater Sci Eng R Rep 101:1–38. https://doi.org/10.1016/j.mser.2015.12.001

Xiao J-W, Ma S, Yu S, Zhou C, Liu P, Chen Y, Zhou H, Li Y, Chen Q (2018) Ligand engineering on CdTe quantum dots in perovskite solar cells for suppressed hysteresis. Nano Energy 46:45–53. https://doi.org/10.1016/j.nanoen.2018.01.013

Xu M-F, Wu Z, Zhang Y, Zhu PT, Xu T, Wang C-N, Yang X-H (2022) Improving efficiency and stability of perovskite solar cells by SiO_2 layer incorporation. Org Electron 108:106583. https://doi.org/10.1016/j.orgel.2022.106583

Yakusheva A, Saranin D, Muratov D, Gostishchev P, Pazniak H, Di Vito A, Le TS, Luchnikov L, Vasiliev A, Podgorny D (2022) Photo stabilization of p-i-n perovskite solar cells with bathocuproine: MXene. Small 18(30):2201730. https://doi.org/10.1002/smll.202201730

Yang JM, Luo Y, Bao Q, Li YQ, Tang JX (2019) Recent advances in energetics and stability of metal halide perovskites for optoelectronic applications. Adv Mater Interfaces 6(3):1801351. https://doi.org/10.1002/admi.201801351

Yelzhanova Z, Nigmetova G, Aidarkhanov D, Daniyar B, Baptayev B, Balanay MP, Jumabekov AN, Ng A (2022) A morphological study of solvothermally grown SnO_2 nanostructures for application in perovskite solar cells. Nano 12(12):1686. https://doi.org/10.3390/nano12121686

Yin W-J, Shi T, Yan Y (2016) Unusual defect physics in $CH_3NH_3PbI_3$ perovskite solar cell absorber. ACS Energy Lett 1(5):438–444. https://doi.org/10.1021/acsenergylett.6b00211

Zhang Q, Dandeneau CS, Zhou X, Cao G (2009) ZnO nanostructures for dye-sensitized solar cells. Adv Mater 21(41):4087–4108. https://doi.org/10.1002/adma.200803827

Zhang W, Anaya M, Lozano G, Calvo ME, Johnston MB, Míguez H, Snaith HJ (2015) Highly efficient perovskite solar cells with tunable structural color. Nano Lett 15(3):1698–1702. https://doi.org/10.1021/nl504349z

Zhang Q, Xiong S, Ali J, Qian K, Li Y, Feng W, Hu H, Song J, Liu F (2020) Polymer interface engineering enabling high-performance perovskite solar cells with improved fill factors of over 82. J Mater Chem A 8(10):5467–5475. https://doi.org/10.1039/C9TA14168D

Zhao T, Shi W, Xi J, Wang D, Shuai Z (2016) Intrinsic and extrinsic charge transport in $CH_3NH_3PbI_3$ perovskites predicted from first-principles. Sci Rep 6:19968. https://doi.org/10.1038/srep19968

Zhao D, Wu Y, Tu B, Xing G, Li H, He Z (2019) Understanding the impact of cu-in-GaS nanoparticles compactness on holes transfer of perovskite solar cells. Nano 9(2):286. https://doi.org/10.3390/nano9020286

Zhou H, Chen Q, Li G, Luo S, Song T-B, Duan H-S, Hong Z, You J, Liu Y, Yang Y (2014) Interface engineering of highly efficient perovskite solar cells. Science 345(6196):542–546. https://doi.org/10.1126/science.1254050

Chapter 11
Bandgap Modification Strategy in Perovskite Solar Cells

Sakshi Chaudhary, Prachi Jain, Gaurav Pandey, Ritu Rani, and Nadezhda Sergeevna Molokitina

Abstract The perovskite solar cell (PSC) based materials have attracted significant attention from researchers all over the globe due to their charge transport properties, high-conversion efficiency, and facile fabrication process. However, commercial use of PSC cells is still challenging due to their poor stability relative to first-generation solar cells. Several factors affect the PSCs efficiency and stability, such as environmental stability, charge transport layer, facile method, temperature, material, bandgap, etc. In this book chapter, we have thoroughly discussed the importance of the bandgap tuning. The bandgap of the material is crucial for conversion efficiency and light harvesting. Perovskite materials demonstrate a broad range of bandgap relative to the optimal bandgap. However, to minimize the bandgap of the perovskite material, we examine methodologies for bandgap modification, including doping, compositional engineering, interface engineering, and dimensional alterations, which enhance the efficiency and stability of perovskite solar cells.

Keywords Perovskite materials · Bandgap tuning · Doping · Interface engineering

---

S. Chaudhary · R. Rani
School of Advanced Engineering, UPES, Dehradun, Uttarakhand, India

P. Jain
Materials Analysis and Research Laboratory, Department of Physics, Netaji Subhas University of Technology, New Delhi, India

G. Pandey (✉)
School of Advanced Engineering, UPES, Dehradun, Uttarakhand, India

Earth Cryosphere Institute, Tyumen Scientific Center, SB, RAS, Tyumen, Russia
e-mail: gaurav.pandey@ddn.upes.ac.in

N. S. Molokitina
Earth Cryosphere Institute, Tyumen Scientific Center, SB, RAS, Tyumen, Russia

M. Kumar (ed.), *Computational Analysis of Perovskite Solar Cells*, Springer Tracts in Electrical and Electronics Engineering,
https://doi.org/10.1007/978-981-95-6521-4_11

Introduction

The global need for environmentally friendly and sustainable energy sources has significantly increased the demand for renewable energy worldwide. Among these, solar energy stands out because it is abundant, low-cost, and eco-friendly, making photovoltaic technology highly promising (Yang et al. 2018; Noel et al. 2014). Perovskite solar cells (PSCs) have emerged as very promising photovoltaic cells owing to their straightforward production technique, affordability, and efficient performance (Helal Miah et al. 2023). These cells display distinctive properties, such as longer charge-carrier lifetimes, elevated charge-separation, adjustable band gaps, and remarkably high absorption coefficients (Guerrero et al. 2014; Salhi et al. 2018; Rahman et al. 2023; Miah et al. 2023). These characteristics allow perovskite solar cells (PSCs) to attain elevated power conversion efficiencies (PCE), illustrating its promise as a preeminent solar technology. The initial effectiveness achieved a photoconversion efficiency ranging from 3.1% to 3.8% (Kojima et al. 2009). Progress in PSCs has been significant, with recent developments showing PCEs reaching 25%, outpacing the majority of silicon and thin-film solar cells (Min et al. 2021; Green et al. 2024). Nonetheless, there is much opportunity for advancements in PSC technology still.

Figure 11.1 depicts the rapid progress and significant enhancements in the photovoltaic sector. The stability of PSCs determines their ability to maintain efficiency over time in diverse environmental conditions, a crucial factor affecting their performance and practical use. The stability of perovskite solar cells (PSCs) presents a significant problem due to their inherent vulnerability to degradation from heat, light, oxygen, and moisture, resulting in performance deterioration over time (Miah et al. 2023; Chowdhury et al. 2023). Furthermore, factors including crystal flaws, ion migration, hysteresis, and hydrophobicity influence their performance and

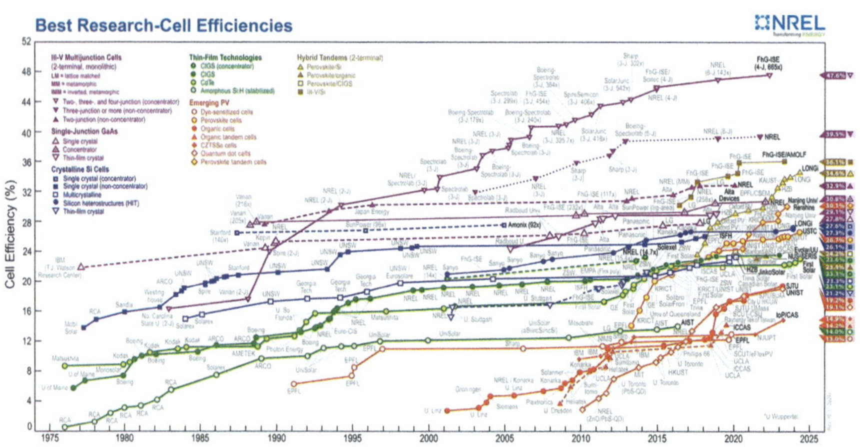

Fig. 11.1 Recent improvements in the photoconversion efficiency of photovoltaic cells, based on data from NREL (National Renewable Energy Laboratory 2025)

stability (Snaith et al. 2014; Azpiroz et al. 2015; Elumalai and Uddin 2016; Pazos-Outón et al. 2018; Chen et al. 2019). For reducing these issues, band gap tuning via structural alterations is essential for optimizing perovskite solar cells for improved performance and stability. The band gap defines the spectrum of light energy that perovskite materials can effectively absorb. The band gap should be optimized to correspond with the solar energy spectrum to enhance light absorption and increase PSC efficiency. Prevalent techniques for band gap tuning encompass compositional engineering, doping, interface engineering, and dimensional alterations (Noh et al. 2013; Eperon et al. 2014; Liu et al. 2019; Ou et al. 2019). The principal solutions for band gap tuning are focused on enhancing the conversion efficiency of perovskite solar cells (shown in Fig. 11.2). Researchers can optimize the optical and electrical properties of perovskite materials by modifying their composition and structure. Furthermore, altering the band gap not only increases photovoltaic efficiency but also promotes stability against external and internal degradation causes (Yuren et al. 2023; Xu et al. 2024).

This chapter provides a comprehensive investigation of the effects of band gap engineering on PSC stability and performance. It highlights how dimensional reduction, impurity-induced bandgap modulation, and ion substitution in the perovskite structure shape material properties. Additionally, the chapter examines computational methods such as SCAPS-1D, COMSOL, ATLAS, SILVACO, and density functional theory (DFT), which provide valuable insights into PSC optimization. By bridging experimental and theoretical studies, this chapter aims to offer a comprehensive understanding of how structural and compositional modifications enhance PSC efficiency and durability. This discussion is intended to address existing challenges and unlock the full potential of perovskite solar cells, contributing to their future development and commercialization.

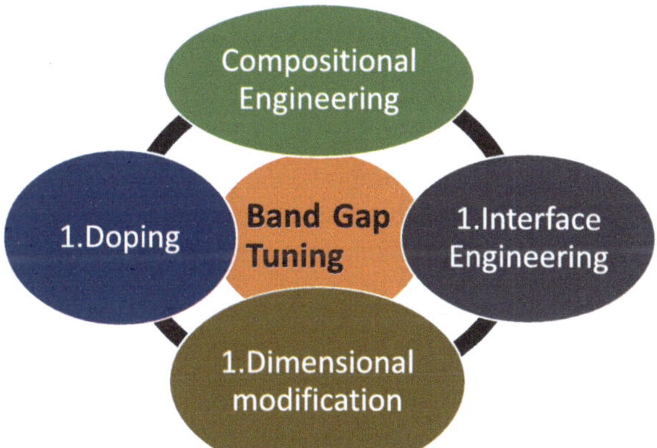

Fig. 11.2 Different strategies for band gap tuning to enhance the power conversion efficiency (PCE) of perovskite solar cells (PSCs)

Band Gap Tuning

Regulation of photon absorption: The interaction between photons and semiconducting materials can lead to one of three outcomes: reflection, absorption, or transmission. Absorption causes electron excitation from the valence to the conduction band, which contributes to power generation in photovoltaic devices, while reflection and transmission are generally considered losses. The formula for estimating the energy required for photon absorption is based on the energy of the photon and the wavelength of the photon. Photons are absorbed if their energy exceeds the material's band gap. Electron-hole pairs cannot be generated by photons with energy levels that are lower than the bandgap, while photons with higher energy levels primarily generate heat, which decreases efficiency.

The absorption range and efficiency are determined by the semiconductor band gap. In order to optimize photon absorption in the 300–800 nm range, the band gap should align with the solar spectrum for efficient solar cells (Goetzberger et al. 2014; Lei et al. 2022; Jung et al. 2023). The majority of materials employ a bandgap of 1.5 eV. The band gap can be optimized through a variety of methods to reduce the generation of heat, which is frequently the result of high-energy photons being underutilized. Numerous researchers have investigated a variety of methods to align material band gaps with the solar spectrum (Haq et al. 2021; Xing et al. 2018). These initiatives have resulted in enhanced performance, increased structural stability, and improved light absorption in solar energy conversion.

Tuning the band gap: The optoelectronic characteristics of materials are directly influenced by the band gap, which is the energy difference between the valence band maximum and the conduction band minimum. It is essential to adjust the band gap in perovskite material, as it determines the wavelengths of light that the material can absorb and convert into electrical signals. Perovskite materials are appealing for applications such as photovoltaics, light-emitting devices, photodetectors, and high-energy particle detection due to their capacity to adjust the band gap, which can range from low to high in value (Miah et al. 2024). Due to their tunable band gaps that are in accordance with solar energy, metal halide perovskites are particularly promising for the development of highly efficient perovskite solar cells (PSCs). In order to optimize the band gap and improve the performance of perovskite-based devices, a variety of strategies have been investigated, including experimental modeling, such as doping, dimensional reduction, compositional, interface engineering as well as computational modeling techniques like SCAPS, COSMOL, ATLAS, and DFT.

Compositional Engineering

Methylammonium (MA)-Based PSCs

Liquid-junction solar cells based on methylammonium were initially introduced. Nevertheless, their stability was restricted by the utilization of liquid-type electrodes (Kojima et al. 2009). In contrast to the liquid-junction versions that were previously used, Kim et al. independently developed solid-state mesoscopic heterojunction PSCs that considerably enhanced stability using $MAPbI_3$ (Kim et al. 2012; Lee et al. 2012). $MAPbI_3$ has been extensively adopted as a standard perovskite material due to its tetragonal phase at room temperature, and PSCs have become a prominent research focus since that time. The interstitial spaces are occupied by MA^+ cations, while PbI_6 octahedra serve as corner-connectors in this structure, as illustrated in Fig. 11.3a. This material has a competitive advantage over other thin-film photovoltaic technologies due to its capacity to generate MA-based perovskite films through a simple solution spin-coating process.

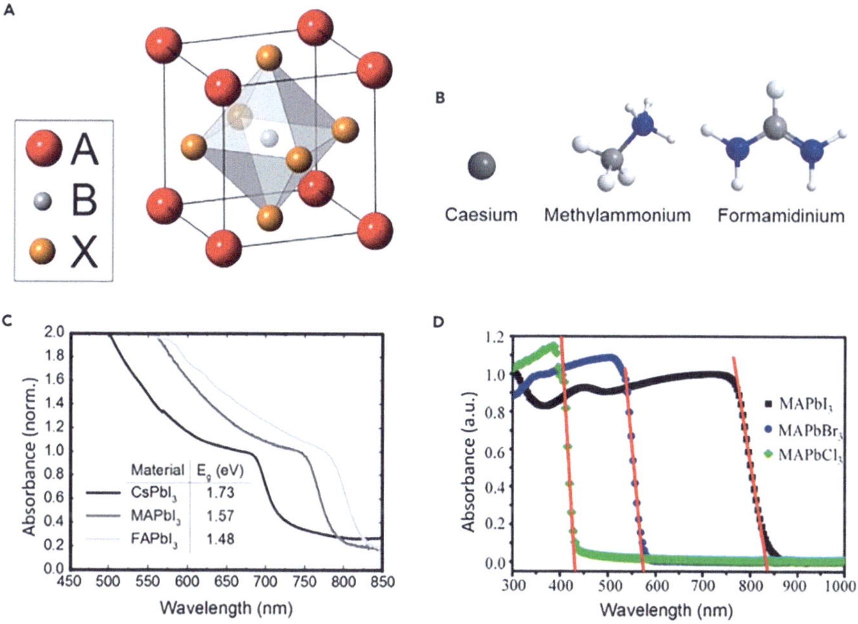

Fig. 11.3 Band gaps and the crystal structure of perovskites: (**a**) Schematic diagram of crystal structure of the ABX_3 perovskite: (**b**) The three A-site cations—Cs, MA, and FA—represented by their atomic structures. (**c**) $APbI_3$ perovskites' UV-vis spectra where A is represented as Cs, MA, or FA. (**d**) $MAPbX_3$ perovskites' UV-vis spectra, where X might be I, Br, or Cl. With permission from Eperon et al. (2014), copyright 2014 Royal Society of Chemistry, panels (**a**–**c**). With permission from Liu et al. (2015), copyright 2015 Wiley-VCH, Panel (**d**) has been modified

The power conversion efficiency of MA-based PSCs was initially less than 10%, primarily as a result of the spin-coated films' weak morphology and crystallinity. Thermal evaporation, sequential deposition, and anti-solvent dipping were among the numerous fabrication techniques that were introduced (Liu et al. 2013; Burschka et al. 2013; Im et al. 2014; Jeon et al. 2014). The PCE of $MAPbI_3$-based PSCs was promptly elevated to 17% as a result of these advancements. Additionally, a halide mixture was identified as a method to improve both efficiency and stability. For example, the $MAPb(I_{1-x}Br_x)_3$ composition achieved a certified PCE of 16.2% by replacing 10–15 mol% of iodine with bromine, thereby enhancing the stability and efficiency of the device (Jeon et al. 2014; Noh et al. 2013). It is intriguing that chlorine does not replace iodine $MAPbI_3$, but it does function as a crystallization assist, which increases the efficiency of the device and increases the length of carrier diffusion (Liu et al. 2013; Stranks et al. 2013). Moreover, the substitution of bromine for iodine enables bandgap tuning within the range of 1.53 eV–2.97 eV, resulting in the formation of colorful PSCs, as illustrated in Fig. 11.3d (Noh et al. 2013; Liu et al. 2015). The method of manufacture is still far from commercialization, although MA-based PSCs have made significant progress in terms of efficiency and stability, with PCEs reaching approximately 20% through molecularly controlled intermediates (Ahn et al. 2015; Cao et al. 2016). The emphasis of researchers has shifted to FA-based PSCs, which exhibit the potential for increased stability and efficacy in comparison to their MA-based counterparts.

Formamidinium (FA)-Based PSCs

With a bandgap of 1.55 eV, $MAPbI_3$ absorbs less of the solar spectrum than an ideal single-junction solar cell, which has an optimum range of 1.1–1.4 eV according to Shockley-Queisser. The bandgap of the perovskite is reduced to 1.48 eV (Fig. 11.3c) when MA is substituted with a slightly larger FA molecule, bringing it closer to the Shockley-Queisser optimum value. Due to better light absorption, this modification could result in a greater power conversion efficiency (PCE) for $FAPbI_3$-based PSCs than MAPbI3. Furthermore, $FAPbI_3$ is a viable option for commercial PSCs because it has an excellent crystal structure with a Goldschmidt factor close to 1, and FA is more stable than the volatile MA cation.

The first efficiency of PSCs based on $FAPbI_3$ was 14.2% (Eperon et al. 2014). They found that $FAPbI_3$ has more thermal stability than $MAPbI_3$, particularly when exposed to heat stress, which is essential for stability over a prolonged period. Phase stability is difficult to maintain since the non-photoactive δ-phase of $FAPbI_3$ is the most stable at room temperature. Because of its high thermodynamic transition temperature, it is challenging to obtain a finely crystalline pure α-phase $FAPbI_3$. Multication or halide mixing techniques have been devised to improve crystallinity and stabilize the α-phase in order to overcome these difficulties, opening the door for stable and effective $FAPbI_3$-based PSCs.

Table 11.1 Performance parameters (V_{OC}, J_{SC}, and PCE) of FA-based PSCs (perovskite solar cells) with various interfacial layers

FA-based PSCs device	V_{OC} (V)	J_{SC} ($mAcm^{-2}$)	PCE (%)	Ref.
$C_{60}/FAPbI_3/CuAlO_2$	1.18	17.39	16.17	Noman et al. (2024)
$PCBM/FAPbI_3/CuAlO_2$	1.13	22.51	20.48	
$PCBM/FAPbI_3/CuI$	1.14	21.64	17.61	
$PCBM/FAPbI_3/Cu_2O$	1.13	22.51	20.33	
$C_{60}/FAPbI_3/CuI$	1.13	16.33	13.57	
$C_{60}/FAPbI_3/Cu_2O$	1.11	17.52	16.13	

$FAPbI_3$ is a viable choice for PSCs because it has a wider absorption spectrum than $MAPbI_3$. Earlier research, however, did not show that $FAPbI_3$-based PSCs' PCE significantly surpassed those of $MAPbI_3$-based PSCs (Eperon et al. 2014; Lee et al. 2014). The low caliber of $FAPbI_3$ films was the main cause of this. Improving the quality of $FAPbI_3$ films was determined to be a primary area of development.

Using an arrangement of MA and FA cations, researchers developed a mixed-cation method to solve problems such poor crystallinity, morphology, and δ-phase impurities (Pellet et al. 2014). The resultant perovskite composition ($FA_{1-x}MA_{x-}PbI_3$) improved crystallinity and eliminated phase impurities while retaining the wide absorption characteristics of $FAPbI_3$. A PCE of 15% was attained using this method (Pellet et al. 2014). Additionally, adding bromide halides to the composition ($FA_{1-x}MA_xPb(I_{1-y}Br_y)_3$) increased stability and efficiency while improving morphology and crystallization. As a result of the enhanced open-circuit voltage, efficiency surpassed 20% (Yang et al. 2015). Overall, as seen in Table 11.1, this composition has achieved significant advances in the generation of $FAPbI_3$-based PSCs, exhibiting both quick progress and high potential for future advancements.

Alkali Metal-Doped ($FA_{1-x}MA_x$) $Pb(I_{1-y}Br_y)_3$-Based PSCs

Utilizing the $FA_{1-x}MA_xPb(I_{1-y}Br_y)_3$ perovskite composition has resulted in notable progress in FA-based PSCs. Comparing these technologies to the most advanced silicon solar cells, their stability and efficiency are still lacking. A triple-cation perovskite composition incorporating cesium, $FA_{0.95-x}MA_xCs_{0.05}Pb(I_{1-y}Br_y)_3$, was developed by Saliba et al. and showed decreased δ-phase impurities and enhanced thermal stability (Saliba et al. 2016a). Additionally, this formulation was found to be less susceptible to fabrication circumstances, which significantly improved repeatability and led to its widespread use. Cesium enhanced perovskite film crystallization, resulting in a stable power output of more than 21% (Saliba et al. 2016a). Its notable achievement of steady performance for more than 200 hours under constant 1-sun illumination is a critical turning point in PSC development. When compared to conventional silicon solar cells, the triple-cation perovskite's bandgap is still quite large at 1.63 eV, and its open-circuit voltage (V_{OC}) only reached 1.15 V, suggesting a significant V_{OC} loss (Saliba et al. 2016b). By adding rubidium to the triple-cation mixture, a quadruple-cation perovskite composition is created to address this. This change produced a stable PCE of 21.6% by lowering V_{OC} loss to 0.39 V, which is equivalent to the 0.4 V loss seen in silicon solar cells. Additionally, the quadruple-cation composition showed remarkable longevity, sustaining a steady power output for 500 hours at 85 °C under 1-sun illumination. Potassium-doped triple-cation perovskite shows similar gains, achieving larger PCEs because of greater V_{OC} (Bu et al. 2017; Abdi-Jalebi et al. 2018). The superior perovskite films are the cause of the increased V_{OC} seen with alkali metal cation and bromide doping. Nevertheless, this enhancement results in a blue-shifted absorption spectrum, which confines the total PCE and limits the short-circuit current density to 23 $mAcm^{-2}$. It became clear that bromide ions and other mixed cations must be removed from the perovskite composition in order to further improve J_{SC} (Turren-Cruz et al. 2018). This made it clear that new techniques must be developed in order to produce high-quality FA-based perovskite films without the need for bromide.

Jian et al. showed that a sequential deposition technique may be used to effectively create a stable, non-bromide perovskite layer (Jiang et al. 2019). They obtained a verified efficiency of 23.3% for planar structure PSCs by increasing J_{SC} by 1 $mAcm^{-2}$ by decreasing the bromide component in the perovskite composition. However, because of its volatility, MA was determined to be inadequate for long-term stability. Turren-Curz et al. produced highly crystalline MABr-free perovskite films by using alkali cations (Turren-Cruz et al. 2018). Their PSCs based on $FA_{0.85}Cs_{0.1}Rb_{0.05}PbI_3$ demonstrated exceptional stability throughout 1000 h of maximum power point monitoring (Kim et al. 2012).

According to other research, using MACI as an additive that evaporates during the annealing process may result in highly crystalline pure α-phase $FAPbI_3$ films (Kim et al. 2012; Min et al. 2019). As reported by Kim et al., the final perovskite film's crystallinity and grain size were greatly increased by combining 40% MACI with $FAPbI_3$ (Kim et al. 2012). With a JSC of 25.6 $mAcm^{-2}$ and a confirmed efficiency alongside MACI and $FAPbI_3$, the resultant film-maintained absorption properties comparable to pure $FAPbI_3$ due to the volatile nature of MACI (Min et al. 2019). These results contribute to PSCs' record-breaking efficiency by clearly showing a trend in the compositional progression from $MAPbI_3$ to $FAPbI_3$ perovskites.

Doping

As Table 11.2 illustrates, dopants have been used extensively in perovskite solar cells (PSCs). With an extended potential around the dopant atoms, they produce localized instabilities in the host crystal lattice at low doping levels. This often causes a change in the fermi level, which results in a semiconductor of the n-type or p-type with a slightly different bandgap. However, the perovskite material has to

Table 11.2 Photovoltaic characteristics and initial PCE values reported for various dopants in Spiro-OMeTAD HTM

Dopants	J_{SC} ($mAcm^{-2}$)	V_{OC} (V)	PCE (%)	Ref.
Li-TFSI +tBP + bubbled CO_2	21.2	1.14	19.1	Kong et al. (2021)
Li-TFSI +tBP + CoF_3	24.99	1.17	23.05	Wu et al. (2022)
Li-TFSI +tBP + N_3	24.7	1.19	22.7	Han et al. (2022)
Li-TFSI +tBP + $SOCl_2$	23.09	1.16	20.76	Li et al. (2020)
LiPFSI +tBP	22.5	1.12	19.47	Seo et al. (2015)
Li-TFSI +tBP + CuSCN	22.01	1.06	18.02	Li et al. (2016b)
Li-TFSI +tBP + CuI	21.52	1.06	16.67	Zhang et al. (2022)
Spiro-OMeTAD+(TFSI–)2+TBMP+TFSI-	25.52	1.17	25.1	Aberle (2000)

have enough dopants (a process known as alloying) to allow interactions between host states and localized impurity states in order to dramatically alter the bandgap. In order to modify the electrical structure, alloying has been used a lot. The resulting bandgap typically exhibits a roughly quadric connection with the molar ratio of components (Van Vechten and Bergstresser 1970). Thompson et al. first defined this phenomenon as x(1-x). Vab et al. further elaborated on this phenomenon, attributing the bending effect to structural aperiodicity, virtual-crystal effects, and fluctuation in crystal potential (Thompson and Woolley 1967). The B-site atom has a major effect on the band structure in perovskites, as shown in $APbX_3$, where the valence band maximum and conduction band minimum are defined by the leads 6 s and 6p orbitals (Thompson and Woolley 1967). Charge-carrier recombination may rise when foreign atoms are added for bandgap tuning; nevertheless, B-site alloying efficiently lowers bandgaps with little interference on host characteristics. With a composition of $(MAPbBr_3)_{0.15}(FAPbI_3)_{0.85}$, alloying also improves structural stability, producing stable PSCs with increased efficiency (Wang et al. 2017). The stable alloy $FA_{0.85}Cs_{0.15}PbI_3$ was produced by stabilizing unstable $FAPbI_3$ and $CsPbI_3$ by adjusting their ionic radii to maximize the Goldschmidt tolerance factor (Jeon et al. 2015). This method has been used to double perovskites made of lead-free oxide. It is appropriate for absorber layers in PSCs because it is doped with Ba_2SbTaO_6 with tantalum to generate $Ba_2SbV_xTa_{1-x}O_6$, which narrows the bandgap from 2.45 eV to a range of 1.0–1.6 eV (Li et al. 2016a). Additionally, $MAPbI_3$ and $MAPbBr_3$ have effectively reduced their bandgaps to ranges of 2.17–1.89 eV and 1.51–1.28 eV, respectively, by doping with hetrovalent and substituting Bi^{3+} for Pb^{2+} (Abdelhady et al. 2016). High doping levels may nevertheless result in diseases, flaws, and voids in spite of these developments. $FA_{0.75}MA_{0.25}Sn_{1-x}Ge_xI_3$, for example, shows elevated Urbach energy as a result of deep alloying, highlighting the need of defect reduction in further impurity-driven bandgap engineering studies (Ito et al. 2018).

Interface Engineering

As was previously mentioned, compositional engineering of the perovskite absorber layer has greatly increased the efficiency of PSCs by optimizing both the J_{SC} and the V_{OC}. With the exception of efforts to combine lead and tin, which have so far been hampered by low device efficiencies and poor stability, there was no increase in J_{SC} as the perovskite composition neared $FAPbI_3$. At the same time, interface engineering has become a major area of attention for PSCs. As shown by silicon and gallium arsenide semiconductors, surface chemistry has historically been essential to the development of solar cell technologies (Bertness et al. 1994; Sheldon et al. 2012; Chen et al. 2019). Because perovskite films are made using a solution-based technique, crystallographic flaws are often concentrated on the surface. High-quality polycrystalline perovskite films with monolithic grains, where bulk flaws are minimal in comparison to surface defects, may now be produced thanks to modern processes. As a result, interface engineering has grown in significance to achieve record-high stability and efficiency in PSCs. The observation that extra PbI_2 increases device efficiency by decreasing carrier recombination inside the perovskite layer and at the interfaces between the perovskite absorber and electron/hole-transport layers is the origin of the surface passivation idea in PSCs (Bi et al. 2016). Since then, it has been commonplace to include excess PbI_2 in perovskite compositions (Saliba et al. 2016a, b; Turren-Cruz et al. 2018). Later, Bi et al. showed that record V_{OC} loss reductions and improved external electroluminescence quantum efficiency may be achieved by carefully adjusting the PbI_2 level in the perovskite composition (Tsai et al. 2016). Under-coordinated Pb^{2+} ions, Pb clusters, iodine vacancies, organic A cation vacancies, and under-coordinated iodine ions are among the defects found on perovskite surfaces. These surface flaws speed up PSC degeneration and reduce device performance by facilitating non-radiative charge recombination. Because molecular passivation is made possible by the ionic nature of the perovskite lattice, interface engineering provides an efficient method of reducing these defects. Many compounds with well-studied functional groups have been reported for surface passivation. Several popular passivation techniques for enhancing PSC performance are examined in this chapter.

Passivation Via Low-Dimensional Perovskites

The presence of long-chain organic cations in low-dimensional perovskites, like 2D perovskites, contributes significantly to their increased stability and water resistance as compared to their 3D counterparts. However, since these long-chain organic cations come from insulating layers that prevent charge transfer along the out-of-plane direction, the PCE of 2D PSCs is often lower (Grancini et al. 2017). Researchers developed a hybrid 3D/2D interface design to take advantage of the durability of 2D perovskites while preserving the effectiveness of 3D PSCs (Cho et al. 2018; Kim et al. 2019). This method involves depositing a thin layer of 2D perovskite on top of a layer of 3D perovskite. A 2D perovskite layer made of PEA_2PbI_4, for example, may act as a barrier to prevent ion transport and provide moisture protection (Cho et al. 2018). A heterojunction was created at the grain boundaries when the narrow-bandgap 3D perovskite grains were encircled by the wide-bandgap 2D perovskite (Fig. 11.4a). Defect density was greatly decreased by this arrangement, which also enhanced hole transport and converted defect-prone grain boundaries into defect-free areas. The creation of reliable and effective 3D/2D PSCs was made possible by these advancements (Grancini et al. 2017; Cho et al. 2018). A better comprehension of the electrical characteristics and moisture resistance of the 2D structure is necessary for future developments in 3D/2D interface engineering (Kim et al. 2019). A systematic investigation into the creation of a 2D layer of top 3D perovskites employing several long-chain alkylammonium iodides, including butylammonium iodide, octylammonium iodide, and diodecylammonium iodide. The alkyl chain's length affected its ability to block electrons as well as its resistance to dampness. The certified efficiency after octyl ammonium iodide treatment was 22.9% (Kim et al. 2019).

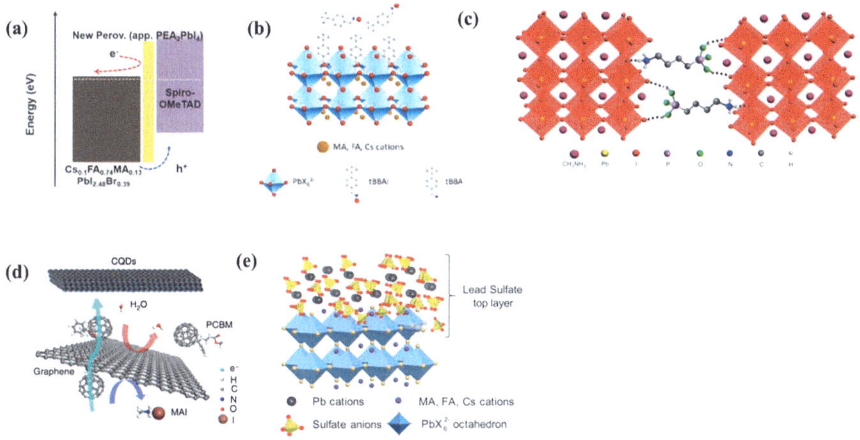

Fig. 11.4 Techniques for PSC passivation. (**a**) An explanation of how the 2D perovskite capping layer improves PCE. Excited electrons cannot go to the HTM layer at the back of the perovskite film due to the 2D perovskite's large bandgap. Permission to adapt (Cho et al. 2018). Royal Society of Chemistry, copyright 2017. (**b**) A simplified figure showing the surface of a perovskite passivated with tBBAI. Permission to adapt (Zhu et al. 2020). Wiley-VCH, © 2020. (**c**) Diagram of nearby grain structures that are cross-linked by butylphosphonic acid 4-ammonium chloride hydrogen bonding interactions. Permission to adapt (Li et al. 2015). Nature Publishing Group, 2015. (**d**) A diagram showing the diffusion processes in an ETL based on nanocarbon. Permission to adapt (Bi et al. 2017). Nature Publishing Group, copyright 2017. (**e**) A surface layer of lead sulfate formed in situ serves as an example of perovskite protection. Permission to adapt (Yang et al. 2019). © 2019 AAAS

Passivation Via Lewis Bases

Lewis base passivation has emerged as a crucial strategy for interface engineering in PSCs. Lewis basses are known to donate lone-pair electrons, which coordinate with neutralized defects such as under-coordinated Pb^{2+} ions or iodine vacancies, forming Lewis adducts. This method, which involves coating the perovskite surface with a thin layer of pyridine or thiophene, was initially shown by Noel et al. (2014). Strong connections were formed between the nitrogen atoms in pyridine and the sulfur atoms in thiophene and Pb^{2+} defects, which improved device stability under maximum power point circumstances and greatly extended the photoluminescence lifespan of perovskite films (Noel et al. 2014). Similarly, $MAPbI_3$-based PSCs performed better when passivated with polyvinyl pyridine, a pyridine-functionalized polymer (Yavari et al. 2018). Other amine-containing Lewis bases, such as with -NH_2 groups, have also proven effective. A dilute solution of phenylalkylamines (using spin coating) on perovskite surfaces could effectively passivate defects, increasing PCE from 14.2% to 19.2% in $FAPbI_3$-based PSCs (Wang et al. 2016). Because there were fewer trap states, the passivated devices had better diode properties and longer carrier lifetimes.

Introducing p-conjugated small molecules, such as indacenodithiophene capped with dicyanomethylene-indanone, for trap passivation. These molecules, with carbonyl and cyano groups, effectively passivated surface traps and grain boundary defects like under-coordinated Pb^{2+} ions (Lin et al. 2017). Additionally, their n-type semiconducting properties enhanced electron extraction and transport, providing multifunctionality. De et al. investigated trio-n-octyl phosphine oxide, 1-octadecanethiol, and triphenylphosphine, finding that trio-n-octyl phosphine oxide significantly increased photoluminescence quantum field and TRPL (time-resolved photoluminescence) lifetime, emphasizing the role of phosphorous and oxygen in mitigating non-radiative recombination (deQuilettes et al. 2016). However, solvents like isopropanol or chlorobenzene might easily remove these molecules due to their weak connection with the perovskite surface. In order to overcome this restriction, 2-mercaptopurine, a bidentate molecule, has a sulfur atom in its mercapto group and a nitrogen atom in its pyridine ring (Zhang et al. 2019). These two functional groups coordinate with Pb^{2+} ions, providing stronger anchoring and adhesion. Compared to monodentate analogs like pyridine and toluene thiol, 2-mercapto pyridine doubled the TRPL lifetime, enhanced resistance to solvent washing and thermal treatment, and improved PCE 18.35% to 20.28%, with an open-circuit voltage nearing 1.18 V. additionally, 2-MP treated films showed exceptional humidity resistance.

The efficiency of passivation could be greatly influenced by the molecular structure along with the Lewis base groups that function, especially when multifunctional groups are present. Small compounds, notably theophylline, caffeine, and theobromine, are used to address surface defects in order to get a better understanding of how the chemical environment of a functional group affects surface defect passivation (Wang et al. 2019). Their study found that an ideal arrangement of N-H and C=O groups within the molecules facilitated hydrogen bonding between N-H and iodine, which in turn helped the carboxyl group bind more effectively with the anti-site Pb defect. In the PSC, this method produced a stable PCE of 22.6%, and the theophylline therapy also produced considerable gains in device stability.

Passivation Via Organic Ammonium Halide Salts

Organic ammonium salts containing ammonium functional groups have been demonstrated as effective agents for surface passivation. The ammonium cation mostly neutralizes negatively charged defects by electrostatic interactions, such as ionic and hydrogen bonding, as opposed to amines, which coordinate to positively charged Lewis acid defects via their lone electron pairs. PSC performance is improved and surface trap flaws are greatly decreased when $MAPbI_3$ is surface passivated with linear diamonium iodide (Zhao et al. 2016). $NH_3(CH_2)_4NH_3I$, on the other hand, produced a 3D-to-2D perovskite phase transition without enhancing performance.

$FA_{0.92}MA_{0.08}PbI_3$ Perovskite was successfully passivated using PEAI salt, with a verified efficiency of 23.32% (Jiang et al. 2019). Notably, this treatment cannot withstand further annealing, as it results in the loss of the passivation effect due to the formation of a 2D perovskite phase. More recently, 4-tert-butyl-bezylammonium iodide (shown in Fig. 11.4b), an amphiphilic molecule, is an efficient passivation material for perovskites, achieving a treated PSC efficiency of 23.5% (Zhu et al. 2020). Because 4-tert-butyl-bezylammonium iodide has tert-butyl groups, which seem to prevent perovskite and increase compatibility with the spiro-MeOTAD, it permitted improved charge extraction from the perovskite layer to the spiro-OMeTAD hole transporter when compared to PEAI-based therapies. Similar ammonium compounds have been shown to be used as dopants in perovskite films by Li et al. (2015). Their study found that alkyl phosphonic acid u-ammonium chlorides form hydrogen bonds with surface iodine atoms, enhancing grain cohesivity, device performance, and stability (Fig. 11.4c). However, a deeper understanding of the role of organic ammonium salts in perovskite surface passivation is needed to optimize future applications.

Passivation Via Thin Protective Capping Layer

Unlike traditional passivation techniques, a technique using a nanostructured carbon layer made up of fullerene carbon quantum dots and N-doped graphene placed between the Ag electrode and the perovskite layer in inverted PSCs (Fig. 11.4d) (Bi et al. 2017). This nanostructured carbon layer effectively prevents ion and molecule diffusion, a major cause of device degradation and defect generation enhancing stability. Additionally, the carbon layer forms a compact electron transport while obstructing ion and moisture penetration. The dual functionality resulted in enhanced efficiency and stability in PSCs employing this method.

In order to ensure long-term stability under thermal stress, a graphene interlayer is used between the perovskite and copper thiocyanate hole transporter (Arora et al. 2017). A new approach based on this idea uses chlorinated graphene oxide as an interlayer between the poly(triarylamine) hole transporter and the perovskite (Arora et al. 2017). This setup operated at maximum power point under AM1.5G solar irradiation at 60 °C for 1000 hours while maintaining 90% of the starting efficiency. In contrast to other approaches that included graphene directly into the ETL, the chlorinated graphene oxide established strong Pb-Cl and Pb-o bonds with the perovskite's Pb-rich surface (Wang et al. 2019). The Pb-rich surface of the perovskite successfully extracts these heterostructures. By minimizing the loss of disintegrated components from the soft perovskite layer and efficiently extracting photogenerated charge carriers, these heterostructures lessened the harm to organic charge-transporting materials. The passivation strategy for lead halide perovskite uses an in situ reaction with sulfate or phosphate ions to form a water-insoluble layer of lead oxysalt on the perovskite surface (Fig. 11.4e) in place of graphene-based capping layers (Yang et al. 2019). By forming solid chemical connections with the perovskite, these lead oxysalt capping layers greatly improved moisture resistance. By passivating under-coordinated lead centers, which serve as defect nucleation sites, the wide bandgap lead oxysalt layers also decreased the density of surface defects. After 1200 h of operation at MPP under simulated AM1.5G irradiation at 65 °C, this method allowed the PSCs to maintain 96.8% of their original efficiency and increased device efficiency to 21.1%.

Dimensional Modification

Dimensionality reduction of perovskite materials presents significant opportunities for bandgap engineering, facilitating alterations in their electrical and optical characteristics (Ou et al. 2019). Quantum confinement effects intensify when the perovskite structure shifts to 2D, 1D, or 0D configurations, leading to tunable bandgaps and distinctive electrical properties (Kahwagi et al. 2020). Structures like quantum dots, nanowires, and two-dimensional perovskite layers demonstrate bandgaps that are contingent upon their size and thickness, enabling meticulous regulation of light absorption and emission spectra (Hoffman et al. 2019). The ability to modify the bandgap by dimensionality reduction creates new opportunities for applications in light detectors, solar cells, and light-emitting diodes, where regulation of optical and electrical characteristics is essential for device efficacy.

In perovskite structures, the traditional 3D framework may be compromised when the A-site cation exceeds the ideal size determined by the tolerance factor (Li et al. 2023). This discrepancy results in the formation and configuration of a two-dimensional layered structure, whereby substantial A-site cations function as spacers, delineating the crystal along designated axes. The octahedral BX_6^{4-} units constitute inorganic layers that alternate with layers of substantial organic cations in the two-dimensional structure, maintained by electrostatic interactions (Stoumpos et al. 2016). These substantial organic ions improve the stability of two-dimensional layered perovskites. While these substantial layers may impede carrier transit, they are essential in regulating the electrostatic forces influencing electron-hole couples. The quantum and dielectric confinement effects in these devices yield increased exciton binding energy (Katan et al. 2019). The organic cation layers function as dielectric barriers in 2D perovskites, while the inorganic layers function as quantum wells (Peng et al. 2022; Pedesseau et al. 2016). The disparity in dielectric constants between these barriers and wells induces a dielectric confinement effect, affecting the band gaps and optical characteristics of the materials according to the quantity of perovskite layers. 2D-3D mixed perovskite heterojunctions have been investigated to mitigate the instability of 3D perovskites while capitalizing on their high efficiency (Li et al. 2023; Katan et al. 2019). This method integrates the advantages of both dimensions, with the 3D aspect facilitating effective light absorption and the 2D configuration offering improved stability. For the perovskite and charge transport layers to align at compatible energy levels and increase the efficiency of charge transport and collection, the right organic cations must be chosen. Studies have shown that two-dimensional layers may alter the bandgap at the interface of three-dimensional perovskites, facilitating energy level modulation. By reducing its conduction band energy below that of the three-dimensional layer, a two-dimensional capping layer could minimize electron recombination. Additionally, by enhancing energy level alignment, conjugated cations added to the two-dimensional capping layer increase the efficiency of carrier extraction (Ma et al. 2021).

In three-dimensional perovskites, grain boundaries and surfaces function as passageways for ion migration, which may lead to device degradation due to interactions with charge transport layers or metallic electrodes. Incorporating a 2D perovskite layer with substantial organic ligands into a 3D perovskite framework has demonstrated a reduction in ion migration (Hu et al. 2020; Zhang et al. 2020). Quantum confinement phenomena in low-dimensional perovskites lead to elevated bandgap energies relative to their bulk equivalents. Altering the dimensions of perovskite nanostructures additionally impacts the bandgap. Studies indicate that diminishing the dimensions of perovskite nanocrystals results in alterations in photoluminescence (PL) emission wavelengths and an elevation in bandgap energy. Recent research on 2D perovskites has demonstrated considerable modulation of photoluminescence spectra and bandgap energy contingent upon the quantity of perovskite layers (Li et al. 2018). Novel methodologies have further improved the stability and efficiency of PSCs. The use of 3-(trifluoromethyl) benzylammonium iodide (3-TFMBAI) as a fluorinated spacer in quasi-2D perovskites yielded devices exhibiting remarkable thermal and moisture stability (Ou et al. 2019). Incorporating tributyl(methyl)phosphonium iodide (TPI) into perovskite precursor solutions resulted in the creation of water-insoluble, mechanically robust one-dimensional structures with improved stability (Jiao et al. 2022). These developments underscore the promise of dimensionality reduction and novel material design in improving the stability, efficiency, and durability of perovskite-based devices for many applications.

Computational Tools for Bandgap Engineering

SCAPS-1D Simulations

Advancing the knowledge and optimization of thin-film solar cells has been significantly facilitated by SCAPS-1D. Through analysis of many charge transport layers and device topologies, researchers have used this modeling technique to enhance the performance of perovskite solar cells (PSCs). Using SCAPS-1D, an investigation aimed at modeling a lead-free, totally inorganic, non-toxic Cs_2TiBr_6-based double perovskite solar cell. In this work, $CuSbS_2$ was used as the hole transport layer (HTL), whereas La-doped $BaSnO_3$ (LBSO) was first introduced as the electron transport layer (ETL) in such simulations. The work also used first-principle density functional theory (DFT) computations to corroborate the bandgap of 1.6 eV of the Cs_2TiBr_6 absorber, therefore ensuring the correctness of the SCAPS-1D findings.

Using SCAPS-1D, further optimization found the perfect characteristics to attain high efficiency. Identified as 1000 nm, 10^{11} cm^{-1}, and 1.4 eV respectively were the ideal absorber thickness, defect density, and bandgap. The best-performing HTL also exhibited an electron affinity of 4.1 eV, a 400 nm optimal thickness, and hole mobility of 10^2 cm^2V^{-1} s^{-1}. Fascinatingly, it was discovered that general device efficiency was not much affected by the ETL's thickness. Determined to be 10^{15}, 10^{20}, and 10^{21} cm^{-1}, respectively, were the ideal doping densities for the absorber layer, HTL, and ETL. The work found $CuSbS_2$ as the most effective HTL material by comparing many HTLs depending on built-in voltage (V_{BI}) and open-circuit voltage (V_{OC}). With an amazing power conversion efficiency (PCE) of 29.13%, the optimal device was finally attained (Shivesh et al. 2022).

Another work investigated many electron and hole transport layers in PSCs based on $CsPbI_3$ using SCAPS-1D. Although $CsPbI_3$ has attracted much interest as a potential absorber material for perovskite solar cells, its efficiency still falls below that of hybrid PSCs. This work investigated the structural, electrical, and optical characteristics of the cubic $CsPbI_3$ absorber via density functional theory (DFT) calculations performed with the Cambridge Serial Total Energy Package (CASTEP). The GGA-PBE method produced a computed electronic bandgap (Eg) of 1.484 eV. Pb-5d orbitals dominated further density of states (DOS) research; electronic charge density mapping indicated significant charge buildup around Pb atoms. While optical property simulations gave information on $CsPbI_3$'s optical response, a Fermi surface study verified its multiband character.

Then many electron and hole transport layers were evaluated using SCAPS-1D simulations. The research assessed a range of ETLs including IGZO, SnO_2, WS_2,

CeO_2, PCBM, TiO_2, ZnO, and C_{60} alongside HTLs including Cu_2O, CuSCN, $CuSbS_2$, Spiro-MeOTAD, V_2O_5, CBTs, CFTS, P_3HT, PEDOT: PSS, NiO, CuO, and CuI. Finding the most effective HTL/$CpbI_3$/ETL combo was the aim. With an efficiency of 17.9%, ITO/TiO_2/$CsPbI_3$/CBTS/Au ranked highest among the 96 tested designs. The effects of absorber and ETL thickness, series and shunt resistance, and operating temperature on device performance were also investigated in this work. Additionally, important properties like generation rate, recombination rate, capacitance-voltage behavior, current density-voltage response, and quantum efficiency were examined. WxAMPS was used to further verify the SCAPS-1D simulation results, therefore guaranteeing the accuracy of the conclusions (Hossain et al. 2022).

Selama et al. integrated experimental methods with SCAPS-1D simulations in another important work to examine the efficiency potential of methylammonium lead triiodide ($MAPbI_3$) PSCs. Their studies produced a power conversion efficiency of more than 31%. The work consisted of creating $MAPbI_3$ films by one-step spin-coating. While atomic force microscopy (AFM) and profilometry showed homogeneous film surfaces with an ideal thickness for effective light absorption and charge transfer, structural analysis using X-ray diffraction (XRD) verified the development of a tetragonal crystalline phase. Further verification of a direct bandgap of 1.53 eV, which lies within the optimal range for photovoltaic applications, was obtained using ultraviolet–visible (UV–Vis) spectroscopy.

Then an n-i-p planar PSC structure including ITO/SnO_2/$MAPbI_3$/Spiro-OMeTAD/Au was modeled using SCAPS-1D simulations. Realistic factors like series and shunt resistances, front contact reflection, and many recombination systems—including Shockley-Read-Hall (SRH), radiative, and Auger recombination—were included into the models. Optimizing defect concentrations, layer thicknesses, and interface characteristics produced a maximum PCE of 31.44%.

The research examined substitute materials for traditional devices aspects in order to improve stability and performance even further. With better stability and efficiency than Spiro-OMeTAD, researchers found D-PBTTT-14 as a possible substitute in the HTL, furthermore suggested as affordable substitutes for traditional electrodes like gold (Au) and indium tin oxide (ITO) were low graphene oxide (RGO) and aluminum-doped zinc oxide (AZO). Comprising AZO/SnO_2/$MAPbI_3$/D-PBTTT-14/RGO, the optimized PSC structure showed a simulated PCE of 28.39%, thereby highlighting the significance of material developments on solar cell efficiency. This work emphasizes the significant cooperation between experimental research and modeling instruments like as SCAPS-1D in driving PSC technology toward better efficiencies (Rabhi et al. 2024).

COMSOL Multiphysics Modeling

Particularly in the realm of bandgap engineering, COMSOL Multiphysics has been very essential in expanding the knowledge and optimization of perovskite solar cells (PSCs). Comprehensive multiphysics simulations made possible by this program let scientists examine the complicated interplay of optical, electrical, and thermal events in PSCs. Optimizing bandgap characteristics depends on such a combined strategy, hence improving the general performance of these solar cells.

COMSOL Multiphysics has been successfully used recently to simulate PSCs including two-dimensional $Ti_3C_2T_x$ MXene materials. Because of their unusual surface-terminating functional groups (T_x), which provide remarkable electrical qualities, these MXenes have attracted much interest. $Ti_3C_2T_x$ MXene is a potential material for solar cell developments as its incorporation into PSCs shows increases in efficiency and optoelectronic performance.

MXene-doped perovskite solar cells were simulated in this study using COMSOL Multiphysics. Three major layers comprised the simulated device: a hole transport layer (HTL), an absorber layer made of perovskite ($MAPbI_3$) combined with MXene ($Ti_2C_2T_x$), and an electron transport layer (ETL). The general arrangement was intended as ETL/$MAPbI_3$ + MXene/HTL. Whereas the HTL was made of Spiro-OMeTAD with a thickness of 140 nm, the ETL utilized was TiO_2 with a thickness of 120 nm.

The influence of absorber layer thickness and MXene content was methodically investigated in this work to maximize solar cell efficiency. The ideal arrangement produced a solar cell structure with a fill factor of 0.57, an open-circuit voltage (V_{OC}) of 1.10 V, and a short-circuit current density (J_{SC}) of 31.97 mA/cm^2. Peak efficiency of 19.87% was obtained. To the best of our knowledge, this is the first time COMSOL Multiphysics has been utilized to replicate perovskite solar cells with 2D $Ti_3C_2T_x$ MXene, therefore offering insightful information for further manufacturing and study in this field (Beepat et al. 2024).

The development of a mathematical device model for meso-structured PSCs was another important breakthrough in PSC study. Using meso-structured PSCs, Xue et al. developed a model looking at device performance and recombination processes. This model optimizes device shape for maximum efficiency by combining experimental data with numerical simulations to identify main recombination routes.

Six layers composed the meso-structured PSC: a gold anode, a Spiro-MeOTAD hole-transporting layer, a $CH_3NH_3PbI_3$ capping layer, a mesoporous layer comprising $CH_3NH_3PbI_3$ and TiO_2, a compact TiO_2 electron-transporting layer, and an FTO cathode (as shown in Fig. 11.5a). Crucially for validating the mathematical model, cross-sectional scanning electron microscopy (SEM) pictures offered exact measurements of the thickness of every layer (Fig. 11.5b) (Xue et al. 2017).

Combining charge carrier movement, recombination processes (radiative, Auger, and Shockley-Read-Hall [SRH] recombination), and optical production using a transfer matrix technique, a steady-state model was built. Under different light intensities, calibration was accomplished by fitting the model to experimentally

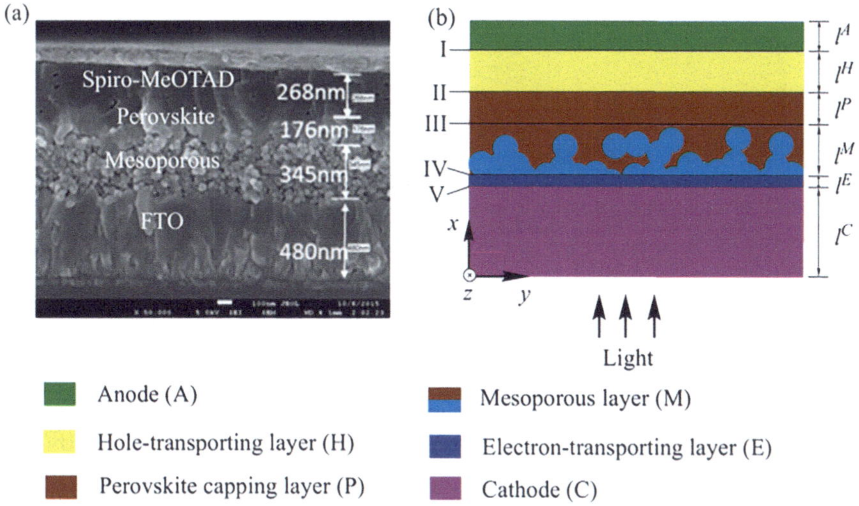

Fig. 11.5 (**a**) Cross-sectional SEM image of a fabricated meso-structured perovskite solar cell (PSC) showing different layers with their respective thicknesses: Spiro-MeOTAD (268 nm), perovskite (176 nm), mesoporous layer (345 nm), and FTO (480 nm) (Xue et al. 2017). (**b**) Schematic representation of the PSC structure, illustrating six distinct layers and five interfaces. The layers include: (A) gold anode (green), (H) Spiro-MeOTAD hole-transporting layer (yellow), (P) $CH_3NH_3PbI_3$ perovskite capping layer (brown), (M) mesoporous layer composed of $CH_3NH_3PbI_3$ and mesoporous TiO_2 (dark brown), (E) compact TiO_2 electron-transporting layer (blue), and (C) FTO cathode (magenta). Light enters from the bottom side of the structure (Xue et al. 2017)

recorded current-voltage characteristics, therefore allowing the exact determination of unknown parameters (Xue et al. 2017).

The results exposed that the main recombination losses happened at the junction between the mesoporous TiO_2 layer and the perovskite layer. At maximum power output, this interface explained around 46% of the overall recombination losses. These findings highlight the need of interfacial engineering in reducing recombination losses and improving PSC efficiency. Within a 50 nm–500 nm range, further simulations investigated the effects of changing the thicknesses of both the perovskite capping layer and the mesoporous layer. The findings showed that device performance is much influenced by the optimal layer thicknesses. More specifically, whereas changes in the mesoporous layer thickness altered interfacial recombination, increasing the perovskite capping layer thickness enhanced light absorption and charge transfer. These results provide useful design rules for the construction of high-efficiency meso-structured PSCs. To provide better understanding of the performance of meso-structured PSCs, this work combines experimental results with sophisticated mathematical modeling. The work provides useful approaches for raising PSC efficiency by spotting prominent recombination paths and maximizing layer geometries. The use of two-dimensional MXene ($Ti_3C_2T_x$) doping in perovskite materials to change the bandgap and improve PSC performance was investigated

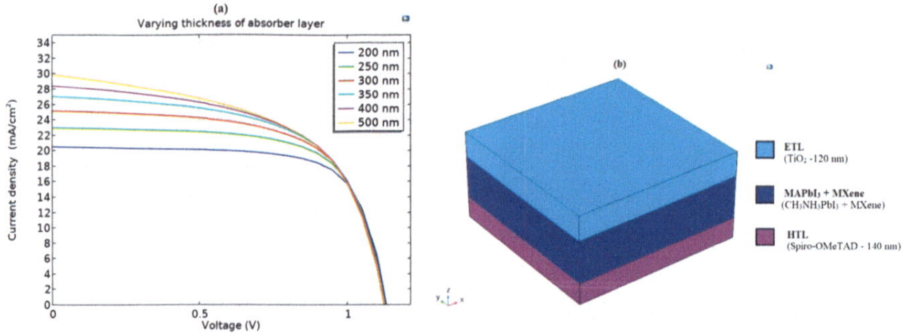

Fig. 11.6 (**a**) JV characteristics for different absorber layer thicknesses (Beepat et al. 2025). (**b**) Schematic representation of the simulated multijunction solar cell structure, consisting of an electron transport layer (ETL) made of TiO_2 (120 nm), an absorber layer of $MAPbI_3$ + MXene ($CH_3NH_3PbI_3$ + MXene), and a hole transport layer (HTL) of Spiro-OMeTAD (140 nm) (Beepat et al. 2025)

too. $Ti_3C_2T_x$ MXenes have considerable promise to increase the efficiency and optoelectronic behavior of PSCs because to their unique electrical characteristics resulting from surface-terminating functional groups (T_x) (Fig. 11.6).

MXene-doped PSCs with an ETL/absorber layer/HTL arrangement were modeled in this work using COMSOL Multiphysics. The HTL was Spiro-OMeTAD (140 nm) and the ETL was TiO_2 (120 nm). The impact of changing absorber layer thickness—between 200 nm and 500 nm—on current density and voltage characteristics was investigated using a JV curve. Furthermore generated was a schematic depiction of the simulated solar cell construction to show the layer configuration used in the research (Beepat et al. 2025).

Through methodically investigating absorber layer thickness and doping concentration, researchers found an optimal solar cell design with a maximum efficiency of 19.46%, a fill factor of 0.61, an open-circuit voltage (V_{OC}) of 1.11 V, and a short-circuit current density (J_{SC}) of 29.03 mA/cm^2. This work shows the first known use of COMSOL Multiphysics in modeling PSCs doped with 2D $Ti_3C_2T_X$ MXene, therefore offering important information for further work and manufacturing of such devices.

Particularly in bandgap engineering, COMSOL Multiphysics remains a strong tool for simulating and optimizing PSCs overall. Its multiphysics properties help to clarify the intricate interactions within PSCs, therefore opening the path for the creation of solar cells with custom bandgaps providing increased efficiency and stability.

ATLAS and SILVACO Simulations

Because of their remarkable efficiency and low cost of manufacture, perovskite solar cells (PSCs) have become somewhat well-known in photovoltaic research bandgap adjustment is a major component in optimizing PSC performance as it directly influences light absorption and general efficiency. Analyzing the results of bandgap engineering depends much on advanced simulation tools as Silvaco's ATLAS. These instruments let scientists maximize material qualities and device architectures and replicate varying bandgap values.

In one study, Wang et al. constructed and investigated the optimal configuration of a $Cs_2AgBiBr_6$ double perovskite/silicon heterojunction tandem structure using Silvaco ATLAS. Their computer computations foresaw a 27.25% theoretical efficiency. The efficiency rose to 37.14%, above that of individual sub-cells by substituting Cs_2AgBi_6 with $Cs_2AgBi_{0.6}$ in the top cell. Highly fit for solar uses, lead-free double perovskites provide structural stability, a configurable bandgap, and environmental friendliness. Density functional theory (DFT) computations with CASTEP investigated crystal structure and electrical characteristics of mixed halide $Cs_2AgB_{1-x}Sb_xBr_6$ (x = 0, 0.25). The findings showed that, with a bandgap of 1.8 eV, Sb-doped Cs_2AgB_6 is more fit for the top cell than $Cs_2AgBiBr_6$, which has a bandgap of 2.05 eV. This adjusted bandgap guarantees better spectrum absorption, enhanced current matching, and greater theoretical efficiency than traditional perovskite tandem cells (Wang et al. 2023).

Using the SILVACO program, Fakhri et al. performed a 3D simulation investigation on the optical and electrical characteristics of PSCs. They obtained an efficiency of 28.064% by using an inverted planar construction with inorganic transport materials and well-chosen anti-reflective coatings and back contact layers, hence improving stability without the requirement for silicone composites. To maximize performance, several anti-reflective materials—CaF_2, SiO_2, and Al_2O_3—were tried in different thicknesses. Under a CaF_2 anti-reflective coating at 110 nm, the optimal absorber layer thickness was found to be 500 nm.

While ZnO was utilized as an electron transport material (ETM), Spiro-OMeTAD and inorganic materials such as Cu_2O and NiOx were used as hole transport materials (HTMs), hence optimizing efficiency. The results were efficiencies of 29.261% for Spiro-OMeTAD/ZnO, 28.064% for Cu_2O/ZnO, and 27.325% for NiOx/ZnO. Although Spiro-OMeTAD showed the best efficiency, its complicated synthesis, great cost, and instability presented difficulties. Though it somewhat reduced efficiency by 1.197%, Cu_2O was suggested as a reasonably affordable substitute. Additionally investigated was the use of graphene as a connection, which provides affordability, flexibility, and low weight. The results imply that paired with an inverted planar layout and inorganic transport materials, graphene-based PSCs show considerable potential for future developments (Fakhri et al. 2021).

The photovoltaic sector is progressively focusing on advanced perovskite–silicon tandem solar cells to further enhance power conversion efficiency. Combining the benefits of perovskite and silicon, these tandem cells transcend single-junction cell efficiency constraints. Still, problems with material compatibility and deposition methods persist.

Recent work employing Silvaco ATLAS's Victory Process and Device Simulator enhanced simulation techniques to recreate real-world manufacturing environments. Three main scenarios—shunting, planarization, and conformal deposition—were used in the study to replicate experimental configurations. Results showed how well process models could precisely forecast and improve tandem solar cell photovoltaic performance. Process simulations allowed two separate perovskite-silicon tandem cell designs to be designed with respective conversion efficiencies of 27.51% and 29.08% (Kumar et al. 2025).

With comprehensive simulations maybe directing the production of optimal perovskite/silicon tandem cells, this work emphasizes the importance of simulation tools in promoting tandem solar cell technology.

Density Functional Theory (DFT) Calculations

Our knowledge of bandgap tuning in perovskite solar cells (PSCs) has been much advanced in great part by density functional theory (DFT) computations. These computational techniques enable researchers to methodically study the interactions between atomic composition, electronic characteristics, and optoelectronic performance by offering thorough understanding of the basic electronic structure of perovskite materials. These insights help DFT studies to be rather important in directing the design and optimization of PSCs to reach increased performance, more efficiency, and better stability. DFT has become a crucial instrument in the creation of next-generation solar cells as it allows one to forecast and study material characteristics at the quantum-mechanical level.

Focusing on how changes in B-site atom composition affect light-matter interactions, Hasan et al. investigated the electronic characteristics of CsSn1 − xGexI thoroughly using density functional theory (DFT). The work sought to investigate how changing the elemental makeup of the perovskite absorber layer influences important optoelectronic properties like bandgap tuning and absorption characteristics. Using the extensively used computational tool Solar Cell Capacitance Simulator (SCAPS-1D), the researchers assessed the device performance of these modified perovskite structures. But SCAPS-1D approximates rather than provides a completely correct quantum-mechanical model by estimating the absorption spectrum using a simpler mathematical formula. Derived from DFT computations, essential characteristics including absorbance spectra and bandgap values were included into SCAPS-1D to improve the accuracy of these simulations and close the gap between theoretical predictions and practical performance. The results of this work showed that widening of the bandgap and improvement in the absorption profile resulting from increasing the concentration of Ge within the perovskite structure increases the efficiency of solar energy conversion. Moreover, utilizing the finite-difference time-domain (FDTD) framework, further studies concerning temperature and field distribution were conducted in order to get better understanding of the effect of these changes. The study team obtained an exceptional power conversion efficiency (PCE) of 23.55% by precisely adjusting the absorber layer composition to include a 75% Ge concentration. These results provide important direction for further studies aiming at the synthesis of lead-free, high-performance halide PSCs. The findings highlight even more the possibilities of atomic doping techniques in customizing the characteristics of perovskite absorber layers, thereby supporting the construction of reasonably affordable, stable, and very efficient solar cells able to satisfy the increasing need for sustainable energy solutions (Hasan et al. 2024).

The capacity to concurrently lower the bandgap and raise the general quality of the perovskite material is one of the most important elements in improving the performance of inorganic perovskite solar cells. Maximizing light absorption, reducing charge recombination, and enhancing PSC long-term stability all depend on reaching this equilibrium. In order to change the bandgap and concurrently improve the quality of the perovskite film, scientists investigated the inclusion of Sn^{2+} into inorganic $CsPbBr_3$ perovskites. Calculations using density functional theory (DFT) revealed important new perspectives on how the presence of Sn^{2+} affects the electronic structure of the material, namely that the insertion of Sn^{2+} causes a significant downward shift in the conduction band minimum. As the Sn^{2+} concentration rises, this change causes the bandgap to gradually decrease, hence improving the material's fit for effective solar energy absorption. Moreover, it was shown that the suitable replacement of Pb^{2+} with Sn^{2+} generates high-quality doped perovskite films with better structural and optoelectronic qualities than undoped films. These developments support important performance advancements including a decrease in charge recombination losses and a rise in light absorption efficiency in the constructed PSCs. These optimizations resulted in a much improved power conversion efficiency (PCE) of 8.04%, compared to the 5.93% efficiency recorded for the control device. The $CsPb_{0.95}Sn_{0.05}Br_3$-based PSC, built with a device architecture comprising an FTO/TiO2/perovskite/carbon layer structure, showed this greatly enhanced PCE. Furthermore extensively evaluated under ambient circumstances was the stability of this optimal perovskite solar cell. Specifically, the $CsPb_{0.95}Sn_{0.05}Br_3$ PSC effectively maintained over 91% of its starting efficiency for 720 hours in ambient storage even without any encapsulation. This amazing stability emphasizes the strength of Sn^{2+}-incorporated perovskite solar cells and their possible use in actual solar energy systems. A major advance in the evolution of next-generation, long-lasting perovskite solar systems is the capacity of these materials to sustain great efficiency over extended periods without deterioration (Wang et al. 2022).

Because of their better environmental stability and reduced toxicity, halide double perovskite semiconductors—such as $Cs_2AgBiBr_6$—have attracted a lot of interest as possible substitutes for traditional lead-halide perovskites. Photovoltaics and photoredox catalysis are only two of the many light conversion uses for which these materials have great promise. Nevertheless, a main drawback of $Cs_2AgBiBr_6$ is its very broad and indirect bandgap, which limits its capacity to effectively absorb and use sunlight for energy conversion. Researchers investigated the regulated replacement of Bi^{3+} with Fe^{3+} using a mechanochemical synthesis method in order to solve this problem and showed that this focused change allows an amazing degree of tunability in the absorption beginning of the material. The researchers greatly expanded the light absorption capacity of the material by precisely modifying the Fe^{3+} concentration within the perovskite structure, therefore obtaining an absorption start range from 2.1 eV down to almost 1 eV. First-principles density functional theory (DFT) calculations revealed important new directions for the mechanism underlying this bandgap reduction: the introduction of Fe^{3+} mostly results in a lowering of the conduction band energy, which in turn promotes more absorption of lower-energy photons. Furthermore underlining the possible uses of Fe^{3+}-substituted halide double perovskites in light-driven catalytic processes, the work showed that this tunability in conduction band energy is strongly connected with the photoredoxin activity of the material. DFT computations especially showed that the material moves from an indirect bandgap to a direct bandgap structure when more than half of Bi^{3+} is replaced with Fe^{3+}. This change is especially important as materials with direct bandgaps usually show improved optical absorption and charge carrier dynamics, which qualifies them more for effective solar energy uses. These revolutionary results not only provide fresh opportunities for raising the light absorption efficiency of double perovskite semiconductors but also help to shape the direction of next-generation sustainable energy technologies development. Researchers may fully exploit these materials for use in environmentally friendly, high-performance solar cells and other optoelectronic applications by carefully engineering their electrical characteristics using atomic-scale changes (Jöbsis et al. 2024).

Conclusion

The performance of perovskite solar cells is investigated in this chapter with reference to band gap engineering. By changing the structure, interfaces, doping techniques, and material composition, researchers have found ways to increase the stability and efficiency of these devices. These methods help to modify material properties so that they would be more useful in solar energy applications. While dimension reduction improves material endurance, doping improves the charge transport property of the perovskite material; interface engineering improves the surface property of the perovskite materials. Further validation of these approaches comes from the simulation results using SCAPS-1D, COMSOL Multiphysics, ATLAS SILVACO, and DFT computations. Higher power conversion efficiency is shown by SCAPS-1D simulations to follow from better material composition and contact attributes. COMSOL Multiphysics simulations show how important charge transport and thermal stability are to preserving constant device performance. Customized doping techniques improve charge carrier mobility and lower recombination losses, according to ATLAS SILVACO modeling. DFT computations provide an understanding of the electrical structure and defect states, thereby verifying the effect of band gap adjustment on device stability and efficiency. High-performance and durable PSC devices are achieved in great part by these approaches and simulation tools. Future studies will continue to improve these methods to solve the last difficulties and fully use perovskite solar cells in the context of renewable energy.

Acknowledgments We would like to acknowledge UPES Dehradun for the support and also like to acknowledge this work under National Project (124022900010-9) from TSC SB-RAS at Earth Cryosphere Institute, Tyumen, Russia.

References

Abdelhady AL, Saidaminov MI, Murali B, Adinolfi V, Voznyy O, Katsiev K, Bakr OM (2016) Heterovalent dopant incorporation for bandgap and type engineering of perovskite crystals. J Phys Chem Lett 7(2):295–301. https://doi.org/10.1021/acs.jpclett.5b02681

Abdi-Jalebi M, Andaji-Garmaroudi Z, Cacovich S, Stavrakas C, Philippe B, Richter JM, Pearson AJJ (2018) Maximizing and stabilizing luminescence from halide perovskites with potassium passivation. Nature 555(7697):497–501

Aberle AG (2000) Surface passivation of crystalline silicon solar cells: a review. Prog Photovolt Res Appl 8(5):473–487. https://doi.org/10.1002/1099-159X(200009/10)8:5<473::AID-PIP337>3.0.CO;2-D

Ahn N, Son D-Y, Jang I-H, Kang SM, Choi M, Park N-G (2015) Highly reproducible perovskite solar cells with average efficiency of 18.3% and best efficiency of 19.7% fabricated via Lewis Base adduct of Lead(II) iodide. J Am Chem Soc 137(27):8696–8699. https://doi.org/10.1021/jacs.5b04930

Arora N, Dar MI, Hinderhofer A, Pellet N, Schreiber F, Zakeeruddin SM, Grätzel M (2017) Perovskite solar cells with CuSCN hole extraction layers yield stabilized efficiencies greater than 20%. Science 358(6364):768–771. https://doi.org/10.1126/science.aam5655

Azpiroz JM, Mosconi E, Bisquert J, De Angelis F (2015) Defect migration in methylammonium lead iodide and its role in perovskite solar cell operation. Energy Environ Sci 8(7):2118–2127. https://doi.org/10.1039/C5EE01265A

Beepat K, Sharma D, Mahajan A, Pathak D, Kumar V (2024) Mxene doped perovskite solar cell simulation for enhanced efficiency

Beepat KG, Sharma DP, Mahajan A, Pathak D, Kumar V (2025) MXene doped perovskite solar cell simulation for enhanced efficiency COMSOL conference 2025

Bertness KA, Kurtz SR, Friedman DJ, Kibbler AE, Kramer C, Olson JM (1994) 29.5%-efficient GaInP/GaAs tandem solar cells. Appl Phys Lett 65(8):989–991. https://doi.org/10.1063/1.112171

Bi D, Tress W, Dar MI, Gao P, Luo J, Renevier C, Hagfeldt A (2016) Efficient luminescent solar cells based on tailored mixed-cation perovskites. Sci Adv 2(1):1. https://doi.org/10.1126/sciadv.1501170

Bi E, Chen H, Xie F, Wu Y, Chen W, Su Y, Han L (2017) Diffusion engineering of ions and charge carriers for stable efficient perovskite solar cells. Nat Commun 8:15330. https://doi.org/10.1038/ncomms15330

Bu T, Liu X, Zhou Y, Yi J, Huang X, Luo L, Zhong J (2017) A novel quadruple-cation absorber for universal hysteresis elimination for high efficiency and stable perovskite solar cells. Energy Environ Sci 10(12):2509–2515. https://doi.org/10.1039/c7ee02634j

Burschka J, Pellet N, Moon S-J, Humphry-Baker R, Gao P, Nazeeruddin MK, Grätzel M (2013) Sequential deposition as a route to high-performance perovskite-sensitized solar cells. Nature 499(7458):316–319. https://doi.org/10.1038/nature12340

Cao J, Jing X, Yan J, Hu C, Chen R, Yin J, Zheng N (2016) Identifying the molecular structures of intermediates for optimizing the fabrication of high-quality perovskite films. J Am Chem Soc 138(31):9919–9926. https://doi.org/10.1021/jacs.6b04924

Chen B, Rudd PN, Yang S, Yuan Y, Huang J (2019) Imperfections and their passivation in halide perovskite solar cells. Chem Soc Rev 48(14):3842–3867. https://doi.org/10.1039/C8CS00853A

Cho KT, Grancini G, Lee Y, Oveisi E, Ryu J, Almora O (2018) Selective growth of layered perovskites for stable and efficient photovoltaics. Energy Environ Sci 11(4):952–959

Chowdhury TA, Bin Zafar MA, Sajjad-Ul Islam M, Shahinuzzaman M, Islam MA, Khandaker MU (2023) Stability of perovskite solar cells: issues and prospects. RSC Adv 13(3):1787–1810. https://doi.org/10.1039/D2RA05903G

deQuilettes DW, Koch S, Burke S, Paranji RK, Shropshire AJ, Ziffer ME, Ginger DS (2016) Photoluminescence lifetimes exceeding 8 μs and quantum yields exceeding 30% in hybrid perovskite thin films by ligand passivation. ACS Energy Lett 1(2):438–444. https://doi.org/10.1021/acsenergylett.6b00236

Elumalai NK, Uddin A (2016) Hysteresis in organic-inorganic hybrid perovskite solar cells. Sol Energy Mater Sol Cells 157:476–509. https://doi.org/10.1016/j.solmat.2016.06.025

Eperon GE, Stranks SD, Menelaou C, Johnston MB, Herz LM, Snaith HJ (2014) Formamidinium lead trihalide: a broadly tunable perovskite for efficient planar heterojunction solar cells. Energy Environ Sci 7(3):982–988. https://doi.org/10.1039/C3EE43822H

Fakhri N, Salay Naderi M, Gholami Farkoush S, SaeidNahaei S, Park S-N, Rhee S-B (2021) Simulation of perovskite solar cells optimized by the inverse planar method in SILVACO: 3D electrical and optical models. Energies 14(18):5944. https://doi.org/10.3390/en14185944

Goetzberger A, Knobloch J, Voß B (2014) Photovoltaics. In: Crystalline silicon solar cells. Wiley, New York, pp 1–4

Grancini G, Roldán-Carmona C, Zimmermann I, Mosconi E, Lee X, Martineau D, Graetzel M, Nazeeruddin MK (2017) One-Year stable perovskite solar cells by 2D/3D interface engineering. Nat Commun Sci 8(1):15684

Green MA, Dunlop ED, Yoshita M, Kopidakis N, Bothe K, Siefer G, Hao X (2024) Solar cell efficiency tables (version 63). Prog Photovoltaics Res Appl 32(1):3–13. https://doi.org/10.1002/pip.3750

Guerrero A, Juarez-Perez EJ, Bisquert J, Mora-Sero I, Garcia-Belmonte G (2014) Electrical field profile and doping in planar lead halide perovskite solar cells. Appl Phys Lett 105(13):133902. https://doi.org/10.1063/1.4896779

Han Y, Zhang G, Xie H, Kong T, Li Y, Zhang Y, Bi D (2022) Azide additive acting as a powerful locker for Li+ and TBP in spiro-OMeTAD toward highly efficient and stable perovskite solar cells. Nano Energy 96:107072. https://doi.org/10.1016/j.nanoen.2022.107072

Haq MA, Saiduzzaman M, Asif TI, Shuvo IK, Hossain KM (2021) Ultra-violet to visible band gap engineering of cubic halide KCaCl3 perovskite under pressure for optoelectronic applications: insights from DFT. RSC Adv 11(58):36367–36378. https://doi.org/10.1039/D1RA06430D

Hasan N, Anik MHK, Hasan MM, Islam S, Kabir A (2024) Comprehensive numerical analysis of doping controlled efficiency in lead-free CsSn1−xGexI3 perovskite solar cell. Appl Phys A 131(1):34. https://doi.org/10.1007/s00339-024-08125-y

Helal Miah M, Bulu Rahman M, Khatun F, Khandaker MU, Wan Muhammad Hatta SF, Soin NB, Islam MA (2023) Optimization and detail analysis of novel structure Pb-free CsGeI3-based all-inorganic perovskite solar cells by SCAPS-1D. Optik 281:170819. https://doi.org/10.1016/j.ijleo.2023.170819

Hoffman, J. M., Che, X., Sidhik, S., Li, X., Hadar, I., Blancon, J.-C., .Kanatzidis, M. G. (2019). From 2D to 1D electronic dimensionality in halide perovskites with stepped and flat layers using Propylammonium as a spacer. J Am Chem Soc, 141(27), 10661–10676. https://doi.org/10.1021/jacs.9b02846

Hossain MK, Rubel MHK, Toki GI, Alam I, Rahman MF, Bencherif HJAO (2022) Effect of various electron and hole transport layers on the performance of CsPbI3-based perovskite solar cells: a numerical investigation in DFT, SCAPS-1D, and wxAMPS frameworks. ACS Omega 7(47):43210–43230

Hu J, Wang C, Qiu S, Zhao Y, Gu E, Zeng L, Guo F (2020) Spontaneously self-assembly of a 2D/3D Heterostructure enhances the efficiency and stability in printed perovskite solar cells. Adv Energy Mater 10(17):2000173. https://doi.org/10.1002/aenm.202000173

Im J-H, Jang I-H, Pellet N, Grätzel M, Park N-G (2014) Growth of CH3NH3PbI3 cuboids with controlled size for high-efficiency perovskite solar cells. Nat Nanotechnol 9(11):927–932. https://doi.org/10.1038/nnano.2014.181

Ito N, Kamarudin MA, Hirotani D, Zhang Y, Shen Q, Ogomi Y, Hayase S (2018) Mixed Sn–Ge perovskite for enhanced perovskite solar cell performance in air. J Phys Chem Lett 9(7):1682–1688. https://doi.org/10.1021/acs.jpclett.8b00275

Jeon NJ, Noh JH, Kim YC, Yang WS, Ryu S, Seok SI (2014) Solvent engineering for high-performance inorganic–organic hybrid perovskite solar cells. Nat Mater 13(9):897–903. https://doi.org/10.1038/nmat4014

Jeon NJ, Noh JH, Yang WS, Kim YC, Ryu S, Seo J, Seok SI (2015) Compositional engineering of perovskite materials for high-performance solar cells. Nature 517(7535):476–480. https://doi.org/10.1038/nature14133

Jiang Q, Zhao Y, Zhang X, Yang X, Chen Y, Chu Z, You J (2019) Surface passivation of perovskite film for efficient solar cells. Nat Photonics 13(7):460–466. https://doi.org/10.1038/s41566-019-0398-2

Jiao H, Ni Z, Shi Z, Fei C, Liu Y, Dai X, Huang J (2022) Perovskite grain wrapping by converting interfaces and grain boundaries into robust and water-insoluble low-dimensional perovskites. Sci Adv 8(48):eabq4524. https://doi.org/10.1126/sciadv.abq4524

Jöbsis HJ, Fykouras K, Reinders JW, van Katwijk J, Dorresteijn JM, Arens T, Hutter EMJAFM (2024) Conduction band tuning by controlled alloying of Fe into Cs2AgBiBr6 double perovskite powders. Adv Funct Mater 34(50):2306106

Jung S-K, Park N-G, Lee J-W (2023) Light management in perovskite solar cells. Mater Today Energy 37:101401. https://doi.org/10.1016/j.mtener.2023.101401

Kahwagi RF, Thornton ST, Smith B, Koleilat GI (2020) Dimensionality engineering of metal halide perovskites. Front Optoelectron 13(3):196–224. https://doi.org/10.1007/s12200-020-1039-6

Katan C, Mercier N, Even J (2019) Quantum and dielectric confinement effects in lower-dimensional hybrid perovskite semiconductors. Chem Rev 119(5):3140–3192. https://doi.org/10.1021/acs.chemrev.8b00417

Kim HS, Lee CR, Im JH, Lee KB, Moehl T, Marchioro A, Park NG (2012) Lead iodide perovskite sensitized all-solid-state submicron thin film mesoscopic solar cell with efficiency exceeding 9%. Sci Rep 2:591. https://doi.org/10.1038/srep00591

Kim H, Lee SU, Lee DY, Paik MJ, Na H, Lee J, Seok SI (2019) Optimal interfacial engineering with different length of Alkylammonium halide for efficient and stable perovskite solar cells. Adv Energy Mater 9(47):1902740. https://doi.org/10.1002/aenm.201902740

Kojima A, Teshima K, Shirai Y, Miyasaka T (2009) Organometal halide perovskites as visible-light sensitizers for photovoltaic cells. J Am Chem Soc 131(17):6050–6051. https://doi.org/10.1021/ja809598r

Kong J, Shin Y, Röhr JA, Wang H, Meng J, Wu Y, Taylor AD (2021) CO2 doping of organic interlayers for perovskite solar cells. Nature 594(7861):51–56. https://doi.org/10.1038/s41586-021-03518-y

Kumar R, Kanjariya P, Abu-Jrai A, Patil N, Ab Yajid MS, Kaur J, Khairandish MI (2025) Addressing fabrication challenges in perovskite-silicon tandem solar cells with advanced simulation techniques. Mater Renew Sustain Energy 14(1):18. https://doi.org/10.1007/s40243-024-00284-5

Lee MM, Teuscher J, Miyasaka T, Murakami TN, Snaith HJ (2012) Efficient hybrid solar cells based on meso-superstructured organometal halide perovskites. Science 338(6107):643–647. https://doi.org/10.1126/science.1228604

Lee J-W, Seol D-J, Cho A-N, Park N-G (2014) High-efficiency perovskite solar cells based on the black polymorph of HC(NH2)2PbI3. Adv Mater 26(29):4991–4998. https://doi.org/10.1002/adma.201401137

Lei Y, Li Y, Jin Z (2022) Photon energy loss and management in perovskite solar cells. Energy Rev 1(1):100003. https://doi.org/10.1016/j.enrev.2022.100003

Li X, Ibrahim Dar M, Yi C, Luo J, Tschumi M, Zakeeruddin SM, Grätzel M (2015) Improved performance and stability of perovskite solar cells by crystal crosslinking with alkylphosphonic acid ω-ammonium chlorides. Nat Chem 7(9):703–711. https://doi.org/10.1038/nchem.2324

Li Z, Yang M, Park J-S, Wei S-H, Berry JJ, Zhu K (2016a) Stabilizing perovskite structures by tuning tolerance factor: formation of formamidinium and cesium Lead iodide solid-state alloys. Chem Mater 28(1):284–292. https://doi.org/10.1021/acs.chemmater.5b04107

Li M, Wang Z-K, Yang Y-G, Hu Y, Feng S-L, Wang J-M, Liao L-S (2016b) Copper salts doped Spiro-OMeTAD for high-performance perovskite solar cells. Adv Energy Mater 6(21):1601156. https://doi.org/10.1002/aenm.201601156

Li M, Begum R, Fu J, Xu Q, Koh TM, Veldhuis SA, Sum TC (2018) Low threshold and efficient multiple exciton generation in halide perovskite nanocrystals. Nat Commun 9(1):4197. https://doi.org/10.1038/s41467-018-06596-1

Li Z, Wu J, Liu X, Zhu Q, Yang Y, Dou Y, Lin J-Y (2020) Highly efficient and stable perovskite solar cells using thionyl chloride as a p-type dopant for spiro-OMeTAD. J Alloys Compd 847:156500. https://doi.org/10.1016/j.jallcom.2020.156500

Li D, Xing Z, Meng X, Hu X, Hu T, Chen Y (2023) Selection of functional spacer cations for efficient 2D/3D perovskite. Solar Cells 5(4):781–801. https://doi.org/10.31635/ccschem.023.202202409

Lin Y, Shen L, Dai J, Deng Y, Wu Y, Bai Y, Huang J (2017) π-Conjugated Lewis Base: efficient trap-passivation and charge-extraction for hybrid perovskite solar cells. Adv Mater 29(7):1418513. https://doi.org/10.1002/adma.201604545

Liu M, Johnston MB, Snaith HJ (2013) Efficient planar heterojunction perovskite solar cells by vapour deposition. Nature 501(7467):395–398. https://doi.org/10.1038/nature12509

Liu Y, Yang Z, Cui D, Ren X, Sun J, Liu X, Liu S (2015) Two-inch-sized perovskite CH3NH3PbX3 (X = Cl, Br, I) crystals: growth and characterization. Adv Mater 27(35):5176–5183. https://doi.org/10.1002/adma.201502597

Liu G, Kong L, Yang W, Mao H-k (2019) Pressure engineering of photovoltaic perovskites. Mater Today 27:91–106. https://doi.org/10.1016/j.mattod.2019.02.016

Ma K, Atapattu HR, Zhao Q, Gao Y, Finkenauer BP, Wang K, Dou L (2021) Multifunctional conjugated ligand engineering for stable and efficient perovskite solar cells. Adv Mater 33(32):2100791. https://doi.org/10.1002/adma.202100791

Miah MH, Rahman MB, Nur-E-Alam M, Das N, Soin NB, Hatta SFWM, Islam MA (2023) Understanding the degradation factors, mechanism and initiatives for highly efficient perovskite solar cells. ChemNanoMat 9(3):e202200471. https://doi.org/10.1002/cnma.202200471

Miah MH, Khandaker MU, Aminul Islam M, Nur-E-Alam M, Osman H, Ullah MH (2024) Perovskite materials in X-ray detection and imaging: recent progress, challenges, and future prospects. RSC Adv 14(10):6656–6698. https://doi.org/10.1039/D4RA00433G

Min H, Kim M, Lee S-U, Kim H, Kim G, Choi K, Seok SI (2019) Efficient, stable solar cells by using inherent bandgap of α-phase formamidinium lead iodide. Science 366(6466):749–753. https://doi.org/10.1126/science.aay7044

Min H, Lee DY, Kim J, Kim G, Lee KS, Kim J, Il Seok S (2021) Perovskite solar cells with atomically coherent interlayers on SnO_2 electrodes. Nature 598(7881):444–450. https://doi.org/10.1038/s41586-021-03964-8

National Renewable Energy Laboratory (2025) *Best research-cell efficiencies* [chart]. NREL. https://www.nrel.gov/pv/cell-efficiency.html

Noel, N. K., Stranks, S. D., Abate, A., Wehrenfennig, C., Guarnera, S., Haghighirad, A.-A., .Snaith, H. J. (2014). Lead-free organic–inorganic tin halide perovskites for photovoltaic applications. Energy Environ Sci, 7(9), 3061–3068. https://doi.org/10.1039/C4EE01076K

Noh JH, Im SH, Heo JH, Mandal TN, Seok SI (2013) Chemical management for colorful, efficient, and stable inorganic–organic hybrid nanostructured solar cells. Nano Lett 13(4):1764–1769. https://doi.org/10.1021/nl400349b

Noman M, Shahzaib M, Jan ST, Khan Z, Ismail M, Khan AD (2024) Optimizing band gap, electron affinity, & carrier mobility for improved performance of formamidinium lead tri-iodide perovskite solar cells. Mater Sci Eng B Solid-State Mater Adv Technol 300:117114. https://doi.org/10.1016/j.mseb.2023.117114

Ou Q, Bao X, Zhang Y, Shao H, Xing G, Li X, Bao Q (2019) Band structure engineering in metal halide perovskite nanostructures for optoelectronic applications. Nano Mater Sci 1(4):268–287. https://doi.org/10.1016/j.nanoms.2019.10.004

Pazos-Outón LM, Xiao TP, Yablonovitch E (2018) Fundamental efficiency limit of Lead iodide perovskite solar cells. J Phys Chem Lett 9(7):1703–1711. https://doi.org/10.1021/acs.jpclett.7b03054

Pedesseau, L., Sapori, D., Traore, B., Robles, R., Fang, H.-H., Loi, M. A., .Kepenekian, M. (2016). Advances and promises of layered halide hybrid perovskite semiconductors. ACS Nano, 10(11), 9776–9786. https://doi.org/10.1021/acsnano.6b05944

Pellet N, Gao P, Gregori G, Yang TY, Nazeeruddin MK, Maier J, Grätzel M (2014) Mixed-organic-cation Perovskite photovoltaics for enhanced solar-light harvesting. Ang Chem Int Ed 53(12):3151–3157
Peng S, Ma J, Li P, Zang S, Zhang Y, Song Y (2022) Regulation of quantum Wells width distribution in 2D perovskite films for photovoltaic application. Adv Funct Mater 32(43):2205289. https://doi.org/10.1002/adfm.202205289
Rabhi S, Sekar K, Kalna K, Bouderbala YI, Bouri N, Oueldna N, Attaf N (2024) Experimental findings and SCAPS-1D simulations for high-efficiency MAPbI3 perovskite solar cells beyond 31%. Opt Quant Electron 56(8):1372. https://doi.org/10.1007/s11082-024-07282-x
Rahman MB, Noor EA, Miah MH, Khandaker MU, Islam MA (2023) Selection of a compatible electron transport layer and hole transport layer for the mixed perovskite FA0.85Cs0.15Pb (I0.85Br0.15)3, towards achieving novel structure and high-efficiency perovskite solar cells: a detailed numerical study by SCAPS-1D. RSC Adv 13(25):17130–17142. https://doi.org/10.1039/D3RA02170J
Salhi B, Wudil YS, Hossain MK, Al-Ahmed A, Al-Sulaiman FA (2018) Review of recent developments and persistent challenges in stability of perovskite solar cells. Renew Sust Energ Rev 90:210–222. https://doi.org/10.1016/j.rser.2018.03.058
Saliba M, Matsui T, Domanski K, Seo J-Y, Ummadisingu A, Zakeeruddin SM, Grätzel M (2016a) Incorporation of rubidium cations into perovskite solar cells improves photovoltaic performance. Science 354(6309):206–209. https://doi.org/10.1126/science.aah5557
Saliba M, Matsui T, Seo JY, Domanski K, Correa-Baena JP, Nazeeruddin MK, Grätzel M (2016b) Cesium-containing triple cation perovskite solar cells: improved stability, reproducibility and high efficiency. Energy Environ Sci 9(6):1989–1997. https://doi.org/10.1039/c5ee03874j
Seo J, Jeon NJ, Yang WS, Shin H-W, Ahn TK, Lee J, Seok SI (2015) Effective Electron Blocking of CuPC-Doped Spiro-OMeTAD for Highly Efficient Inorganic–Organic Hybrid Perovskite Solar Cells. Adv Energy Mater 5(20):1501320. https://doi.org/10.1002/aenm.201501320
Sheldon MT, Eisler CN, Atwater HA (2012) GaAs passivation with Trioctylphosphine sulfide for enhanced solar cell efficiency and durability. Adv Energy Mater 2(3):339–344. https://doi.org/10.1002/aenm.201100666
Shivesh K, Alam I, Kushwaha AK, Kumar M, Singh SV (2022) Investigating the theoretical performance of Cs2TiBr6-based perovskite solar cell with La-doped BaSnO3 and CuSbS2 as the charge transport layers. Int J Energy Res 46(5):6045–6064
Snaith HJ, Abate A, Ball JM, Eperon GE, Leijtens T, Noel NK, Zhang W (2014) Anomalous hysteresis in perovskite solar cells. J Phys Chem Lett 5(9):1511–1515. https://doi.org/10.1021/jz500113x
Stoumpos CC, Cao DH, Clark DJ, Young J, Rondinelli JM, Jang JI, Kanatzidis MG (2016) Ruddlesden–popper hybrid lead iodide perovskite 2D homologous semiconductors. Chem Mater 28(8):2852–2867. https://doi.org/10.1021/acs.chemmater.6b00847
Stranks SD, Eperon GE, Grancini G, Menelaou C, Alcocer MJP, Leijtens T, Snaith HJ (2013) Electron-hole diffusion lengths exceeding 1 micrometer in an organometal trihalide perovskite absorber. Science 342(6156):341–344. https://doi.org/10.1126/science.1243982
Thompson AG, Woolley JC (1967) Energy-gap variation in mixed iii–v alloys. Canadian J Phys 45:255–261
Tsai H, Nie W, Blancon J-C, Stoumpos CC, Asadpour R, Harutyunyan B, Mohite AD (2016) High-efficiency two-dimensional Ruddlesden–Popper perovskite solar cells. Nature 536(7616):312–316. https://doi.org/10.1038/nature18306
Turren-Cruz S-H, Hagfeldt A, Saliba M (2018) Methylammonium-free, high-performance, and stable perovskite solar cells on a planar architecture. Science 362(6413):449–453. https://doi.org/10.1126/science.aat3583
Van Vechten JA, Bergstresser TK (1970) Electronic structures of semiconductor Alloys. Phys Rev B 1(8):3351–3358. https://doi.org/10.1103/PhysRevB.1.3351
Wang F, Geng W, Zhou Y, Fang H-H, Tong C-J, Loi MA, Zhao N (2016) Phenylalkylamine passivation of organolead halide perovskites enabling high-efficiency and air-stable photovoltaic cells. Adv Mater 28(45):9986–9992. https://doi.org/10.1002/adma.201603062

Wang T, Daiber B, Frost JM, Mann SA, Garnett EC, Walsh A, Ehrler B (2017) Indirect to direct bandgap transition in methylammonium lead halide perovskite. Energy Environ Sci 10(2):509–515. https://doi.org/10.1039/C6EE03474H

Wang, R., Xue, J., Wang, K.-L., Wang, Z.-K., Luo, Y., Fenning, D., .Yang, Y. (2019). Constructive molecular configurations for surface-defect passivation of perovskite photovoltaics. Science, 366(6472), 1509–1513. https://doi.org/10.1126/science.aay9698

Wang G, Bi J, Chang JR, Lei M, Zheng H, Yan YJICF (2022) Bandgap tuning of CsPbBr3 perovskite with synergistically improved quality via Sn2+ doping for high-performance carbon-based inorganic perovskite solar cells. Inorg Chem Front 9:4359

Wang Y, Tian H, Zhang D, Liu W, Ma X, Wang J (2023) Study on High-Efficiency Double Perovskite/Silicon Heterojunction Tandem Cells with Sb-Doped Cs2AgBiBr6. J Electron Mater 52(11):7728–7739. https://doi.org/10.1007/s11664-023-10692-4

Wu J, Ma Z, Huang S, Lei Y, Guo H, Fang Y, Lin Y (2022) Directly purifiable pre-oxidation of Spiro-OMeTAD for stability enhanced perovskite solar cells with efficiency over 23%. Chem Eng J 437:135457. https://doi.org/10.1016/j.cej.2022.135457

Xing J, Zhao Y, Askerka M, Quan LN, Gong X, Zhao W, Sargent EH (2018) Color-stable highly luminescent sky-blue perovskite light-emitting diodes. Nat Commun 9(1):3541. https://doi.org/10.1038/s41467-018-05909-8

Xu J, Castriotta LA, Di Carlo A, Brown TM (2024) Air-stable Lead-free antimony-based perovskite inspired solar cells and modules. ACS Energy Lett 9(2):671–678. https://doi.org/10.1021/acsenergylett.3c02409

Xue H, Fu K, Wong LH, Birgersson E, Stangl R (2017) Modelling and loss analysis of meso-structured perovskite solar cells. J Appl Phys 122(8):083105. https://doi.org/10.1063/1.4986115

Yang WS, Noh JH, Jeon NJ, Kim YC, Ryu S, Seo J, Seok SI (2015) High-performance photovoltaic perovskite layers fabricated through intramolecular exchange. Science 348(6240):1234–1237. https://doi.org/10.1126/science.aaa9272

Yang Z, Guo Y, Li H, Zhou Y, Zuo X, Yu Y, Rafailovich MH (2018) Roles of interfacial tension in regulating internal organization of low Bandgap polymer bulk heterojunction solar cells by polymer additives. Adv Mater Interfaces 5(15):1800435

Yang S, Chen S, Mosconi E, Fang Y, Xiao X, Wang C, Huang J (2019) Stabilizing halide perovskite surfaces for solar cell operation with wide-bandgap lead oxysalts. Science 365(6452):473–478

Yavari M, Mazloum-Ardakani M, Gholipour S, Tavakoli MM, Taghavinia N, Hagfeldt A, Tress W (2018) Reducing surface recombination by a poly(4-vinylpyridine) interlayer in perovskite solar cells with high open-circuit voltage and efficiency. ACS Omega 3(5):5038–5043. https://doi.org/10.1021/acsomega.8b00555

Yuren X, Mengfei Z, Lina Q, Cheng Z, Daocheng H, Yuxi T, Zhong J (2023) Organic-inorganic hybrid quasi-2D perovskites incorporated with fluorinated additives for efficient and stable four-terminal tandem solar cells. Energy Mater 3(1):300004. https://doi.org/10.20517/energymater.2022.55

Zhang H, Wu Y, Shen C, Li E, Yan C, Zhang W, Zhu WH (2019) Efficient and stable chemical passivation on perovskite surface via bidentate anchoring. Adv Energy Mater 9(13):1803573. https://doi.org/10.1002/aenm.201803573

Zhang C, Wu S, Tao L, Arumugam GM, Liu C, Wang Z, Mai Y (2020) Fabrication strategy for efficient 2D/3D perovskite solar cells enabled by diffusion passivation and strain compensation. Adv Energy Mater 10(43):2002004. https://doi.org/10.1002/aenm.202002004

Zhang T, Wang F, Kim HB, Choi IW, Wang C, Cho E, Gao F (2022) Ion-modulated radical doping of spiro-OMeTAD for more efficient and stable perovskite solar cells. Science 377(6605):495–501. https://doi.org/10.1126/science.abo2757

Zhao T, Chueh C-C, Chen Q, Rajagopal A, Jen AKY (2016) Defect Passivation of Organic–Inorganic Hybrid Perovskites by Diammonium Iodide toward High-Performance Photovoltaic Devices. ACS Energy Lett 1(4):757–763. https://doi.org/10.1021/acsenergylett.6b00327

Zhu H, Liu Y, Eickemeyer FT, Pan L, Ren D, Ruiz-Preciado MA, Grätzel M (2020) Tailored amphiphilic molecular Mitigators for stable perovskite solar cells with 23.5% efficiency. Adv Mater 32(12):1907757. https://doi.org/10.1002/adma.201907757

Chapter 12
Role of Machine Learning in the Perovskite Solar Cell Investigations

Ankur Taya, Shweta Dhakla, Parvesh K. Deendyal, Sarvesh Kumar, and Manish K. Kashyap

Abstract Perovskite solar cells (PSCs) have emerged as game changers in photovoltaics domain owing to their ease of fabrication and impressive high power conversion efficiencies. However, the challenges like performance optimization, scalability and long-term stability continue to be substantial barriers to commercialization. Machine learning (ML) approach has proven to be an indispensable tool for overcoming these difficulties by offering predictive insights and accelerating the development of innovative materials and device architectures. This chapter emphasizes the pivotal role of ML in PSC research, focusing on data-driven approaches to material discovery and device optimization. Moreover, the comparative analysis of ML techniques including various models such as regression, ensemble, and neural network, have been presented in this chapter. By offering a comprehensive overview of the integration of ML and PSC research, the potential of ML in revolutionizing the photovoltaic technologies and paving a path for next-generation, robust and highly efficient solar cells have been highlighted.

Keywords Perovskite · Photovoltaics · Machine learning · Material discovery

A. Taya
Department of Physics, Mukand Lal National College, Yamunanagar, Haryana, India

S. Dhakla · M. K. Kashyap (✉)
Renewable Energy Laboratory, School of Physical Sciences, Jawaharlal Nehru University, New Delhi, India
e-mail: mkkashyap@jnu.ac.in

P. K. Deendyal
Renewable Energy Laboratory, School of Physical Sciences, Jawaharlal Nehru University, New Delhi, India

Government Polytechnic for Women, Faridabad, Haryana, India

S. Kumar
Inter-University Accelerator Centre (IUAC), Aruna Asaf Ali Marg, New Delhi, India

Department of Physics & Astrophysics, University of Delhi, Delhi, India

M. Kumar (ed.), *Computational Analysis of Perovskite Solar Cells*, Springer Tracts in Electrical and Electronics Engineering,
https://doi.org/10.1007/978-981-95-6521-4_12

Introduction

The changing global energy scenario along with the imminent need for climate change mitigation have shifted the focus from nonrenewable to renewable energy technologies. The sustainability and abundance of solar energy have made it stand out as one of the most promising and effective solutions. With a quest for efficient solar energy conversion, the foundation of modern photovoltaics was laid with the development of Si-based first-generation solar cells in 1950s. With high efficiency and long durability, these solar cells dominated the solar industry for long but their high manufacturing and materials costs underlined the requirement of alternative technologies, which resulted in the development of thin-film technologies based second-generation solar cells (Green 2002; Sharma et al. 2015). This generation registered improvements on some fronts but at the same time accompanied by the challenges like moderate efficiencies, inclusion of toxic materials, and limited availability of deployed materials. The entire focus of third-generation solar cells intensified on developing novel materials and architectures to attain higher efficiencies but at lower costs (Goetzberger et al. 2002). This category is exemplified by the quantum dot solar cells, dye-sensitized solar cells (DSSCs), and organic solar cells (Brown and Wu 2009) etc. as shown in Fig. 12.1. From these, DSSCs garnered significant attention owing to its simple fabrication techniques alongside use of inexpensive materials (Gong et al. 2017). Nevertheless, the challenges of limited efficiencies and long-term stability issues further drove the exploration of alternative new materials. Though known for a very long time, 2009 marked the turning point for organometal halide perovskites when Miyasaka and his team (Kojima et al. 2009) introduced methylammonium lead iodide ($CH_3NH_3PbI_3$) as a light absorber in DSSCs and successfully obtained an initial power conversion efficiency

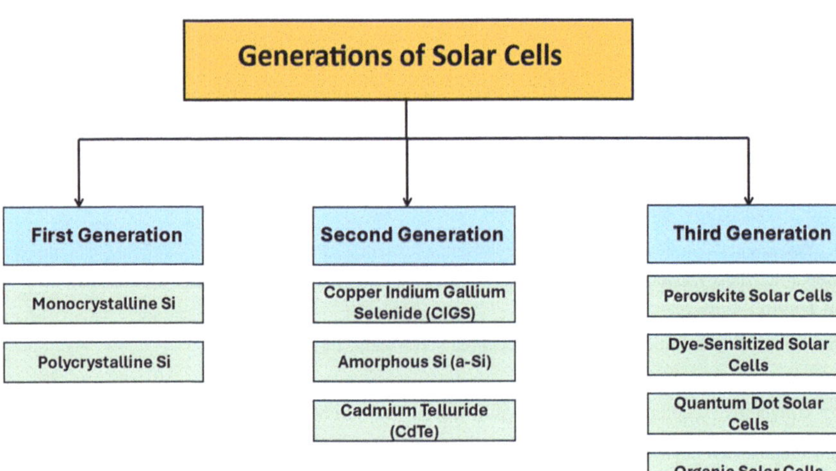

Fig. 12.1 Different generations of solar cells

(PCE) of 3.8%, demonstrating the remarkable potential of perovskites and their derivatives as efficient optoelectronic materials.

Building on these initial developments and findings, the researchers swiftly transitioned from DSSC-inspired architectures to more refined mesoporous and fully planar perovskite solar cell (PSC) designs which further expanded their potential (Zhao et al. 2014; Jiang et al. 2017). Due to these relentless efforts worldwide, PCE first surpassed 10% and then reached 27.3% over the past fifteen years, rivaling the efficiencies of Si solar cells. Notably, the most propelling factor in the success of PSC was the ease of solution-based processing and good compatibility with tandem and flexible device configurations (Bailie and McGehee 2015).

At present, within the category of third generation solar cells, PSCs are the most transformative technologies, and throughout these years their journey from exploratory research to high-performance devices witnessed various important conjectures on theoretical, experimental, and computational fronts (Lopez-Varo et al. 2018). New innovations, technological advancements, and infrastructural growth over the centuries are intrinsically linked with the materials discovery, and intuition, serendipity, and trial-and-error based Edisonian approaches have played an instrumental role in that. But over time, these paradigms of material discovery are shifting from empirical explorations to model based theoretical science. The latter half of the twentieth century witnessed the growth of computational science with computer-based simulations, assuming pivotal role with each passing decade. Theoretical and computational studies have greatly contributed to the evolution of PSCs by providing the deeper insights in the fundamental properties and working mechanisms of many perovskites, guiding numerous experimental efforts and assisting in the development of high-performance PSCs. For instance, much of the computational work concerning PSCs focus on studying the effect of strategies such as doping and substitution on electronic structure and bandgap engineering to provide the most optimized perovskite material (Todorov et al. 2015; Jiang et al. 2022). The influence of orientation and movement of MA molecules along its central C-N axis on the H-I bonding in the crystal and its relation to the variation in electronic structure and bandgap values was explored (Quarti et al. 2014; Liu et al. 2022a). Theoretical studies were employed to probe the effect of defect states within perovskite materials on photovoltaic properties as the nature of defects has a direct impact on charge recombination (Duan and Uddin 2022; Lei et al. 2021). This understanding provides a systematic guide for the development of various passivation techniques that can further enhance the performance of the PSCs. Another important aspect of theoretical studies pertained to the investigation of the stability of perovskite materials under different environmental stressors (Rolston et al. 2018; Li et al. 2020). These studies have highlighted the possible degradation pathways and provided a framework for making informed choices for designing stable and efficient perovskite compositions.

Material modeling has also provided insights into the impact of interface of perovskite and adjacent electron transport layer (ETL)/hole transport layer (HTL) by studying the factors like density of states (DOS), dielectric constant, and spatial distribution of recombination losses (Shao and Loi 2020; Fakharuddin et al. 2017).

Interestingly, it was found that interfacial dipoles, formed due to energy level alignments at interface, greatly influence the open circuit voltage and the built-in potential in PSCs (Ma et al. 2023). Deciphering the factors affecting carrier recombination dynamics, bandgap tunability is one of the most crucial contributions of the theoretical studies that the researchers have used for benchmarking experimental efforts directed towards achieving maximum PCE.

Experimental studies have successfully studied and identified several promising candidates for PSCs by exploring the choices of anions and cations in perovskite structure and by incorporating the dopants and additives to optimize the optoelectronic properties and improve film formation and stability. Numerous synthesis techniques like vapor deposition, solution processing, and co-evaporation are typically used to fabricate perovskite layers and the choice of synthesis method is crucial for deciding the morphology, crystallinity, and overall performance of the solar cells (Kumar and Naidu 2021). Another aspect that holds importance is related to device architecture of PSC. While planar configurations are easier to fabricate, mesoporous architectures are more commonly used in efficient and stable PSCs due to their superior charge transport properties (Shi et al. 2015). Experimental evaluations of PCE, short circuit current density (J_{sc}), open circuit voltage (V_{oc}), fill factor (FF) under standard and varied conditions of temperature, and light intensity provide understanding of potential operational challenges and device performance. For the understanding of degradation mechanism, stability testing of PSC is done by exposing them to environmental stressors like varied temperature, humidity, and photon influx (Wang et al. 2022). To analyze the perovskite materials and devices, various characterization techniques such as impedance spectroscopy, UV-Vis spectroscopy, photoluminescence, and electron microscopy are used to evaluate charge recombination and transport dynamics, optical properties, crystallography, and morphology (Pazoki et al. 2019). For the advancement of this promising technology, more experimental studies addressing the challenges of uniformity, reproducibility, and large scale-large area fabrication are going on.

Although these traditional approaches of experiment, theory, and computation for perovskite discovery are notably time consuming and poorly efficient, the extensive data generated from these approaches have laid the foundation for fourth paradigm of materials discovery. The four paradigms of materials discovery are shown in Fig. 12.2.

Inspired by the success of the cognitive computing and artificial intelligence (AI) in flourishing AI ecosystems in domains like automation, gaming, finance, and cybersecurity (Chan and Stolfo 1998; Silver et al. 2016; Liu and Tian 2010) in last two decades, the researchers started exploring the promises that AI and machine learning (ML) hold for materials science and engineering. While data-driven research in chemistry and biology has been here for a while, its prominence in material science is rather recent (Silver et al. 2016; Noordik 2004). The ML is accelerating the material designing by utilizing the present knowledge and the knowledge generated while designing the materials to discover the new materials.

Many independent developments in the domain of materials science, especially on the availability of data and incessant developments and improvements in the ML

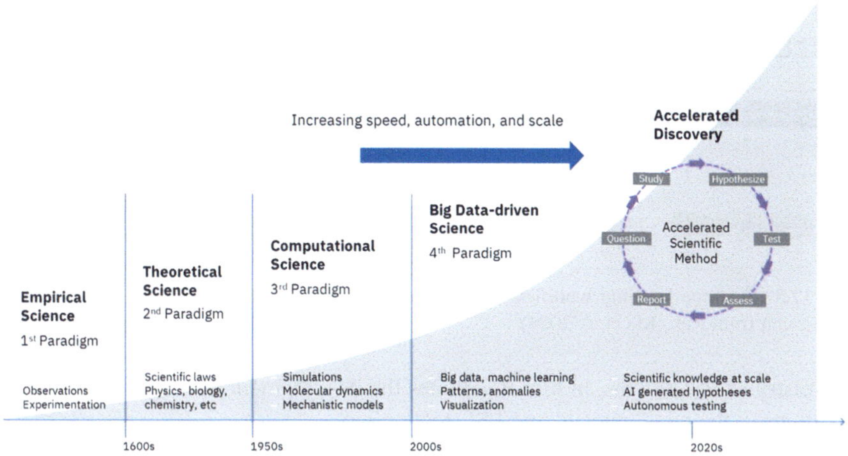

Fig. 12.2 The four paradigms for material discovery: empirical, theoretical, computational, and data-driven. Adapted with permission from Ref. (Pyzer-Knapp et al. 2022)

algorithms and models to utilize that data, has spurred a lot of excitement and expectations. The ML techniques are being utilized to discover new materials and to determine the properties and functions of existing materials for various applications such as superconductivity, glasses, polymers, and renewable energy (Baldi and Brunak 2001). The targeted design approach of ML is hugely benefiting the energy research in photovoltaics, piezoelectrics, rechargeable alkali-ion batteries, and thermoelectrics. In this context, ML has emerged as a game changer, accelerating the pace of discovery and innovation in PSC research. By leveraging vast datasets and sophisticated algorithms, ML has demonstrated unparalleled capabilities in identifying patterns, making predictions and optimizing processes that were previously beyond human reach. The ability of ML to process complex and high-dimensional data has found applications in various domains of PSC research, including materials discovery, device optimization, performance prediction, and defect analysis.

The introduction of ML into PSC research marks a paradigm shift in the way scientific investigations are conducted. Traditional experimental and computational methods, while effective, are often time-consuming and resource-intensive. The ML, on the other hand, offers a data-driven approach that can significantly enhance the efficiency and precision of research efforts. For instance, ML algorithms can predict the properties of novel materials, guide the synthesis of high-performance perovskite structures, and optimize device architectures with minimal experimental trials. This capability is particularly valuable in the context of PSCs, where the material composition and device design space are vast and complex. Moreover, ML enables the researchers to extract meaningful insights from experimental and simulation data that might otherwise remain obscured, using a workflow which typically flows the process shown in Fig. 12.3. By identifying hidden correlations and trends, the ML models can provide a deeper understanding of the underlying physics and

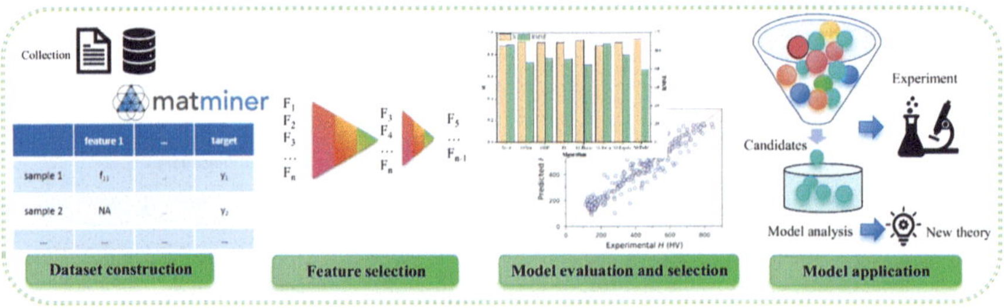

Fig. 12.3 Machine learning workflow for materials discovery and optimization. Adapted with permission from Ref. (Ma et al. 2024)

chemistry of PSCs. This, in turn, facilitates the development of robust and scalable solutions to the challenges facing PSC technology.

Comparative Analysis of Machine Learning Techniques

Regression Models

Regression models refer to the statistical techniques that are employed to access the relationships of one or more independent variables with a dependent variable. In materials science, these models are typically used to predict behavior, performance metrics, and properties based on observed data (Maulud and Abdulazeez 2020). For instance, regression models assist the researchers in optimizing the PSC materials for better scalability and performance by identifying and quantifying the relationship of various perovskite materials with their performance parameters like operational efficiency, lifespan, and stability. Depending on the type of relationships between the dependent and independent variables, regression models are typically characterized as linear, polynomial, and multivariate regression models. Under linear regression models, the dependent variable y (e.g. PCE of a PSC) is assumed to be linearly linked with the independent variables $x_1, x_2, \ldots, x_n$ (e.g., material properties such as electron mobility, bandgap, etc.) following the general Eq. (12.1):

$$y = \beta_0 + \beta_1 x_1 + \beta_2 x_2 + \ldots + \beta_n x_n + \epsilon \tag{12.1}$$

where, β_i and ϵ are the regression coefficients and error term, respectively (Su et al. 2012).

However, polynomial regression is the extension of the linear regression as it includes the nonlinear relationship between the dependent and independent variables for modeling as shown in Eq. (12.2):

$$y = \beta_0 + \beta_1 x_1 + \beta_2 x_1^2 + \beta_3 x_2 + \beta_4 x_2^2 + \ldots + \beta_n x_n^k + \epsilon \tag{12.2}$$

where, x_i^k are the higher-order polynomial terms that permit the model to capture nonlinear and complex relationships between the properties of the materials and their performance parameters (Ostertagová 2012).

Multivariate regression comes into the picture for cases where multiple independent variables called features influence the dependent variable simultaneously in the perspective of PSC. This regression model comes in handy in PSC research where there is confluence of several material properties directly influencing the device's performance and this model can easily capture that interaction. Eq. (12.3) represents governing equation for the interaction:

$$y = \beta_0 + \sum_{1=1}^{n} \beta_i x_i + \epsilon \tag{12.3}$$

It can be noted that apart part from inclusion of multiple predictors, Eq. (12.3) resembles that of linear regression (Alexopoulos 2010).

Sometimes overfitting of the data can result in deviated predictions of materials; to overcome this, a regularized version of linear regression is developed by incorporating a penalty term to the loss function. When the added penalty term is proportional to the square of the coefficients, then the resulting regression is called Ridge regression (Marquardt and Snee 1975), whereas the Lasso regression (Hans 2009) incorporates a penalty that is proportional to the coefficients' absolute values. This approach significantly reduces the model's complexity by eliminating and shrinking the lesser significant coefficients. Notably, the addition of such techniques is particularly useful for the cases having strong multicollinearity among independent variables.

Ensemble Models

The basic idea behind ensemble models is rooted in the idea that predictive task obtained after combining the prediction of multiple individual models/base learners is more accurate, reliable, and robust as a group of models synergizing together is ought to outperform a single model by minimizing the bias/variance/both. This technique is particularly helpful in the contexts where individual models possess complementary weaknesses and strengths. The ensemble methods are broadly categorized using three primary approaches, namely, bagging, boosting, and stacking. With their applications and characteristics, each of these approaches are well-suited for multiple tasks in PSC research. The bagging technique in ensemble models involves the training of multiple instances of the same model type on different subsets of the training data, which are typically obtained through the process of bootstrapping. The predictions obtained from these models are then either voted on or averaged out after considering the nature of task at hand. Ultimately, the aim is to diminish the variance by averaging out the individual model error (Ganaie et al. 2022).

However, the boosting in ensemble techniques pertains to the sequential training of the models with each new model formed upon the errors obtained from the previous ones and the resultant prediction is based on a weighted sum of predictions obtained from the individual models. Thus, by correcting error of the base models, the boosting primarily aims to reduce both variance and bias in the resulting model (González et al. 2020). Stacking approach is an ensemble technique that first entails the training of different types of the models like support vector machines (SVM), decision trees, and neural networks on the same dataset and then uses their output prediction as input to a higher level "meta-model" which finally gives the decision combining them all. By leveraging the strengths of multiple base models, this approach works effectively in scenarios where each model excels in different aspects of the given task (Rashid et al. 2022).

The typical ensemble models used in materials science are:

1. Random Forest (RF): It is a bagging-based ensemble model that consists of multiple decision trees, where each tree is trained on a random subset of the data. The final prediction is obtained through majority voting for classification tasks or by averaging the outputs for regression tasks (Liu et al. 2012).
2. Gradient Boosting Machines (GBM): It is an ensemble model connected sequentially in a manner that each model corrects the errors reported by the previous one. The training of the models is done in a way that iteratively optimizes the loss function (Natekin and Knoll 2013).
3. Extreme Gradient Boosting (XGBoost): Known for its performance and speed, XGBoost is an optimized version of gradient boosting as it not only permits the parallelization during training but also incorporates the regularization to avoid overfitting of the data. It is used extensively in materials science, typically for regression and classification problems (Chen 2015).
4. Adaptive Boosting (AdaBoost): For the problems like material phase transitions and classification in crystalline structures, i.e., where the precision is crucial in prediction, AdaBoost works adequately by accommodating the weights of incorrectly classified instances, assigning more importance to difficult cases for subsequent models. Thus, the basic idea working behind is only to combine weak learners to form a strong learner (Margineantu and Dietterich 1997).
5. Light Gradient Boosting Machine (LightGBM): When dealing with the large-scale prediction tasks like polymer/semiconductor properties prediction or prediction of energy storage capacity in battery materials, a gradient boosting framework is optimized for efficiency and speed for large datasets is required (Lin et al. 2022). Such framework is provided by LightGBM, which uses histogram-based methods to efficiently handle continuous features.

These ensemble models are frequently used in studying the materials that often contain nonlinear relationships between different features as these can improve predictive accuracy, handle high-dimensional data, and manage the complexities and intricacies involved in the perovskite systems.

Neural Network Models

The neural network (NN) models have emerged as transformative tools for identifying hidden patterns and optimizing various processes in the ways that are almost impossible with traditional methods for complex material systems such as PSCs. By analyzing the larger datasets via mimicking the biological neural systems functioning, NNs can assist in discovery of new materials, prediction of their performance parameter, and in the optimization of numerous fabrication processes. The NN is made by interconnecting the layers of artificial neurons/nodes and works such that the input data is processed in a hierarchical manner (Wilamowski 2009). The architecture of a typical neural network is described below:

1. Input layer to receive the raw data like processing conditions or material properties.
2. After attaching appropriate weights and nonlinear activation function to the input, the task of doing intermediate computations is assigned to the middle layers. Depending upon the complexity in relationships for a given problem, the deep NNs can have numerous hidden layers.
3. For producing the final output such as performance prediction (stability or efficiency), output layer is also used in NNs.

The mapping of input data (x) to output predictions ($\hat{y}$) in NNs is done through forward propagation. To minimize the error between predicted $\hat{y}$ and actual output y, the weights in the network are adjusted at the time of the training process. Generally, optimization techniques like stochastic gradient descent (SGD) (Bottou 2012) are utilized to minimize the loss function ($L(\hat{y}, y)$) to learn the relationship between input and output variables. The training of a NN has two main steps. First is the forward propagation, which involves the propagation of input data through the network, layer by layer, to compute the output. This is followed by the backpropagation in which the difference of the actual and predicted output is passed back through the network to adjust the weights using optimization algorithms. Moreover, the weights of the model are updated iteratively to minimize the error in its predictions.

Depending upon the specific task and the nature of the data, various types of NNs are employed and the commonly used types are briefed below.

1. Feedforward Neural Networks (FNNs): The simplest form of NNs in which information is moved unidirectionally from input to the output. Generally, FNNs are used in tackling regression tasks like performance forecasting or material property prediction for PSCs (Ojha et al. 2017).
2. Convolutional Neural Networks (CNNs): The main use of CNN models pertains to the processing of spatial and image data. For PSCs, CNNs can be used to correlate morphological features with PSC performance or for analyzing the microstructure of perovskite films (O'Shea 2015).

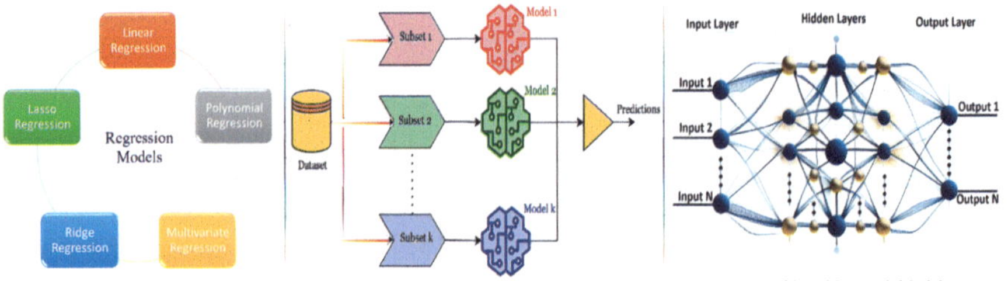

(a) Regression Models (b) Ensemble Models (c) Neural Network Model

Fig. 12.4 Different types of models in machine learning

3. Recurrent Neural Networks (RNNs): Based on the sequential data, the RNNs are particularly useful for the modeling of various time-dependent processes. For instance, these can be applied to access and examine the stability or degradation of PSCs over time (Salehinejad et al. 2017).

Thus, NNs are highly versatile for various applications in the domain of PSCs as these can be chosen and designed to carter different data types and relationships. The schematics of different models are shown in Fig. 12.4.

Applications of Machine Learning in Perovskite Solar Cell Research

Materials Discovery

A large number of papers focusing on the use of ML in perovskite research have been published in the past 5 years, and this number is going up exponentially due to increasing interest of researchers in both PSCs and ML. Although, greater part of the research pertaining to ML for PSCs seems to be aligned toward discovering novel and efficient perovskite-like material having high stability, less toxicity, and appropriate band gap values. The typical steps involved in the study of materials science using ML are shown in Fig. 12.5.

The screening space for perovskites is generally curated by combining the elements from periodic table, extracted from existing databases and repositories, or sometimes targeted data is also generated using high-throughput computations and experiments. The next crucial step is the representation of perovskite data in machine readable form, often known as featurization or fingerprinting. It involves the representation of material in terms of a spectrum of numbers that can be automatically generated or handcrafted. With this available information at hand, progress can be made in predicting new materials or for their property prediction. A spectrum of numerous old and new algorithms is used for training the model that then establishes a mapping between the required property/outcome with the representations of the perovskite material.

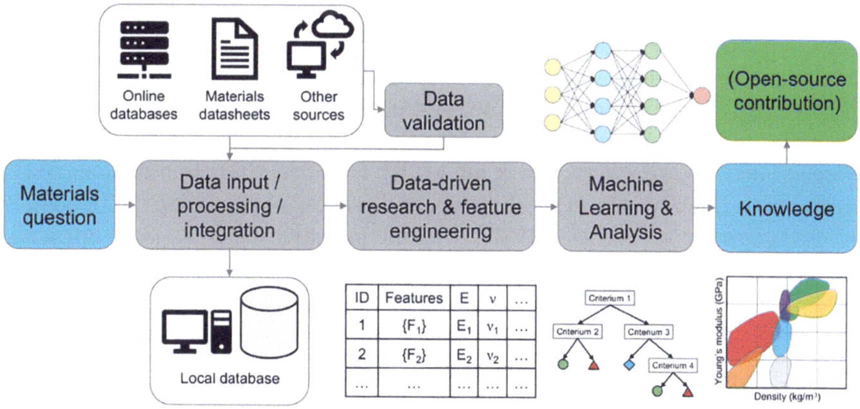

Fig. 12.5 Schematic of a machine learning study in materials science. Adapted with permission from Ref. (Wang et al. 2020)

In this context, several studies using ML on computationally and experimentally generated data for PSCs have been conducted. For instance, using the data generated using DFT, Lu et al. (2018) illustrated how ML can be effectively used for screening suitable perovskite materials for PSCs by using the band gap as a key output parameter. With a training dataset of 212 perovskites, first descriptors were analyzed using the gradient boosting regression (GBR) algorithm, reducing the number from 30 potential descriptors to 14 key descriptors, which included ionic charge (ICB), ionization energy (IEB), tolerance factor (T_f), octahedral factor (O_f), p-orbital electron count (Xp-electron), ionic polarizability (PB, PA), electron affinity (EAB), etc. Subsequently, several ML techniques like GBR, decision tree regression (DTR), Gaussian process regression (GPR), kernel ridge regression (KRR), support vector regression (SVR), and multilayer perceptron regression (MPR) were tested. Out of these, GBR provided the best fit for the training dataset by accurately predicting band gap values that lie near to the values evaluated by DFT computations. The band gaps of 5158 potential perovskite materials, generated from the combination formed by 32 A-site and 43 B-site elements in ABX_3 configuration, were predicted. Thereafter, six perovskites were finally screened on other factors like stability, band gap suitability, experimental accessibility, and toxicity. Similarly, clustering-based similarity analysis was done by Kim et al. (2017) to form a database of 1346 hybrid halide perovskites and used DFT to analyze their properties. Other studies like that by Xu et al. (2018) and Takahashi et al. (2018) also aimed to predict band gaps and formability of perovskites using algorithms like SVM and random forests. Other significant contributions include the work by Kanno et al. (2019) that involved a high-throughput material search scheme based on the ML analysis via screening of over 28,125,225 double perovskite materials to predict 24 potential candidate materials as cheaper and toxic-free alternatives of conventional $MAPbX_3$. Jain et al. (2019) integrated DFT with ML by using SVM and convex hull analysis to screen and study ABX_3 compounds and predict formability and band gaps of 454 materials. Also, the materials with formation probability (P) ≥ 0.8 were further checked for thermodynamic stability using a combination of open materials databases like MP (Jain et al. 2013), AFLOW (Curtarolo et al. 2012), and OQMD (Saal et al. 2013), and provided a detailed insight into the electronic and surface properties of $RbSnCl_3$ and $RbSnBr_3$. These studies alongside such similar studies (Odabaşı and Yıldırım 2019; Liu et al. 2022b; Tao et al. 2021; Bansal et al. 2023) highlight the power of ML in accelerating the discovery of promising PSC materials by predicting the properties and selecting the potential materials in significantly less time.

Device Optimization

A series of studies have also successfully used many ML techniques to enhance the performance of PSCs. Li et al. (2019) targeted multiple ML techniques like Bayesian ridge regression and SVR to investigate the transport of electrons in $MAPbI_3$/metal electrode junctions and to predict their decomposition energies. A correlation analysis on the partial substitution of Pb in $MAPbBr_3$ was done by Mannodei-Kanakkithodi et al. (2019) and the defect formation energies were evaluated for 53 elements using DFT which were then further validated experimentally. On the similar lines, Park et al. (2019) used RF, generalized linear models (GLM), and deep learning (DL) with key descriptors like lone pair numbers and cationic radii to estimate the phase stability of different perovskites. In another study, they predicted the relationship between octahedral deformation in halide perovskites and chalcogenide and band gaps using models like RF, KNN, and clustered them based on their structural features. In a crucial study, Ren et al. (2019) adopted a novel approach for perovskite optimization by using an autoencoder NN. They used 50,000 artificially generated current-voltage (JV) curves to train the material parameters for optimization. In another study, Yu et al. (2018) combined both computational and experimental data to evaluate the compatibility of 50 amines for post-treatment of $MAPbI_3$. To access the reactivity of various amines, they measured the UV – Vis absorption of $MAPbI_3$ films before and after the treatment and depending on the value of their residue index, their reactivity was classified as nonreactive (1) and reactive (0). 53 chemical descriptors were determined from JChem for Office software, from which 31 were selected for SVM classification. With an accuracy of 86%, the model validated 5 previously unseen amines and correctly predicted their reactivity. In another study, Sun et al. (2018) fabricated and characterized 75 perovskite-inspired materials deposited on glass substrates and analyzed their XRD and UV – Vis spectra using fully connected deep NN to classify these materials into 0D, 2D and 3D structures. Due to small size of this experimental dataset to augment the data, they incorporated 164 powder XRD patterns from the ICSD database (Belsky et al. 2002) and used a data augmentation technique to simulate the differences between simulated powder XRD patterns and experimental thin-films. Thus, the inclusion of ML with both computational and experimental data is significantly advancing PSC research by augmenting small datasets. Apart from these, by employing the text mining and leveraging the recent data analytics techniques, ML have become an indispensable tool for identifying trends, understanding the factors that drive their efficiency and stability by optimizing material choices. With the refinement of current methods and advancement of research, the new avenues for applying ML in this rapidly evolving field are growing manifolds with each passing day.

Challenges and Limitations

The accelerated progress in perovskite research has paralleled the burgeoning interest for usage of ML in the materials science. This trend is expected to increase, given the incessant advancements in computational and experimental tools alongside data management techniques and ML methodologies. Data frameworks like Findable, Accessible, Interoperable, Reusable (FAIR) for sharing of data and the trend of open-source publications are further contributing to this momentum. Despite these promises, several hurdles persist in fully utilizing the ML in perovskite research. Most notable challenge is the insufficient availability of the good experimental and highly accurate computational data on PSCs as they are expensive in terms of time and cost. This inherently induces limitations in the model by reducing their generalization capability and making them susceptible to noise, outliers and imbalanced data. To overcome this hurdle, different experimental approaches involving combinatorial can be developed (Schmidt et al. 2019). Moreover, the ML techniques targeting small datasets may also improve the predictability of the model. Additionally, model reliability can also be enhanced by removing the outliers that pose distortion for model performance in small datasets. Significantly reduced dimensionality achieved by feature engineering can also aid in making ML models more powerful with fewer yet effective parameters. Furthermore, synthetic data can also be generated to balance the imbalance in datasets. The practical and valuable approach of using both low and high accuracy data along with combining computational and experimental data can be helpful in expanding the dataset and improving the model performance by strengthening its predictive power.

With the constant shifting of research trends from the material discovery to device fabrication and practical implementation, it became imminent to integrate of data from diverse sources so that the physical reality can be sufficiently represented in the given data as the stability of the material alone can never guarantee the cell's stability because the factors like operational conditions, storage and manufacturing impact it tremendously. These issues also present the requirement of standardized testing and reporting protocols in PSC domain.

Another key concern is the intrinsic challenge pertaining to the ML itself, arising because of lack of complete analysis and reporting of uncertainties in model's prediction. The difference between the aleatoric and epistemic uncertainty is often overlooked in ML applications. The aleatoric uncertainty arises due to inherent variability in data whereas uncertainty arising from lack of knowledge constitutes epistemic uncertainty. It is noteworthy that epistemic uncertainties can be reduced by improving the ML models while aleatoric uncertainties are irreducible (Hüllermeier and Waegeman 2021). With the maturity of this domain, the demand for more rigorous and effective models is going to be greater with an increasing awareness among both model users and developers for these uncertainties.

Since the performance of PSCs relies on multiple variables, the resulting feature space becomes high-dimensional, making the accurate modeling of all possible combinations. It is also seen that at many times, the ML models successfully learn specific nuances about the given data but fail tremendously when it comes to generalization of results to new and unseen perovskite compositions. Many a time, the employed ML techniques, specifically deep learning techniques, act just like a black box, as the learner cannot clearly interpret the nature of dependence of input parameters like composition or processing conditions with predicted outcomes. The lack of such understanding clearly impedes the researcher's ability to gain insights into the elementary mechanisms that drives the perovskite performance.

Future Perspectives

Given the pace of rapid advancement and convergence in the domains of PSC and ML, the perspectives for the usage of ML in PSCs are highly promising, transforming and crucial. The concept of inverse design, i.e., predicting the perovskite materials with desired properties is going to be a reality with the advancement in techniques like generative adversarial networks (GANs) (Goodfellow et al. 2020) and variational autoencoders (VAEs) (Doersch 2016). The ML models can be designed for the analysis and prediction of degradation pathways that can significantly reduce the need for experimental testing by guiding the researchers for long-term degradation mechanism simulations. Futuristic autonomous close-loop system can be designed for scaling up the PSC manufacturing by integrating the trained ML model with process control systems that can enable the continuous adjustment of processing parameters to adequate values, ensuring uniform and high-quality perovskite films and solar cells. Doing the multifidelity modeling by combining the high-fidelity experimental data with low-fidelity computational data can be proven to be more efficient for exploration of perovskite structural space. The integration of ML with robotics can transform the laboratory landscape by automating the designing, synthesizing, and testing of PSCs via guiding the experimental setups, optimizing the synthesis protocols and presenting the real-time analysis of the results. This would also enable the evolution in feedback loops, which pertains to immediate incorporation of new insights from experiments into the model, further refining the future predictions. More emphasis on developing the models that work on transfer learning can also boost the perovskite development process by utilizing the knowledge gain in one material class in another's predictions. With enhanced capabilities, ML can also be used for fabricating customized perovskite devices catering to the needs like transparency, flexibility, and bifacial PSCs production.

Conclusion

The rapid pace with which ML is marking its impact in the perovskite field clearly indicates the profound influence that it is going to leave in times to come. The way, it has accelerated the material discovery, device optimization, improvement in stability, and efficiency predictions besides facilitating stable manufacturing has already catalyzed the transformation in this field. The ML models will persistently evolve with the availability of highly accurate and abundant data, giving deeper and valuable insights into the perovskite materials and solar cells' behaviors, enabling efficient and faster development of next-generation solar technologies. The unique confluence of ML with perovskite research can unlock the emergence of novel materials, innovations, and technologies that could offer a sustainable and cost-effective solution to soaring global energy demands.

References

Alexopoulos EC (2010) Introduction to multivariate regression analysis. Hippokratia 14(Suppl 1):23

Bailie CD, McGehee MD (2015) High-efficiency tandem perovskite solar cells. MRS Bull 40(8):681–686

Baldi P, Brunak S (2001) Bioinformatics: the machine learning approach. The MIT Press, Cambridge, MA

Bansal NK, Mishra S, Dixit H, Porwal S, Singh P, Singh T (2023) Machine learning in perovskite solar cells: recent developments and future perspectives. Energ Technol 11(12):2300735

Belsky A, Hellenbrandt M, Karen VL, Luksch P (2002) New developments in the inorganic crystal structure database (ICSD): accessibility in support of materials research and design. Acta Crystallogr Sect B: Struct Sci 58(3):364–369

Bottou L (2012) Stochastic gradient descent tricks. In: Neural networks: tricks of the trade: Second Edition. Springer Berlin Heidelberg, Berlin, Heidelberg, pp 421–436

Brown GF, Wu J (2009) Third generation photovoltaics. Laser Photonics Rev 3(4):394–405

Chan PK, Stolfo SJ (1998) Toward scalable learning with non-uniform class and cost distributions: a case study in credit card fraud detection. In: Agrawal R, Stolorz P, Piatetsky G (eds) KDD'98 proceedings of fourth international conference on knowledge discovery and data mining. AAAI Press, New York, NY, pp 164–168

Chen T (2015) Xgboost: extreme gradient boosting R package version 0.4-2, 1(4)

Curtarolo S, Setyawan W, Hart GL, Jahnatek M, Chepulskii RV, Taylor RH, Morgan D (2012) AFLOW: an automatic framework for high-throughput materials discovery. Comput Mater Sci 58:218–226

Doersch C (2016) Tutorial on variational autoencoders. arXiv preprint arXiv:1606.05908

Duan L, Uddin A (2022) Defects and stability of perovskite solar cells: a critical analysis. Mater Chem Front 6(4):400–417

Fakharuddin A, Schmidt-Mende L, Garcia-Belmonte G, Jose R, Mora-Sero I (2017) Interfaces in perovskite solar cells. Adv Energy Mater 7(22):1700623

Ganaie MA, Hu M, Malik AK, Tanveer M, Suganthan PN (2022) Ensemble deep learning: a review. Eng Appl Artif Intell 115:105151

Goetzberger A, Luther J, Willeke G (2002) Solar cells: past, present, future. Sol Energy Mater Sol Cells 74(1–4):1–11

Gong J, Sumathy K, Qiao Q, Zhou Z (2017) Review on dye-sensitized solar cells (DSSCs): advanced techniques and research trends. Renew Sust Energ Rev 68:234–246

González S, García S, Del Ser J, Rokach L, Herrera F (2020) A practical tutorial on bagging and boosting based ensembles for machine learning: algorithms, software tools, performance study, practical perspectives and opportunities. Inf Fusion 64:205–237

Goodfellow I, Pouget-Abadie J, Mirza M, Xu B, Warde-Farley D, Ozair S, Bengio Y (2020) Generative adversarial networks. Commun ACM 63(11):139–144

Green MA (2002) Third generation photovoltaics: solar cells for 2020 and beyond. Physica E 14(1–2):65–70

Hans C (2009) Bayesian lasso regression. Biometrika 96(4):835–845

Hüllermeier E, Waegeman W (2021) Aleatoric and epistemic uncertainty in machine learning: an introduction to concepts and methods. Mach Learn 110(3):457–506

Jain A, Ong SP, Hautier G, Chen W, Richards WD, Dacek S, Persson KA (2013) Commentary: the materials project: a materials genome approach to accelerating materials innovation. APL Mater 1(1):011002

Jain D, Chaube S, Khullar P, Srinivasan SG, Rai B (2019) Bulk and surface DFT investigations of inorganic halide perovskites screened using machine learning and materials property databases. Phys Chem Chem Phys 21(35):19423–19436

Jiang Q, Chu Z, Wang P, Yang X, Liu H, Wang Y, Yin Z, Jinliang W, Zhang X, You J (2017) Planar-structure perovskite solar cells with efficiency beyond 21%. Adv Mater 29(46):1703852

Jiang Q, Tong J, Scheidt RA, Wang X, Louks AE, Xian Y, Zhu K (2022) Compositional texture engineering for highly stable wide-bandgap perovskite solar cells. Science 378(6626):1295–1300

Kanno S, Imamura Y, Hada M (2019) Alternative materials for perovskite solar cells from materials informatics. Phys Rev Mater 3(7):075403

Kim C, Huan TD, Krishnan S, Ramprasad R (2017) A hybrid organic-inorganic perovskite dataset. Sci Data 4(1):1–11

Kojima A, Teshima K, Shirai Y, Miyasaka T (2009) Organometal halide perovskites as visible-light sensitizers for photovoltaic cells. J Am Chem Soc 131(17):6050–6051

Kumar NS, Naidu KCB (2021) A review on perovskite solar cells (PSCs), materials and applications. J Mater 7(5):940–956

Lei Y, Xu Y, Wang M, Zhu G, Jin Z (2021) Origin, influence, and countermeasures of defects in perovskite solar cells. Small 17(26):2005495

Li L, You Y, Hu S, Shi Y, Zhao G, Chen C, Ren W (2019) Electronic transport of organic-inorganic hybrid perovskites from first-principles and machine learning. Appl Phys Lett 114(8):083102

Li N, Luo Y, Chen Z, Niu X, Zhang X, Lu J, Zhou H (2020) Microscopic degradation in formamidinium-cesium lead iodide perovskite solar cells under operational stressors. Joule 4(8):1743–1758

Lin R, Liu Z, Han P, Lin R, Lu Y, Cao H, Li X (2022) A machine learning study on superlattice electron blocking layer design for AlGaN deep ultraviolet light-emitting diodes using the stacked XGBoost/LightGBM algorithm. J Mater Chem C 10(46):17602–17610

Liu S-S, Tian Y-T (2010) Facial expression recognition method based on Gabor wavelet features and fractional power polynomial kernel PCA. In: Zhang L, Lu B-L, Kwok J (eds) Advances in neural networks—ISNN 2010. Springer, Berlin, Heidelberg, pp 144–151

Liu Y, Wang Y, Zhang J (2012) New machine learning algorithm: random forest. In: Information computing and applications: third international conference, ICICA 2012, Chengde, China, September 14–16, 2012. proceedings 3. Springer, Berlin Heidelberg, pp 246–252

Liu S, Guo R, Xie F (2022a) The effects of organic cation rotation in hybrid organic-inorganic perovskites: a critical review. Mater Des 221:110951

Liu Y, Yan W, Han S, Zhu H, Tu Y, Guan L, Tan X (2022b) How machine learning predicts and explains the performance of perovskite solar cells. Sol RRL 6(6):2101100
Lopez-Varo P, Jiménez-Tejada JA, García-Rosell M, Ravishankar S, Garcia-Belmonte G, Bisquert J, Almora O (2018) Device physics of hybrid perovskite solar cells: theory and experiment. Adv Energy Mater 8(14):1702772
Lu S, Zhou Q, Ouyang Y, Guo Y, Li Q, Wang J (2018) Accelerated discovery of stable lead-free hybrid organic-inorganic perovskites via machine learning. Nat Commun 9(1):3405
Ma Y, Gong J, Zeng P, Liu M (2023) Recent progress in interfacial dipole engineering for perovskite solar cells. Nano-Micro Lett 15(1):173
Ma Y, Xu P, Li M, Ji X, Zhao W, Lu W (2024) The mastery of details in the workflow of materials machine learning. NPJ Comput Mater 10(1):141
Mannodi-Kanakkithodi A, Park JS, Jeon N, Cao DH, Gosztola DJ, Martinson AB, Chan MK (2019) Comprehensive computational study of partial lead substitution in methylammonium lead bromide. Chem Mater 31(10):3599–3612
Margineantu DD, Dietterich TG (1997) Pruning adaptive boosting. ICML 97:211–218
Marquardt DW, Snee RD (1975) Ridge regression in practice. Am Stat 29(1):3–20
Maulud D, Abdulazeez AM (2020) A review on linear regression comprehensive in machine learning. J Appl Sci Technol Trends 1(2):140–147
Natekin A, Knoll A (2013) Gradient boosting machines, a tutorial. Front Neurorobot 7:21
Noordik, J. H. Cheminformatics developments: history, reviews and current research (IOS Press, Amsterdam, 2004)
Odabaşı Ç, Yıldırım R (2019) Performance analysis of perovskite solar cells in 2013–2018 using machine-learning tools. Nano Energy 56:770–791
Ojha VK, Abraham A, Snášel V (2017) Metaheuristic design of feedforward neural networks: a review of two decades of research. Eng Appl Artif Intell 60:97–116
O'Shea K (2015) An introduction to convolutional neural networks. arXiv preprint arXiv:1511.08458
Ostertagová E (2012) Modelling using polynomial regression. Procedia Eng 48:500–506
Park H, Mall R, Alharbi FH, Sanvito S, Tabet N, Bensmail H, El-Mellouhi F (2019) Learn-and-match molecular cations for perovskites. Chem A Eur J 123(33):7323–7334
Pazoki M, Hagfeldt A, Edvinsson T (2019) Characterization techniques for perovskite solar cell materials. Elsevier
Pyzer-Knapp EO, Pitera JW, Staar PW, Takeda S, Laino T, Sanders DP, Curioni A (2022) Accelerating materials discovery using artificial intelligence, high performance computing and robotics. NPJ Comput Mater 8(1):84
Quarti C, Mosconi E, De Angelis F (2014) Interplay of orientational order and electronic structure in methylammonium lead iodide: implications for solar cell operation. Chem Mater 26(22):6557–6569
Rashid M, Kamruzzaman J, Imam T, Wibowo S, Gordon S (2022) A tree-based stacking ensemble technique with feature selection for network intrusion detection. Appl Intell 52(9):9768–9781
Ren Z, Oviedo F, Xue H, Thway M, Zhang K, Li N, Buonassisi T (2019) Physics-guided characterization and optimization of solar cells using surrogate machine learning model. In: 2019 IEEE 46th photovoltaic specialists conference (PVSC). IEEE, pp 3054–3058
Rolston N, Bush KA, Printz AD, Gold-Parker A, Ding Y, Toney MF, Dauskardt RH (2018) Engineering stress in perovskite solar cells to improve stability. Adv Energy Mater 8(29):1802139
Saal JE, Kirklin S, Aykol M, Meredig B, Wolverton C (2013) Materials design and discovery with high-throughput density functional theory: the open quantum materials database (OQMD). JOM 65:1501–1509
Salehinejad H, Sankar S, Barfett J, Colak E, Valaee S (2017) Recent advances in recurrent neural networks. arXiv preprint arXiv:1801.01078
Schmidt J, Marques MR, Botti S, Marques MA (2019) Recent advances and applications of machine learning in solid-state materials science. NPJ Comput Mater 5(1):83

Shao S, Loi MA (2020) The role of the interfaces in perovskite solar cells. Adv Mater Interfaces 7(1):1901469

Sharma S, Jain KK, Sharma A (2015) Solar cells: in research and applications—a review. Mater Sci Appl 6(12):1145–1155

Shi Y, Xing Y, Li Y, Dong Q, Wang K, Du Y, Ma T (2015) CH3NH3PbI3 and CH3NH3PbI3–x cl x in planar or mesoporous perovskite solar cells: comprehensive insight into the dependence of performance on architecture. J Phys Chem C 119(28):15868–15873

Silver D et al (2016) Mastering the game of go with deep neural networks and tree search. Nature 529:484–489

Su X, Yan X, Tsai CL (2012) Linear regression. Wiley Interdiscip Rev Comput Stat 4(3):275–294

Sun Y, Szűcs G, Brandt AR (2018) Solar PV output prediction from video streams using convolutional neural networks. Energy Environ Sci 11(7):1811–1818

Takahashi K, Takahashi L, Miyazato I, Tanaka Y (2018) Searching for hidden perovskite materials for photovoltaic systems by combining data science and first principle calculations. ACS Photonics 5(3):771–775

Tao Q, Xu P, Li M, Lu W (2021) Machine learning for perovskite materials design and discovery. NPJ Comput Mater 7(1):23

Todorov T, Gershon T, Gunawan O, Lee YS, Sturdevant C, Chang LY, Guha S (2015) Monolithic perovskite-CIGS tandem solar cells via in situ band gap engineering. Adv Energy Mater 5(23):1500799

Wang AYT, Murdock RJ, Kauwe SK, Oliynyk AO, Gurlo A, Brgoch J, Sparks TD (2020) Machine learning for materials scientists: an introductory guide toward best practices. Chem Mater 32(12):4954–4965

Wang Y, Ahmad I, Leung T, Lin J, Chen W, Liu F, Djurišić AB (2022) Encapsulation and stability testing of perovskite solar cells for real life applications. ACS Mater Au 2(3):215–236

Wilamowski BM (2009) Neural network architectures and learning algorithms. IEEE Ind Electron Mag 3(4):56–63

Xu Q, Li Z, Liu M, Yin WJ (2018) Rationalizing perovskite data for machine learning and materials design. J Phys Chem Lett 9(24):6948–6954

Yu J, Wang Z, Majumdar A, Rajagopal R (2018) DeepSolar: a machine learning framework to efficiently construct a solar deployment database in the United States. Joule 2(12):2605–2617

Zhao Y, Nardes AM, Zhu K (2014) Mesoporous perovskite solar cells: material composition, charge-carrier dynamics, and device characteristics. Faraday Discuss 176:301–312

Chapter 13
Summary and Future Outlook on Computational Analysis of Perovskite Solar Cells

Manish Kumar and Khushboo Gupta

Abstract Computational analysis offers a cost-effective approach to exploring the doping concentration, device architecture, operating conditions, and material composition of perovskite solar cells (PSCs). A key aspect of computational research focuses on unravelling the electronic structure and optoelectronic behavior of perovskite materials. Density Functional Theory (DFT) calculations play a crucial role in determining energy levels, band structure, and density of states. Additionally, simulation-based modeling provides valuable insights into the influence of defects and interfaces on PSC performance, as factors like vacancies and impurities can significantly impact charge transport, recombination dynamics, and overall efficiency. With the rising popularity of multi-physics computational techniques, researchers can now analyze diverse electrical and thermal characteristics of PSCs in greater detail. This book presents a comprehensive exploration of advanced computational tools, including DFT, SCAPS-1D, and Multi-physics COMSOL, to examine critical properties such as optoelectronic behavior, electrical performance, band gap tuning strategies, doping effects, and defect densities. Notably, this edited volume introduces discussions on grading effects, multivalent defect analysis, scripting methodologies, and multi-terminal/multi-junction tandem configurations—topics integrated for the first time based on the latest research findings.

Keywords Perovskite solar cell · DFT · SCAPS · COMSOL · Future Outlook

---

M. Kumar (✉)
Experimental Research Laboratory, Department of Physics, ARSD College, University of Delhi, Dhaula Kuan, New Delhi, India
e-mail: mkumar2@arsd.du.ac.in

K. Gupta
Department of Physics, School of Sciences, IILM University, Greater Noida, Uttar Pradesh, India

M. Kumar (ed.), *Computational Analysis of Perovskite Solar Cells*, Springer Tracts in Electrical and Electronics Engineering,
https://doi.org/10.1007/978-981-95-6521-4_13

Introduction

Perovskite solar cells have gained significant attention in recent years due to their remarkable optoelectronic properties, low-cost fabrication, and high-power conversion efficiencies (PCEs) (Hossain et al. 2021; Rajagopal et al. 2017). The unique characteristics of perovskite materials, such as their tunable bandgap, strong light absorption, and long carrier diffusion lengths, make them highly promising for next-generation photovoltaic applications (Kahandal et al. 2024). Despite their rapid progress, several challenges, including material stability, defect generation, and optimization of device architectures, remain crucial in determining their commercial viability. Addressing these challenges requires a deep understanding of the physical and electronic properties of perovskite materials, which is where computational studies play an essential role.

Computational analysis has revolutionized the way PSCs are studied, offering a cost-effective and efficient method to investigate material properties, device performance, and potential design optimizations. Various computational techniques, such as DFT, SCAPS-1D, VASP simulations (Gupta et al. 2024; Baro and Borgohain 2023; Hafner 2008), and multi-physics modeling, provide valuable insights into different aspects of perovskite materials and devices. These techniques help researchers explore the effects of doping concentration, defect generation, charge transport mechanisms, and thermal stability, all of which contribute to the overall performance and longevity of PSCs. This chapter presents an overview of the computational methodologies employed in the study of PSCs and their role in advancing the field.

A critical aspect of computational research on perovskite materials is the study of defect generation and its impact on device performance. Defects in perovskite absorbers can arise due to various factors, including chemical doping, annealing, ion bombardment, and intrinsic material variations. Chemical doping is widely used to modify the electronic structure of perovskite materials by replacing some of the lattice elements with dopant atoms (Nie et al. 2023). This can influence stability, trap state density, and charge transport properties. For instance, doping with cesium or rubidium can create lead vacancies (Pb vacancies), which help in reducing trap states and enhancing material stability. Annealing, another common technique, introduces vacancy-related and grain boundary defects through controlled heating and cooling cycles. These defects can influence charge carrier dynamics and recombination rates, ultimately affecting device efficiency. Ion bombardment, which exposes the perovskite surface to high-energy ions, creates lattice imperfections and allows precise control over defect density. Additionally, intrinsic material variations occurring during synthesis can lead to non-uniformities, grain boundaries, and structural imperfections that impact charge transport and recombination mechanisms. Understanding these defects and their effects is crucial for optimizing PSC performance, and computational modeling provides the necessary framework for their detailed analysis.

Another important computational technique used in the study of PSCs is DFT, which is instrumental in investigating the structural and electronic properties of perovskite materials. DFT calculations provide insights into the stability, bonding, and energetics of perovskite structures by optimizing atomic positions and lattice parameters to find the lowest energy configuration. This method is particularly useful for predicting defect formation energies, charge carrier mobility, and band structure modifications resulting from the introduction of vacancies, interstitials, or substitutional dopants. DFT is also employed in studying surface and interface properties, which play a significant role in charge extraction and transport in PSCs (Uddin et al. 2024; Fang et al. 2020). By analyzing adsorption energies, charge transfer mechanisms, and band alignments at interfaces, researchers can design better-performing perovskite solar cells. Furthermore, DFT-based simulations are used to explore band structure and optical properties, providing valuable information on bandgap characteristics, density of states, absorption spectra, and exciton binding energies. These insights guide the development of novel perovskite compositions with enhanced efficiency and stability.

In addition to electronic structure calculations, computational models are widely used for assessing the physical parameters of PSCs. These studies offer several advantages, such as cost-effectiveness, predictive insights, parameter sensitivity analysis, design optimization, and a deeper mechanistic understanding of device operation. Computational techniques enable researchers to study the impact of various material properties, interface effects, and device configurations without the need for expensive and time-consuming experimental setups (Kumar et al. 2022). By analyzing different parameters, such as layer thickness, doping concentrations, and interface engineering, computational models help optimize device designs for maximum efficiency. However, these studies also have certain limitations, including the need for approximations and simplifications, accuracy constraints of computational models, material and process variability, limited experimental validation, and the requirement for significant computational resources. Despite these challenges, computational studies remain a powerful tool in advancing the understanding and development of PSCs.

The integration of multi-physics simulations in PSC research has further expanded the capabilities of computational analysis. PSCs operate in a complex environment where optical, electrical, and thermal properties interact dynamically. Multi-physics modeling allows researchers to simulate the interaction of light with the perovskite material, optimize light absorption, and analyze charge carrier dynamics. These simulations consider optical properties, such as the absorption coefficient and refractive index, to determine the ideal active layer thickness and composition for efficient light harvesting. Additionally, multi-physics studies help investigate charge transport mechanisms, including carrier diffusion, drift, and recombination processes. By simulating current-voltage characteristics and electrical losses, researchers can better understand the impact of defects and interfaces on device performance. Thermal management is another crucial aspect explored through multi-physics simulations, as heat generation and dissipation can significantly influence the stability and efficiency of perovskite solar cells. This chapter discusses various aspects of multi-physics modeling and its implications for PSC optimization.

Theoretical studies have also been extensively applied to all-inorganic and organic-inorganic mixed halide PSCs. While perovskite solar cells have demonstrated impressive efficiency, their long-term stability and lead toxicity remain major concerns that hinder commercialization. Inorganic perovskites, such as cesium-based compositions, offer improved stability in ambient conditions, while mixed halide perovskites provide tunable bandgaps for better light absorption (Kahandal et al. 2024). Computational studies have been instrumental in evaluating the optoelectronic properties, defect tolerance, and structural stability of these materials. Simulations help researchers understand the influence of various dopants, interface engineering strategies, and material compositions on the performance and durability of PSCs (Raj et al. 2021). This chapter provides an in-depth discussion on the fundamental properties, performance parameters, and challenges associated with all-inorganic and mixed halide PSCs.

Another key area of computational research is the development of perovskite tandem solar cells, which aim to surpass the efficiency limits of single-junction devices. Tandem solar cells combine multiple absorber layers with complementary bandgaps to maximize light absorption across the solar spectrum. Computational modeling plays a vital role in selecting suitable materials, optimizing layer thicknesses, and identifying effective interlayer materials for efficient charge transport. Drift-diffusion simulations coupled with optical models provide insights into key performance parameters, including power conversion efficiency (PCE), open-circuit voltage (Voc), short-circuit current density (Jsc), and fill factor (FF) (Uddin et al. 2024). By leveraging computational studies, researchers can fine-tune device architectures to achieve higher efficiency and stability in perovskite tandem solar cells. This chapter explores the role of computational modeling in advancing tandem solar cell technology and discusses future directions in this field.

In summary, computational studies have significantly contributed to the progress of PSCs by providing a deeper understanding of material properties, device physics, and performance optimization strategies. The ability to simulate defects, electronic structure, charge transport, and thermal management has enabled researchers to design more efficient and stable perovskite solar cells. Despite certain limitations, computational techniques remain an indispensable tool in PSC research, complementing experimental studies and accelerating the development of next-generation photovoltaic technologies. This chapter aims to consolidate key findings from previous chapters, reflecting on the strengths and challenges of computational approaches while highlighting their potential for future advancements in perovskite solar cell technology.

Summary of the Work

PSCs have gained significant attention in recent years due to their exceptional power conversion efficiency and cost-effective fabrication processes. As researchers continue to explore ways to enhance their performance and stability, computational

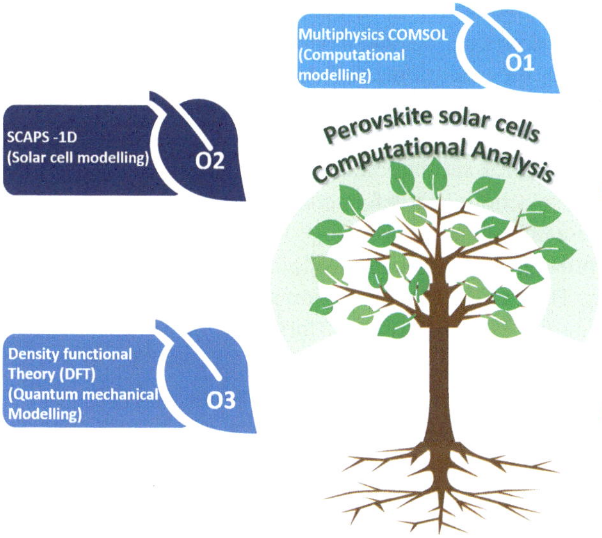

Fig. 13.1 The schematic illustration of software and tools utilized for PSC devices

approaches have emerged as powerful tools for understanding material properties, optimizing device structures, and predicting long-term behavior. This book provides an in-depth analysis of the role of computational modeling in PSC research, covering key topics such as defect analysis, interface engineering, stability studies, and the integration of machine learning techniques. The schematics of all the possible tools used for PSC devices in the present book are shown in Fig. 13.1.

In this book, Chapter 1 introduces the significance of computational analysis in the context of PSCs. The chapter elaborates on how hybrid halide perovskites (HHP) have emerged as a revolutionary technology in the renewable energy sector, offering impressive PCE of 26.7%. Despite the remarkable properties of HHP materials, such as tunable band gaps, low exciton binding energy, high charge carrier mobility, and cost-effective fabrication, PSCs are still far from being commercially viable. Therefore, the importance of computational approaches alongside experimental techniques is highlighted. Various computational methods like DFT, molecular dynamics (MD) simulations, and machine learning (ML) are discussed, all of which have proven to be instrumental in predicting material properties, optimizing device designs, and discovering new perovskite compositions. The chapter also delves into various device simulation and modeling software, such as SCAPS-1D, COMSOL, and AMPS, providing a comprehensive view of the tools that have facilitated advancements in PSC technology. Furthermore, the chapter presents future research directions, underlining how computational methods are expected to significantly influence the development of sustainable energy technologies and the commercialization of PSCs.

Chapter 2 explores the generation of defects in perovskite absorbers, a crucial factor that influences the efficiency, stability, and overall performance of PSCs. The chapter categorizes defects into vacancies, interstitials, and grain boundaries, and examines the mechanisms through which these defects are formed. Key factors such as synthesis conditions, thermal processing, and environmental conditions are discussed in detail to explain how they contribute to defect generation. These defects significantly impact the electronic properties of perovskite materials, as well as charge transport and recombination processes, which in turn affect the efficiency of solar cells. The chapter emphasizes the importance of advanced synthesis techniques, appropriate material selection, and defect control strategies for mitigating the effects of defects. Researchers have been exploring methods to optimize perovskite materials to improve their performance and stability, which is critical for advancing PSC technology. The chapter also underscores the role of computational approaches in understanding the underlying defect formation mechanisms and in developing strategies to minimize defect-induced challenges in perovskite absorbers.

Chapter 3 highlights the importance of interface engineering in improving the efficiency and stability of PSCs. As organic-inorganic halide PSCs have achieved notable PCE, the chapter discusses the factors that still limit their performance, such as poor intrinsic stability under ambient conditions. Interface engineering has emerged as a promising strategy to enhance both PCE and stability. This chapter elaborates on how energy level alignment, interfacial defects, and carrier dynamics can be optimized to facilitate efficient carrier extraction and reduce undesirable degradation pathways. Several types of materials used as interlayers at different interfaces, including electrode/ETL, ETL/perovskite, and HTL/electrode interfaces, are discussed in the context of their role in enhancing device performance. The chapter also covers the application of interface engineering to flexible PSCs, recognizing the growing interest in developing PSCs that can be integrated into various applications. The chapter concludes by offering a future outlook on how interface engineering can shape the next generation of PSCs.

Chapter 4 provides an in-depth analysis of physical impedance models used to study the internal mechanisms of PSCs. Impedance spectroscopy is introduced as a powerful tool for gaining insights into the electrical properties of solar cells, including charge transport, recombination dynamics, and degradation processes. The chapter examines how impedance data, when combined with other analytical methods, can provide a deeper understanding of the operational characteristics of PSCs. It also discusses the challenges posed by the absence of a centralized repository for impedance data, which can make it difficult to track the progress of PSC research effectively. By consolidating the understanding of impedance models and equivalent circuits, the chapter aims to establish a comprehensive knowledge base that will support ongoing efforts to optimize PSC performance and stability.

Chapter 5 focuses on the role of multi-physics studies in understanding and optimizing PSCs. PSCs have gained attention for their high power conversion efficiency and low production costs, but achieving optimal performance and long-term stability requires an understanding of various interconnected physical processes. This chapter discusses key multi-physics phenomena, including electrical charge transport, light absorption, optical behavior, thermal management, and mechanical stability. It also highlights various modeling techniques, such as drift-diffusion models, optical simulations, and electro-thermal-mechanical coupling, which help predict device behavior under real-world operating conditions. By integrating these diverse physical aspects into unified multi-physics models, researchers can gain valuable insights into PSC performance, identify potential issues related to efficiency and degradation, and develop strategies to address scalability concerns. The chapter concludes with an outlook on how the integration of multi-physics simulations will contribute to the continued development of PSC technologies.

Furthermore, Chapter 6 presents a theoretical study of all-inorganic and organic-inorganic mixed halide PSCs. The chapter discusses the structural, electronic, and optical properties of these materials as revealed through computational techniques such as DFT and SCAPS-1D modeling. The findings highlight the unique advantages of all-inorganic perovskites, such as exceptional thermal stability, tunable band gaps, and efficient charge carrier dynamics, which are critical for the development of durable and efficient solar cells. On the other hand, organic-inorganic hybrid perovskites, such as $MAPbI_3$, are shown to benefit from the stabilizing role of organic molecules, as well as the impact of compositional engineering at the A-site and X-site on their performance and stability. The chapter discusses important aspects such as defect formation, exciton dynamics, and degradation mechanisms, which significantly affect the long-term reliability of PSCs. The integration of theoretical studies with experimental data is emphasized, with the aim of facilitating the development of next-generation PSCs with enhanced efficiency and durability.

Chapter 7 delves into the computational investigations of tandem solar cells, a promising advancement in photovoltaic technology. The chapter outlines how tandem solar cells—comprising multiple layers with different band gaps—offer the potential to achieve higher PCE compared to single-junction cells. The chapter provides a detailed overview of the principles and structural advantages of tandem configurations, including both perovskite-on-perovskite and heterojunction tandem architectures. A particular focus is placed on the role of interface engineering in enhancing device stability and reducing energy losses. The chapter also discusses the limitations of current computational models, such as accuracy and resource constraints, and emphasizes the importance of combining computational studies with experimental research to overcome these challenges. In doing so, the chapter outlines the potential of tandem PSCs to revolutionize the field of photovoltaics by enabling higher efficiency and more stable devices.

Chapter 8 addresses the significant challenge of stability in PSCs, which remains a major obstacle to their commercialization. The chapter explores various factors that contribute to the instability of PSCs, including moisture, heat, UV light, oxygen, ion migration, and hysteresis. These factors cause material degradation and negatively impact the performance of PSCs over time. The chapter highlights strategies to improve the stability of PSCs, such as compositional engineering, the use of 2D layered materials, nanoengineering, and interfacial modification techniques. In addition, the chapter emphasizes the role of computational methods, including DFT, MD, and ab initio molecular dynamics (AIMD), in understanding the mechanisms behind material degradation and in guiding the development of more stable PSCs.

Chapter 9 examines the impact of 2D materials on the performance and stability of PSCs. The incorporation of two-dimensional materials, known for their excellent properties such as high charge mobility, conductivity, and tunable band gaps, has garnered significant attention in recent studies. The chapter discusses how 2D materials can be used as ETL, HTL, interfacial layers, and back electrodes in PSCs. The chapter presents evidence showing that 2D materials can improve both the efficiency and stability of PSCs by reducing non-radiative recombination, minimizing defects, and improving energy level alignment. The chapter also acknowledges the challenges that remain in the field, including the need for more comprehensive studies to optimize the integration of 2D materials into PSC devices.

Chapter 10 focuses on interfacial engineering in PSCs using simulation approaches. This chapter builds upon the rapid advancements in PSCs, particularly their impressive PCE and cost-effective fabrication, by examining the significance of optimizing the interfaces within PSC devices. The chapter discusses the role of hybrid organic-inorganic halide perovskites in achieving high efficiency, and how interfacial engineering can further enhance charge-carrier diffusion lengths, bandgap tuning, and interfacial stability. By optimizing energy levels and mitigating losses at different interfaces, researchers can improve the overall efficiency of PSCs. The chapter highlights how simulation-based approaches have been essential in these efforts, providing valuable insights into interface optimization and helping to drive the evolution of PSC technology.

Chapter 11 investigates the strategies for modifying the bandgap of perovskite materials, a crucial factor for improving the efficiency and light harvesting capabilities of PSCs. The chapter discusses various approaches to bandgap modification, including doping, compositional engineering, and interface engineering. It is shown that these methods can effectively tune the bandgap of perovskite materials, thereby enhancing their performance. The chapter also explores the challenges associated with balancing the bandgap for optimal efficiency, as well as how these strategies can contribute to improving the stability and long-term performance of PSCs.

Lastly, Chapter 12 highlights the role of ML in advancing the development of PSCs. Machine learning has emerged as a powerful tool in optimizing PSC performance, accelerating material discovery, and solving scalability challenges. This chapter reviews various ML techniques, including regression models, ensemble methods, and neural networks, and their application to PSC research. By providing predictive insights, ML can streamline the process of material discovery and device optimization, enabling the development of robust, high-efficiency solar cells. The chapter emphasizes the transformative potential of ML in revolutionizing photovoltaic technologies and paves the way for next-generation PSCs. The highlights of all the chapters are tabulated in Table 13.1.

Table 13.1 Highlights of the all 12 chapters of this book

Chapters	Highlights
Chapter 1	• PSCs achieve high efficiency of 26.7% with low-cost fabrication • PSCs need computational analysis for commercialization • DFT, MD, ML, and simulation tools optimize PSC design
Chapter 2	• Perovskites excel in light absorption and charge transport • Vacancies, interstitials, and grain boundaries affect efficiency and stability • Advanced synthesis, material selection, and computational methods improve performance
Chapter 3	• PSCs reach 26.1% efficiency but face stability issues • Enhances efficiency and stability by reducing defects and improving carrier extraction • Focus on interfacial materials, ion migration control, and flexible PSC development
Chapter 4	• Advances in PSC research boost solar energy potential • IS helps analyze PSC internal mechanisms • Lack of centralized data hinders research tracking
Chapter 5	• Understanding charge transport, optics, heat, and mechanics is key to stability • Advanced simulations optimize performance, design, and scalability
Chapter 6	• PSCs offer high efficiency, cost-effectiveness, and tunable properties • DFT and SCAPS-1D reveal structural, electronic, and optical behavior • Simulations aid in improving stability, defect control, and performance
Chapter 7	• Tandem solar cells surpass single-junction cells in PCE • Improved light absorption and reduced energy loss enhance performance • Interface engineering and combined research drive stability and efficiency
Chapter 8	• Moisture, heat, UV light, and ion migration degrade PSCs • Strategies include compositional engineering, nanoengineering, and encapsulation • DFT, MD, and AIMD help analyze and improve PSC stability
Chapter 9	• PSCs are a transformative technology, but commercial application is hindered by stability issues • 2D materials improve PSCs by enhancing efficiency, stability, and reducing defects • While 2D materials boost performance, challenges remain, and future prospects are explored
Chapter 10	• PSCs offer high efficiency, low cost, and versatility, challenging silicon solar cells • Perovskites feature high charge-carrier diffusion, variable bandgaps, and low-temperature processability • Perovskite efficiency surged from 3.8% to over 25%, leading next-gen photovoltaics
Chapter 11	• PSCs are known for high efficiency, charge transport, and easy fabrication • poor stability compared to first-generation solar cells limits PSC commercialization • Methods like doping, compositional engineering, and dimensional alterations are used to optimize bandgap for better efficiency and stability
Chapter 12	• ML helps overcome these challenges by predicting material properties and optimizing device design • ML models like regression, ensemble, and neural networks are key to advancing PSC research and creating next-gen solar cells

Concluding Remarks

Computational and theoretical studies have played a crucial role in advancing the understanding and optimization of key physical parameters in PSCs. Through modeling and simulations, researchers have gained valuable insights into the electronic structure, optoelectronic properties, defect behavior, interfacial interactions, and overall device performance of perovskite materials. These computational approaches allow for the systematic exploration of various material compositions, device configurations, and operating conditions in a cost-effective and efficient manner. By identifying promising material combinations and guiding experimental efforts, computational studies significantly accelerate the development of high-performance and stable perovskite solar cells. Moving forward, these studies will remain an essential tool in complementing experimental research and driving advancements in perovskite solar cell technology.

References

Baro M, Borgohain P (2023) SCAPS-1D device simulation of highly efficient perovskite solar cells using diverse charge transport layers. J Electron Mater 52:7623. https://doi.org/10.1007/s11664-023-10681-7

Fang Y, Kong X, Wang D, Liu J, Shang Q, Cui S (2020) Insights into the interactions of g-C3N4/LaMnO3 hetero-junction to their structures and electronic properties by DFT calculations. J Solid State Chem 292:121727. https://doi.org/10.1016/j.jssc.2020.121727

Gupta K, Thakur OP, Kumar M (2024) Efficient and stable perovskite solar cell through the interfacial engineering: experiment and computational investigations. Opt Mater (Amst) 150:115125. https://doi.org/10.1016/j.optmat.2024.115125

Hafner J (2008) *Ab-initio* simulations of materials using VASP: density-functional theory and beyond. J Comput Chem 29:2044–2078. https://doi.org/10.1002/jcc.21057

Hossain MI, Shahiduzzaman M, Saleque AM, Huqe MR, Qarony W, Ahmed S et al (2021) Improved nanophotonic front contact design for high-performance perovskite single-junction and perovskite/perovskite tandem solar cells. Sol RRL 5:2100509. https://doi.org/10.1002/solr.202100509

Kahandal SS, Tupke RS, Bobade DS, Kim H, Piao G, Sankapal BR et al (2024) Perovskite solar cells: fundamental aspects, stability challenges, and future prospects. Prog Solid State Chem 74:100463. https://doi.org/10.1016/j.progsolidstchem.2024.100463

Kumar M, Raj A, Kumar A, Anshul A (2022) Computational analysis of bandgap tuning, admittance and impedance spectroscopy measurements in lead-free MASnI3 perovskite solar cell device. Int J Energy Res 46:11456–11469. https://doi.org/10.1002/er.7942

Nie J, Zhou B, Fang S, Wang Y, Wang Y, Tian B et al (2023) Chemical doping of lead-free metal-halide-perovskite related materials for efficient white-light photoluminescence. Mater Today Phys 31:100992. https://doi.org/10.1016/j.mtphys.2023.100992

Raj A, Kumar M, Bherwani H, Gupta A, Anshul A (2021) Evidence of improved power conversion efficiency in lead-free CsGeI3 based perovskite solar cell heterostructure via scaps simulation. J Vacuum Sci & Technol B Nanotechnol Microelectr 39:012401. https://doi.org/10.1116/6.0000718

Rajagopal A, Yang Z, Jo SB, Braly IL, Liang PW, Hillhouse HW et al (2017) Highly efficient perovskite–perovskite tandem solar cells reaching 80% of the theoretical limit in photovoltage. Adv Mater 29:1702140. https://doi.org/10.1002/adma.201702140

Uddin MS, Hossain MK, Uddin MB, Toki GFI, Ouladsmane M, Rubel MHK et al (2024) An in-depth investigation of the combined optoelectronic and photovoltaic properties of Lead-free $Cs_2AgBiBr_6$ double perovskite solar cells using DFT and SCAPS-1D frameworks. Adv Electron Mater 10:2300751. https://doi.org/10.1002/aelm.202300751

The manufacturer's authorised representative in the EU is Springer Nature Customer Service Centre GmbH, Europaplatz 3, 69115 Heidelberg, Germany. If you have any concerns regarding our products, please contact ProductSafety@springernature.com

Printed and bound by CPI Group (UK) Ltd, Croydon, CR0 4YY
07/10/2026
02229349-0009